STUDIES IN COGNITIVE SYSTEMS

PROGRAM VERIFICATION

STUDIES IN COGNITIVE SYSTEMS

VOLUME 14

The titles published in this series are listed at the end of this volume.

PROGRAM VERIFICATION

Fundamental Issues in Computer Science

Edited by

TIMOTHY R. COLBURN

Department of Computer Science,
University of Minnesota, Duluth

JAMES H. FETZER

Department of Philosophy,
University of Minnesota, Duluth

and

TERRY L. RANKIN

IBM Health Industry Marketing,
Atlanta, Georgia

SPRINGER SCIENCE+BUSINESS MEDIA, B.V.

Library of Congress Cataloging-in-Publication Data

Program verification : fundamental issues in computer science / edited
  by Timothy R. Colburn, James H. Fetzer, and Terry L. Rankin.
       p.   cm. -- (Studies in cognitive systems ; v. 14)
    Includes bibliographical references (p.   ) and indexes.
    ISBN 978-94-010-4789-0   ISBN 978-94-011-1793-7 (eBook)
    DOI 10.1007/978-94-011-1793-7
    1. Computer software--Verification.   I. Colburn, Timothy R.,
  1952-   . II. Fetzer, James H., 1940-   . III. Rankin, Terry L.
  IV. Series.
  QA76.76.V47P76   1993
  005.1'4--dc20   92-26748

ISBN 978-94-010-4789-0

Printed on acid-free paper

To
C. A. R. Hoare

TABLE OF CONTENTS

SERIES PREFACE

This series includes monographs and collections of studies devoted to the investigation and exploration of knowledge, information, and data-processing systems of all kinds, no matter whether human, (other) animal, or machine. Its scope spans the full range of interests from classical problems in the philosophy of mind and philosophical psychology through issues in cognitive psychology and sociobiology (concerning the mental powers of other species) to ideas related to artificial intelligence and computer science. While primary emphasis is placed upon theoretical, conceptual, and epistemological aspects of these problems and domains, empirical, experimental, and methodological studies will also appear from time to time.

The program verification debate affords a welcome opportunity to contribute to a new area of study, which might appropriately be referred to as the philosophy of computer science. The foundations of a field as technical and as recent as computer science raise special problems that tend to resist analysis by conventional means. The editors of this volume have sought to contribute to our understanding of this discipline by bringing together important papers that clarify and illuminate these underlying issues. The senior editor, Timothy R. Colburn, supplies a framework for following the issues, while the bibliography provides a route to other work that is suitably related to these questions. An assortment of rich and fascinating problems are involved here, which deserve to be explored further.

J. H. F.

ACKNOWLEDGMENTS

'Towards a Mathematical Science of Computation', by John McCarthy originally appeared in C. M. Popplewell (ed.), *Information Processing 1962*, Proceedings of DFIP Congress 62 (Amsterdam, The Netherlands: North-Holland Publishing Company, 1963), pp. 21—28.

'Proof of Algorithms by General Snapshots', © 1966 Peter Naur, originally appeared in *BIT* **6** (1966), pp. 310—316.

'Assigning Meanings to Programs', by Robert W. Floyd, originally appeared in *Mathematical Aspects of Computer Science* (Proceedings of Symposia in Applied Mathematics, Vol. 19), American Mathematical Society, 1967, pp. 19—32.

'An Axiomatic Basis for Computer Programming', by C. A. R. Hoare, originally appeared in *Communications of the ACM* **12**(10) (1969), pp. 576—580 and p. 583. © 1969 Association for Computing Machinery, Inc. Reprinted by permission.

'Mathematics of Programming', by C. A. R. Hoare is reprinted with permission, from the August 1986 issue of *BYTE Magazine*. © McGraw-Hill, Inc., New York, NY. All rights reserved.

'On Formalism in Specifications', by Bertrand Meyer, originally appeared in *IEEE Software*. © 1985 IEEE. Reprinted with permission, from IEEE Software; volume 2; issue 1; pp. 6—26; January 1985.

'Formalization in Program Development', © 1982 Peter Naur, originally appeared in *BIT* **22** (1982), pp. 437—453.

'First Steps Towards Inferential Programming', by William L. Scherlis and Dana S. Scott, originally appeared in R. E. A. Mason (ed.), *Information Processing 83*. © 1983 JFIP and Elsevier Science Publishers, The Netherlands, pp. 199—212.

'Outline of a Paradigm Change in Software Engineering', by Christiane Floyd, originally appeared in G. Bjerknes *et al.* (eds.), *Computers and Democracy: A Scandanavian Challenge* (Brookfield, VT: Gower Publishing Company, 1987), pp. 191—210. (Old Post Road, Brookfield, VT 05036, U.S.A.)

'Limits of Correctness in Computers', by Brian Cantwell Smith, originally appeared as a Center for the Study of Language and Information Report, No. CSLI-85-36 (October 1985), © The Center for the Study of Language and Information, Stanford, CA, 1985.

'Formalism and Prototyping in the Software Process', by Bruce I. Blum, originally appeared in *Information and Decision Technologies* **15** (1989), pp. 327—341. © Elsevier Science Publishers, The Netherlands.

'The Place of Strictly Defined Notation in Human Insight', by Peter Naur, originally appeared in *Computing: A Human Activity*, © 1992 by ACM Press, forthcoming. Reprinted with permission of Addison-Wesley Publishing Company, Inc.

'Social Processes and Proofs of Theorems and Programs', by Richard A. De Millo, Richard J. Lipton, and Alan J. Perlis, originally appeared in *Communications of the ACM* **22**(5) (1979), 271—280. © 1979, Association for Computing Machinery, Inc. Reprinted by permission.

'Program Verification: The Very Idea', by James H. Fetzer, originally appeared in *Communications of the ACM* **31**(9) (1988), 1048—1063. © 1988, Association for Computing Machinery, Inc. Reprinted by permission.

'The Notion of Proof in Hardware Verification', by Avra Cohn, originally appeared in *Journal of Automated Reasoning* **5** (1989), 127—139. Reprinted by permission of Kluwer Academic Publishers.

'Program Verification, Defeasible Reasoning, and Two Views of Computer Science', by Timothy R. Colburn, originally appeared in *Minds and Machines* **1** (1991), 97—116. Reprinted by permission of Kluwer Academic Publishers.

'Philosophical Aspects of Program Verification', by James H. Fetzer, originally appeared in *Minds and Machines* **1** (1991), 197—216. Reprinted by permission of Kluwer Academic Publishers.

PROLOGUE

TIMOTHY R. COLBURN

COMPUTER SCIENCE AND PHILOSOPHY

"Computer science and philosophy? Isn't that sort of an odd combination?" Such is the typical cocktail-party response when learning of my academic training in the discipline Socrates called "the love of wisdom" and my subsequent immersion in the world of bytes, programs, systems analysis, and government contracts. And such might very well be the reaction to the title of this essay. But despite its cloistered reputation, and its literary, as opposed to technological image, the tradition of philosophical investigation, as all of us who have been seduced by it know, has no turf limits. No turf limits, at least, in one direction; while few but the truly prepared venture into philosophy's hard-core "inner circle" of epistemology, metaphysics, (meta)ethics, and logic, literally anything is fair philosophical game in the outer circle in which most of us exist. And so we have the "philosophy of's": philosophy of science, philosophy of art, of language, education. Some of the philosophy of's even have names befitting their integration into vital areas of modern society, for example, medical ethics and business ethics, which we can say are shorter names for the philosophies of ethical decisions in medicine and business. This anthology is intended to be perhaps the first organization of various writings in a yet-to-crystalize philosophy of computer science.

Which is not to say that there has not been a vast amount of work done which can be described as a combination of philosophy *and* computer science. The typical cocktail-party reaction to this juxtaposition notwithstanding, the solutions to many problems in computer science have benefited from what we might call *applied* philosophy. For example, hardware logic gate design would not be possible without Boolean algebra, developed by the 19th century mathematician George Boole, whose work helped lay the foundation for modern logic. Later work in logic, particularly the development of predicate calculus by Gottlob Frege, has been drawn upon extensively by researchers in software engineering who desire a formal language for computer program semantics. Predicate calculus is also the formal model used by many of those who implement automated systems for mechanical

3

Timothy R. Colburn et al. (eds.), Program Verification, 3—31.
© 1993 *Kluwer Academic Publishers*.

theorem proving. These theorem proving techniques have even formed the basis for the complete, general purpose programming language, Prolog, which is now the language of choice for artificial intelligence research in Europe and Japan.

But the application of philosophical methods to computer science is not limited to those in logic. The study of ethics, for example, has found broad application to computer-related issues of privacy, security, and law. While these issues are not regarded as germane to the science of computing *per se*, they have arisen directly as a result of the drastic changes society has undergone due to the ubiquity and power of computers. In 1990, a major U.S. software vendor attempted to openly market a large mailing list compiled from public sources, but was forced to withdraw it when the public outcry over invasion of privacy became too great. While the scaling back of the U.S. Strategic Defense Initiative can be seen as a response to technical feasibility questions, a major underlying moral concern is whether a nation *ought* to entrust its security, to such a large extent, to machines.

Within the field of law, many sticky ethical questions related to computers have arisen: Is unauthorized use of a computer from the privacy of one's own home, without damaging any files or programs (*i.e.* hacking), the same as breaking and entering? Can programs which are experts in fields of medicine or law be sued for malpractice? Should computer programs be copyrightable, or free, like air? Should programmed trading be allowed on the stock exchange? Answers to the last two questions, and others like it, would have significant effects on the conduct of our economy. All of these questions could not have been predicted a mere few decades ago. Today, it would be difficult to find a college curriculum which did not include, in either the computer science or philosophy department, a course entitled 'Computers and Society', 'Values and Technology', or the like.

But besides the application of philosophical tools and the engendering of ethical questions, the relationship of philosophy and computer science is perhaps most intimate in the field of artificial intelligence (AI). Although even the *definition* of the field of AI is fraught with philosophical debate, genuine philosophical questions at the core of epistemology and the philosophy of mind come to the fore as researchers attempt to model human intelligence in computer programs: What is the structure of human knowledge (so that we may represent it in computer memory)? What is the process of human thought (so that we may model reasoning, learning, and creativity in computer programs)? Interestingly, while AI researchers must ask the same sorts of cognitive questions as philosophers do, they must ultimately agree with the pervasive assumption, stated by Hume in the 18th century, that "cognition is computation", a point of view certainly not shared by all philosophers.

Thus we can define at least three areas in which computer science and philosophy come together: the application of logical methods and tools, the treatment of new ethical questions, and debate over traditional philosophical questions in the context of minds and machines. There is no dearth of research and writing in these areas, which — a case can be made — all count to some extent as applied philosophy. It is interesting, however, that comparatively little work can be found which would count as attempting to establish philosophical foundations of computer science itself, *i.e.* a philosophy *of* computer science. The essays collected in this volume either do attempt to establish some such foundations, or, in honored philosophical tradition, criticise ones that do. But besides being presented as participating in a dialectic, what ties these essays together as philosophy? Before answering this, it would help to consider what it is that would make something a *philosophical foundation* of computer science. Or, what *is* the philosophy of computer science?

But even before that, we should ask what makes any philosophy a philosophy *of*. Here is one point of view. Epistemology, metaphysics, and (meta)ethics constitute core philosophical areas in that they take for their subject matter the most 'generic' of concepts, stripped of flavor and color: knowledge (in general, not necessarily knowledge *of* anything in particular), existence (of *kinds* of entities, not anything in particular), and values (the meaning and possibility of value-laden or moral terms, not the morality of acts *per se*). We might even call these areas the foundations of philosophy. See the bottom level in the taxonomy presented in Figure 1, which is by no means intended to be exhaustive, only illustrative.

When philosophical investigation is concerned with any of these concepts laced with a particular slant, it is a philosophy *of*. A philosophy *of* can involve questions and problems in one or several of the foundational areas of philosophy. So, philosophy of mind is concerned with the existence and knowledge of *particular* kinds of entities, namely mental objects and mental events. Philosophy of art analyzes the existence and value of *aesthetic* objects. Normative ethics (philosophy of moral decisions) is concerned with the morality of *particular* acts. Finally, more relevant to our current problem, philosophy of science is concerned with the logic of *scientific* concept formation in the analysis of *scientific* knowledge — how we come by it and what it means. (See the second tier of Figure 1.)

We can describe a third tier of philosophical investigation, essentially a level of specialization in relation to the second tier. This is a level in which questions of knowledge, existence, and value are posed within a quite specific context, demanding expertise at least in the language of a specific field of inquiry. Thus, medical ethics is a specialization of normative ethics concerned with the morality of medical acts, like abortion or euthanasia. Philosophies of, say, cinema or dance are specializations of philosophy of art. Philosophy of psychology poses questions for both philosophy of science and philosophy of mind, while philosophies of other social or natural sciences are restricted studies in the philosophy of science. Such specializations are depicted in the third tier of Figure 1, which illustrates philosophy's appeal for both pure generalists and focused specialists, as well as many falling in between.

Clearly, the philosophy of computer science should take a place alongside other specific philosophies of science in the third tier. Work in this tier is typically characterized by one or more Central Questions. A Central Question in the philosophy of biology, for example, is whether the science of the organic can be reduced to the science of the inorganic (the reduction of biological to physical laws). A Central Question in the philosophy of the social sciences is whether the social sciences can or should be value-free. A Central Question in philosophy of physics concerns the function and ontological status of inferred entities like quarks and strings. What are the Central Questions in the philosophy of computer science, and how do the essays in this book address them?

As in all philosophy, the philosophy of science is sustained by controversy. Following World War II, much of the discussion regarding theory building centered around the role of observation. The view, commonplace at the time, of theory building as the mere generalization of observation came under attack by scientists and philosophers alike bent on discrediting the empiricists' reductive analysis. More specific philosophies *of* are also often born of controversy over reductionism. In biology, it was vitalism, or belief in a unique 'vital force' in organic

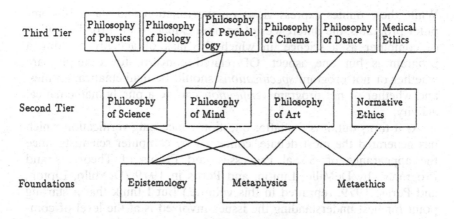

Fig. 1. A taxonomy of philosophy.

life, vs. mechanism, or the belief that all talk about such a force can be reduced to talk about inorganic processes. In psychology, it was behaviorism, or the belief that talk about consciousness can be reduced to talk about stimuli, percepts, and responses, vs. the view that entities such as the self are unique, irreducible, and impossible to study without processes such as introspection.

So it seems that there may be a pattern in the birth and sustenance of philosophy *of*'s involving reductionist attempts. The philosophy of computer science may be no exception, for if the history of philosophy ever looks back and identifies an early Central Question in the philosophy of computer science, it may be this: Can computer science be reduced to a branch of mathematics? This is the broad question around which the essays in this collection are arranged. The range of perspectives from which this question can be addressed is as wide as the difference between the following view, expressed by Hoare:

Computer programs are mathematical expressions. They describe, with unprecedented precision and in the most minute detail, and behavior, intended or unintended, of the computer on which they are executed. (Hoare, 1986, p. 115, reprinted in this volume.)

and this alternative, offered by C. Floyd:

Programs are tools or working environments for people. [They] are designed in processes of learning and communication so as to fit human needs. (Floyd, 1987, p. 196, reprinted in this volume.)

While these quotes express views on the function and status of computer programs, the differences of opinion extend to the broader notion of computer as a science, in which the task of actually creating a program is but one aspect. Of related concern, for example, are whether or not program *specifications* should be mathematical entities, and whether or not program *verification* can be a purely mathematical activity.

As it turns out, it is the latter question concerning verification which has generated the most debate, contested by computer scientists since the appearance of 'Social Processes and Proofs of Theorems and Programs', by DeMillo, Lipton, and Perlis in 1979 (DeMillo, Lipton, and Perlis, 1979, reprinted in this volume.) But I think that a starting point for best understanding the issues involved is at the level of computer programs themselves and what they mean. The view expressed by Hoare above is unequivocal: computer programs *are* mathematical expressions. The quote by Floyd is less precise, but it expresses a view on the *function* of programs for humans in decidedly nonmathematical terms. While these views do not necessarily contradict one another, they can most definitely signal contrasting values as to how computer programs *ought* to be designed, built, and used.

Not until 1988 did these questions draw the attention of a 'pure' philosopher. With 'Program Verification: The Very Idea' (Fetzer, 1988, reprinted in this volume), Fetzer resurrected the program verification/ social process debate of a decade earlier and placed it in the newer light of causal models. Before this time, however, debate on the issues was evidenced mainly by differences in visionary accounts, given by computer science practitioners and teachers, of how the young discipline of computer science ought to proceed. Thus, only three of the 17 essays in this volume are written by practicing philosophers. I definitely do *not*, however, consider this a disadvantage. While the vocabulary and precision accompanying philosophical acumen are an advantage, and possibly a necessity, in the philosophical mainstream of epistemology, metaphysics, ethics, and logic, we are interested here in insights within philosophy's 'third tier', where both experience and the force of unconditioned intuition are as important as academic philosophizing. All of the works by practicing computer scientists here are thoughtful and literate contributions to the budding philosophy of computer science.

Having legitimized both the academic philosophical and the practicing scientific backgrounds for this endeavor, let me say that as a practicing computer scientist trained in philosophy I find myself compelled to frame some of these issues in an historical philosophical perspective. Many philosophers have long held there to be a clear distinction between formal science (mathematics and logic) and factual science (any empirical discipline, including physics, biology, sciology, etc.), offering critieria for the distinction between the two as well as an account of their interplay. Quite possibly we can use this distinction to throw some light on the question of whether computer science can be reduced to a branch of mathematics.

According to one widespread philosophical opinion, the difference between formal and factual sciences is based upon the distinction between *analytic* and *synthetic* statements. For Carnap, for example, an analytic statement is one which is valid according to the syntactic rules of a pertinent language "independently of the truth or falsity of other statements" (Carnap, 1934, p. 124). The idea is that the truth of an analytic statement can be determined by virtue simply of a language's formation and transformation rules, so that "it is a consequence of the null class of statements." Thus, statements of mathematics ('5 is a prime number') and logic ('$F(x) \lor \neg F(x)$') are analytic; they are derivable entirely within the rules of their respective languages. A synthetic statement, on the other hand, is one which is not analytic (and also not contradictory). A synthetic statement therefore is one whose truth or falsity will depend upon statements other than the syntactic transformation rules of a pertinent language. 'Chicago is on Lake Michigan' is synthetic because it is a consequence of geographical statements such as, 'At such-and-such a place there is a conjunction of such-and-such a city and such-and-such a lake'.

Now according to Carnap, "*The distinction between the formal and the factual sciences* consists then in this: the first contains only analytic, while the second also contains synthetic statements" (Carnap, 1934, p. 124). We might try, then, to put this criterion to the test in the case of computer science. We have the claim, championed by Hoare, that "computer programs are mathematical expressions." Presumably, this would be accompanied by the claim that computer programming is a formal (mathematical) science. Are the statements of a computer program all analytic, or are some synthetic? Take a typical statement in a programming language, like

S $A := 13 * 74.$

The problem with analyzing **S** is that it can have either an imperative sense, as in

I   Compute the value 13 times 74 and place it in memory location A,

or a declarative sense, as in

D   Memory location A receives the value of computing 13 times 74.

Imperative statements, as such, are neither true nor false and thus cannot be classified as analytic or synthetic. Let us focus, then, on the declarative sense.

Statement **D** hides an ambiguity concerning the ontological status of the entity described by 'memory location A'. This ambiguity, as Fetzer (1988) observed, arises because A can be either a *physical* memory location or an *abstract* one. If A is a physical memory location in an executing computer, then **D** should read

DP   Physical memory location A receives the value of physically computing 13 times 74.

DP is actually a *prediction* of what will occur when statement **S** is ultimately executed. In reality, **S**, as marks on a piece of paper, or character codes in a computer memory, is not executed *per se*. Instead, it is translated into or interpreted as a set of more primitive operations to be executed when the program is run. But whatever the level of a computer operation, its *representation* in a statement like **S** is at best a static prediction of its later dynamic execution. As such, it cannot be analytic, since the validity of no prediction can be ascertained via syntactic transformation rules alone.

So let us turn to an interpretation of **D** with memory location A in its abstract sense. But what *is* an abstract memory location? In computer science there is the notion of an *abstract* machine, that is, a complete description of the operations of a machine which, were it built to certain specifications and run in certain conditions, would exhibit certain ideal behavior in terms of its state changes and output when given certain input. An abstract machine can be thought of as a specification, realizable either in hardware, through circuit fabrication, or in software, through simulation, of a physically executing machine. As Fetzer points out, physical machines are subject to the problems of reliability and performance that attend any causal systems. Abstract macines, however, do not fail; they are ideal and thus simply *supposed* not to fail. In discourse about programs and their behavior, we tend to unknowingly flip back and forth between reference to abstract and physical machines. When talking about a program and its *intended* behavior, as when predicting that it will run correctly, we assume that the machine on which it runs does not fail its specifications, and thus we are talking about an abstract machine. When talking about a program and its *actual* behavior, say after it is run in analyzing its efficiency, we are talking about a physical machine.

Mathematics, of course, abounds with abstract entities like geometric forms, numbers, structures, and operations and, if Carnap is right, mathematics consists of analytic statements. Does an interpretation of computer programs as running on abstract machines result in their statements being analytic? Consider the statement

DA   Abstract memory location A receives the value of abstractly computing 13 times 74.

How does one analyze the validity of such a statement as **DA**? It seems to be an entirely different breed of statement; one which, like **I**, it would be a category mistake to which to apply an analytic/synthetic determination. Unlike **I**, **DA** has a truth-value, but unlike **DP**, it is true in an innocuous sort of way. Specifying what an abstract memory location shall receive after abstractly performing a computation is rather like a mathematician *stipulating* the contents of a variable, for example 'let $a = 962$'. But such a statement is not true analytically; it is true merely by virtue of its being made, like an Austinian performative utterance. Thus **DA** is very much like a mathematical statement, but not in any interesting way — no more interesting than a mathematician incanting 'let . . .' without a 'therefore . . .'. Any statement in any programming language can be construed in an abstract sense as in **DA**. Thus any program can be regarded as a series of stipulations in an abstract world. As such, a program, like a series of mathematician's 'let's, is not a mathematical expression in any interesting sense.

What is mathematical, however, is *reasoning* about programs in their abstract sense. For example, given a program P consisting of the statements $S_1, \ldots, S_n$, and interpreting each S_i as an abstract machine statement s_i, it is possible to construct statements like

T   Let s_1, let $s_2, \ldots$, and let s_n. Then P has property R,

where R describes some aspect of the execution of P in the abstract sense. By giving precise interpretations to the s_i in a pertinent language and appropriately choosing R, it may be possible that **T** is a theorem in the language and thus analytic. This is in fact the approach taken by modern researchers in software engineering who are concerned with program semantics.

While I have just given a justification of how reasoning *about* programs can be regarded as mathematical, it is yet a much broader claim to say that computer science is, or ought to aspire to be, a branch of mathematics. For there are still the issues of whether the specification, generation, or maintenance of programs (apart from reasoning about completed ones) is or ought to be like a mathematical activity. It is some or all of these issues that the essays in this volume in one way or another address.

In what follows I shall try to describe a thread of issues which, hopefully, stitches the diverse presentations and backgrounds of these essays into a unifying whole. As mentioned, the issue which motivates and underlies much of the tension in philosophical discussion of computer science is formal verification. Although we have collected several of the important papers regarding verification in Part IV, we shall see that verification provides a pervasive undercurrent for discussion in virtually all of them. Consequently, my discussion of the papers in Part IV creeps into what follows slightly out of turn compared to their arrangement in the book.

THE MATHEMATICAL PARADIGM

The essays in Part I of this collection represent some of the earliest attempts to reason mathematically about programs, and characterize what we shall call the mathematical paradigm for computer science. While these authors no doubt would have conceded Fetzer's later point about the causal properties of physical systems vs. the mathematical properties of abstract systems, they probably would not have had a problem with relegating their arguments to programs running on abstract machines. Instead, they found their greatest challenge in somehow establishing an exploitable mathematical link between the *static* properties of programs and the *dynamic* properties of running them (albeit in an abstract sense).

The earliest paper in the collection, 'Towards a Mathematical Science of Computation', was written by McCarthy in 1962. McCarthy is also given credit for coining the term 'artificial intelligence' in the 1950s. While much of McCarthy's work has been driven by a philosophical opinion concerning the relation of human thought and computation, this paper is as much concerned with establishing a basis for a theory of computation which is free of any dependence on a particular computational domain, e.g., integers or character strings. Thus, while his contribution here is not directly concerned with reasoning about programs, it *is* overtly concerned with establishing a language for talk about the abstract activity of computation itself. Because, it might be argued, the most general talk about any computation is always in terms of its bare *function*, the best mathematical framework for discussing computation is that of recursive function theory. Such a framework, however, by itself lacks the essential power of programming languages, namely the ability to conditionally execute a statement based on the outcome of some test. McCarthy, therefore, adds *conditional expressions* to the standard mathematical lexicon and shows how typical programming features can be described in terms of recursive functions. Since recursive functions can be reasoned about, for example, to show that two sets of functions are equivalent, corresponding programs can also be reasoned about. Thus the static-to-dynamic bridging problem is crossed, for McCarthy, through the intermediacy of recursive functions. Not coincidentally, McCarthy is also the designer of a programming language, called Lisp, whose chief programming element is the function, and whose effective use relies heavily on recursion.

Still, McCarthy's motive is not the promotion of a programming language. Indeed, using the techniques he sketches and suggests, the question of which programming language to use becomes a moot one, so long as the language's syntax and semantics are well specified, since the theory of computation he envisions would allow, among other advantages, the automatic translation from one linguistic paradigm to another. We can, in fact, look back now after nearly thirty years and confirm that automatic program translation, with the help of precise language specification, has been accomplished in the case of language compilers. However, that a compiler, however automatically, in all cases *correctly* translates programs into an assembly or machine language, is something that no compiler builder ever guarantees, as every language reference manual's warranty disclaimer demonstrates. Thus, there is the distinction between (1) using mathematical methods during source language translation to produce highly reliable object language code, and (2) using mathematical methods to prove that an object language program behaves on its abstract machine exactly as a source language program behaves on its abstract machine. While the difference between these seems fairly substantial, as the difference between applied and pure mathematics, believers in the mathematical paradigm for computer science, whose goal is the more ambitious (2), remain undaunted.

This distinction between the use of mathematics as an engineering tool vs. the use of mathematics to conclusively prove statements in formal systems fairly accurately characterizes the polarity underlying the contrasting quotes from Floyd and Hoare cited above. It also provides the undercurrent for the program verification debate, in which is contrasted the distinction betwen (1) using mathematical methods during program development and testing to produce highly reliable code, and (2) using mathematical methods to prove that a program's execution on its abstract machine is correct. McCarthy, seeing no obstacle to (2), wrote:

It should be possible almost to eliminate debugging. Debugging is the testing of a program on cases one hopes are typical, until it seems to work. This hope is frequently vain. Instead of debugging a program, one should prove that it meets its specifications, and this proof should be checked by a computer program. (McCarthy, 1963, p. 22)

While McCarthy was one of the first to express this opinion, it was shared by others in the 1960s who strove to describe what such a proof would be like. Naur, in an effort to cross the static-to-dynamic bridge in reasoning about programs ('Proof of Algorithms by General Snapshots', which is reprinted in this volume), recognized that one way to talk about a program both as a static, textual entity and as a dynamic, executing entity, was to conceive of the program as executing, but to from time to time conceptually 'halt' it and make statements about the state of its abstract machine at the time of halting. By making a number of these 'snapshots' of a conceptually executing program, and providing justifications for each on the basis of the previous one, a proof can be constructed about the state of the abstract machine upon termination.

Naur envisioned the use of mathematical thinking not only to prove the correctness of algorithms, but also to aid in the actual construction of program elements, particularly loops. He prescribes a disciplined approach to loop construction which involves the use of snapshots and painstaking attention to the limits and parameters of the loop — all of which are activities actually preparatory to writing the loop, which, it is hoped, simply 'falls out' as a result of the conceptual work involved in the mathematical description of snapshots. That Naur believes the mathematical component of programming to be essential is evident when he writes

It is a deplorable consequence of the lack of influence of mathematical thinking on the way in which computer programming is currently being pursued, that the regular use of systematic proof procedures, or even the realization that such proof procedures exist, is unknown to the large majority of programmers. (Naur, 1966, p. 310)

It should be noted, however, that Naur is also wary of the potential *mis*uses of mathematical thinking in computer science, as evidenced by two other articles of his included in this volume.

Naur's work, and that of R. Floyd, form the basis of much of the contemporary work on proving properties of programs through inductive assertions. Independently of Naur but at about the same time, Floyd investigated the idea of assigning meanings to programs in terms of *interpretations* placed on their flowcharts ('Assigning Meanings to Programs', which is reprinted in this volume). Like Naur's snapshots, Floyd's interpretations are propositions labeling various points in the control flow of a program (called *tags*) which , if proved, may constitute a proof of correctness:

It is, therefore, possible to extend a partially specified interpretation to a complete interpretation, without loss of verifiability, provided that initially there is no closed loop in the flowchart all of whose edges are not tagged and that there is no entrance which is not tagged. This fact offers the possibility of automatic verification of programs, the programmer merely tagging entrances and one edge in each innermost loop; the verifying program would extend the interpretation and verify it, if possible, by mechanical theorem-proving techniques. (Floyd, 1967, p. 25)

Though this idea held much promise for believers in the Mathematical Paradigm, it came under attack in the abovementioned essay by

DeMillo, Lipton, and Perlis, which opens Part IV of this collection, focusing on formal verification. They argue passionately that mechanically produced program verifications, which are long chains of dense logical formulas, are *not* what constitute mathematical proofs. In coming to be accepted, a mathematical proof undergoes social processes in its communication and peer scrutiny, processes which cannot be applied to unfathomable pages of logic. While DeMillo, Lipton, and Perlis do not subscribe to the mathematical paradigm, i.e., that computer science is a branch of formal mathematics, they also do not deny that programming is *like* mathematics. An analogy can be drawn between mathematics and programming, but ". . . the same social processes that work in mathematical proofs doom verifications" (DeMillo, Lipton, and Perlis, 1979, p. 275). Social processes, they argue, are critical:

> No matter how high the payoff, no one will ever be able to force himself to read the incredibly long, tedious verifications of real-life systems, and unless they can be read, understood, and refined, the verifications are worthless. (DeMillo, Lipton, and Perlis, 1979, p. 276)

The main focus of DeMillo, Lipton, and Perlis' attack lies in the practicality and social acceptance of the method of formal verification. They hint, however, at another, nonsocial, reason for deemphasizing verifiability in software design, having to do with conceptual confusion over the ultimate significance of mathematical models:

> For the practice of programming, . . . verifiability must not be allowed to overshadow reliability. Scientists should not confuse mathematical models with reality — and verification is nothing but a model of believability. (DeMillo, Lipton, and Perlis, 1979, p. 279)

Fetzer's (1988) paper, mentioned above, once more brought the debate over formal verification to the forefront, but by elaborating this conceptual confusion in great detail. Fetzer argues, in fact, that the presence or absence of social processes is germane to neither the truth of theorems nor program verifications:

> Indeed, while social processes are crucial in determining what theorems the mathematical ·ommunity takes to be true and what proofs it takes to be valid, they do not thereby make them true or valid. The absence of similar social processes in determining which programs are correct, accordingly, does not affect which programs are correct. (Fetzer, 1988, p. 1049)

DeMillo, Lipton, and Perlis, for example, hit upon the boredom, tedium, and lack of glamor involved in reviewing proofs produced by mechanical verifiers. But for Fetzer, if this is all there is to their criticism of formal verification, it is not substantial. As Fetzer points out, social processes are characterized by transitory patterns of human behavior which, one could imagine, in different circumstances would reserve for program verification the same sort of excitement and collegial collaboration which marks the best mathematical research. Thus DeMillo, Lipton, and Perlis have identified a difference 'in practice' between mathematical research and formal program verification, but not 'in principle':

What this means is that to the extent to which their position depends upon this difference, it represents no more than a contingent, *de facto* state-of-affairs that might be a mere transient stage in the development of program verification within the computer science community. (Fetzer, 1988, p. 1053)

Like DeMillo, Lipton, and Perlis, Fetzer believes that formal program verification cannot fulfill the role that some of its advocates would assign to it within software engineering. But he attacks it from a nonsocial, more strictly philosophical perspective, hinted at but not elaborated by DeMillo, Lipton, and Perlis above. This has to do with the relationship between mathematical models and the causal systems they are intended to describe. Close scrutiny of this relationship reveals, for Fetzer, the relative, rather than absolute, nature of the program correctness guarantee that formal verification can provide, leaving the indispensability of empirical methods (i.e., program testing) in the software development process intact.

ELABORATING THE PARADIGM

Despite the criticisms of DeMillo, Lipton, and Perlis, elaboration of the mathematical paradigm continued throughout the 1980s. Part II of our collection documents three of the more representative attempts, as well as a questioning look in a second paper by Naur. An example of a grand but incompletely specified vision of mathematically basing the process of programming is provided by Scherlis and Scott in their 'First Steps Towards Inferential Programming' (Scherlis and Scott, 1983, reprinted in this volume). In this paper they not only respond to the objections of DeMillo, Lipton, and Perlis with respect to formal verification, but they give a proposal for an overall approach to programming as concerned primarily with a process rather than a product:

> The initial goal of research must be to discern the basic structural elements of the process of programming and to cast them into a precise framework for expressing them and reasoning about them. This understanding must be embodied in a *logic of programming* — a mathematical system that allows us not only to reason about individual programs, but also to carry out *operations* (or transformations) on programs and, most importantly, to reason about the operations. (Scherlis and Scott, 1983, p. 199)

The object of reasoning, in this vision, is not programs themselves, but program *derivations*. Scherlis and Scott are therefore guided by the following analogy (Scherlis and Scott, 1983, p. 207):

Mathematics		*Programming*
problem	...	specification
theorem	...	program
proof	...	program derivation

where a process of program derivation produces a program somewhat like a proof mathematically supports a theorem. A program derivation is thus somewhat akin to a formal derivation in logic and can be reasoned about mechanistically.

Since the derivation is guided by the program specification, it should be possible at all times during the derivation to reason about the process and demonstrate that the development history so far has conformed with the specification. Scherlis and Scott envision this process as a formal one aided by a mechanical theorem prover, and thus they describe the desired process of program development as 'inferential' programming. While they grant that formal verification is not feasible *after* programs have been produced, they claim that this is not a problem under a paradigm of inferential programming, since verification will be built into the very process of programming.

One of the most consistent champions of the mathematical paradigm since the 1960s has been Hoare. In 'An Axiomatic Basis of Computer Programming', included in Part I, he presents the beginnings of a system of axioms and rules which can be used in proofs of program properties. Essentially, these axioms and rules describe principles of general program behavior. By substituting specific statements into the axiom schemata and applying the rules to obtain new statements, proofs of terminating conditions can be obtained, insuring correctness. Hoare realizes, however, that the correctness of a running program depends on more than verification of the program itself:

When the correctness of a program, its compiler, and the hardware of the computer have all been established with mathematical certainty, it will be possible to place great reliance on the results of the program, and predict their properties with a confidence limited only by the reliability of the electronics. (Hoare, 1969, p. 579)

In the 22 years since this pronouncement, no program nearly as large as any production compiler has been mathematically verified. Some progress has been made in the mathematical proving of hardware circuit designs, but even practitioners in the field are aware of the limits of proving hardware, as the essay by A. Cohn, 'The Notion of Proof in Hardware Verification' (Cohn, 1989, reprinted in this volume), attests.

Hoare has persistently believed in "the principle that the whole of nature is governed by mathematical laws of great simplicity and elegance" (Hoare, 1986, p. 122), and the governance of software engineering practices is no exception. In 'Mathematics of Programming' (Hoare, 1986, reprinted in this volume), he develops an analogy between operations on natural numbers and the 'sequential composition' of commands which makes up final programs. Through this analogy, he provides strict mathematical foundations for familiar software engineering practices, for example, procedural abstraction and stepwise refinement. However, his aim is not merely to theoretically base these practices on mathematics, but to inject mathematics into the practices: "I would suggest that the skills of our best programmers will be even more productive and effective when exercised within the framework of understanding and applying the relevant mathematics" (Hoare, 1986, p. 122). Though he gives detailed attention to the arithmetic analogy and the mathematical *justification* of good software engineering practice, he is less specific about what the relevant mathematics is and how it is used in the process of creating programs.

The mathematical paradigm has its adherents in other aspects of programming besides development and verification. In particular, the *specification* of programs is seen by many as an area in which natural language's vagueness and ambiguity have contributed to problems in program development, and thus that a formal approach is preferable.

An example of an actual formally stated program specification is proposed by B. Meyer in 'On Formalism in Specifications' (Meyer, 1985, reprinted in this volume). As evidence of the need for formalism, Meyer describes the deficiencies of a fairly meticulously stated specification, in natural language, of a program implementing a very simple text formatter. In a careful analysis, Meyer uncovers numerous instances of the "seven sins of the specifier", as he calls them, including noise, overspecification, contradiction, and ambiguity. Meyer's recommendation:

> In our opinion, the situation can be significantly improved by a reasoned use of more formal specifications. . . . In fact, we'll show how a detour through formal specification may eventually lead to a better English description. This and other benefits of formal approaches more than compensate for the effort needed to write and understand mathematical notations. (Meyer, 1985, p. 15)

This effort can be considerable. In order to comprehend Meyer's proposed formal specification for the text formatting program, a thorough understanding of functions, relations, sets, sequences, and predicate calculus is required, all in order to make precise the intuitively clear idea of 'filling' text into paragraphs with a given width specification. Meyer makes a point of saying, however, that he is not proposing to replace natural language as a specification vehicle. He shows, for example, how one might go about producing a precisely stated natural language specification from the formal one. Whether the translation process is effective and the resulting product worth the formal effort, however, is open to question.

It is clear then that the elaboration of the mathematical paradigm for computer science has gone far beyond the vision embodied by **T**, in which mathematical reasoning is employed to reason merely about the properties of programs running on abstract machines. In addition, we have the mathematical modeling of computation through recursive functions (McCarthy), deductive reasoning about program derivations with the help of mechanical theorem provers (Scherlis and Scott), mathematical justification of software engineering practices (Hoare), and a strict formal mode for program specification writing (Meyer). Just as clearly, however, a movement away from the mathematical paradigm can be discerned. Naur, in 'Formalization in Program Development' (Naur, 1982, reprinted in this volume), warns of problems in swinging radically to the formal side: "[T]he emphasis on formalization in program development may in fact have harmful effects on the results of the activity" (Naur, 1982, p. 437). For Naur the formal/informal antinomy is not germane to the program development problem:

Instead of regarding the formal mode of expression as an alternative to the informal mode we must view it as a freely introduced part of the basic informal mode, having sometimes great advantages, mostly for the expression of highly specialized assertions, but having also great disadvantages, first and foremost in being limited to stating facts while being inapplicable to the many other shades of expression that can be reached in the informal mode. (Naur, 1982, p. 440)

In a counterpoint predating Meyer, Naur denies the claim that a formal mode of expression in specification writing necessarily removes errors and inconsistencies. As evidence, he takes a very small part of a formal specification for the Algol 60 language and uncovers several problems which the formal mode cannot thereby preclude. But most of all, Naur supports the unique intrinsic value of the informal, intuitive mode: "In fact, an important part of any scientific activity is to express, discuss, criticize, and refine, intuitive notions. In this activity formalization may be very helpful, but it is, at best, just an aid to intuition, not a replacement of it" (Naur, 1982, p. 448).

CHALLENGES, LIMITS, AND ALTERNATIVES

With this paper by Naur, we see a backing away from the treatment of mathematics as a *paradigm* for computer science, and toward its use as a *tool* for those who specify, code, and verify programs. In Part III of this collection we present more reflections on the limits and alternatives to a strict mathematical model for computer science. None of these contributions call for total abandonment of mathematics; to the contrary, all authors recognize the value of formal methods and the advances made possible because of them. As Blum writes in "Formalism and Prototyping in the Software Process" (Blum, 1989, reprinted in this volume):

It should be clear that success in computer science rests on formal methods. Instruction sequences in a computer can only be understood in terms of formal abstractions. Programming languages and their compilers are consistent and reliable to the extent we have good formal models for them. The issue is not one of having formalisms. It is one of identifying when and how they contribute to the software process and of understanding their limitations. (Blum, 1989, p. 10)

The contributions by both Blum and C. Floyd in this part strike a conciliatory tone for the formal/informal polarization in computer science. Blum argues for the acceptance of prototyping within an overall fairly rigorous software process. According to a widely agreed upon model, this process consists of two very general and fundamentally different subprocesses. *Abstraction* is the process of producing a precise specification S of a problem statement from a real world application domain, and *reification* is the process of transforming S into a software product P. The pivotal object in the overall process is therefore the specification S. For Blum, it isn't S being a possibly *formal* object that is objectionable; rather, what can cause problems is that at the start of the reification process, S is normally *incomplete*. For one thing, project sponsors themselves "seldom know enough about the application to define it fully at the start" (Blum, 1989, p. 6), but more importantly, S *should not* be complete or the problem will be overspecified. Overspecification results in S being *isomorphic* to P, in which case S includes details of behavior which are irrelevant to the sponsor, and thus not incumbent upon him/her to provide. "Therefore, during reification it may be necessary to augment S as the problem becomes better understood. ... [T]he software process is not simply one of the logical derivation of a product from a fixed specification." (Blum, 1989, p. 6)

For Blum, the initial incompleteness of S results from the transition from a *conceptual* mode of thinking about the problem to a *formal* mode during the abstraction subprocess. By definition, during an abstraction detail is lost, and it needs to be added to the formal expression of S in a typically iterative process. According to Blum, *rapid prototyping* can facilitate this process, "as a technique to define (or refine) the initial specification" (Blum, 1989, p. 13). Prototyping as a method of software development has often been distinguished from the linear phase method, in which software is conceived as a product arising from a progression of well-defined phases, from requirements analysis to maintenance. Indeed, prototyping is also often presented as incompatible with the linear phase method, since prototyping is not literally one of the phases. Correspondingly, prototyping has been viewed as an alternative paradigm to the mathematical one being studied here. It is Blum's contention, however, that when viewed as part of the specification phase, rapid prototyping is compatible with the linear phase conception and even necessary for determining the problem specification, which is independent of whether formal methods are used to produce a program from the specification. He provides examples to support his resulting view that "prototyping and formal methods can fruitfully coexist in the same paradigm" (Blum, 1989, p. 13).

C. Floyd, in 'Outline of a Paradigm Change in Software Engineering' (Floyd, 1987, reprinted here), also pleas for coexistence, but of two paradigms, one of which she thinks should assume a new prominence in computer science. In doing so, she does not challenge the mathematical/formal paradigm as we have described it here *per se*, but rather a general paradigm for software engineering in which the mathematical model plays the dominating role. This general paradigm comes from what Floyd calls a *product-oriented perspective* which has dominated the discipline of software engineering since its inception: "The product-oriented perspective regards software as a product standing on its own, consisting of a set of programs and related defining texts" (Floyd, 1987, p. 194). According to this perspective, it is assumed that the problem for which a computerized solution is proposed is so well understood as to allow its software requirements to be infallibly determined before software production, allowing the formal approaches to specification and verification to be applied with deductive precision so that software is simply churned out or 'engineered' like complicated electronic hardware components. While a noble goal, it is unrealistic, declares Floyd: "[T]he product-oriented approach . . . does not permit us to treat systematically questions pertaining to the relationship between software and the living human world, which are of paramount importance for the adequacy of software to fulfill human needs" (Floyd, 1987, p. 195).

As an alternative paradigm, Floyd sees the need for a *process-oriented perspective* which "views software in connection with human learning, work, and communication, taking place in an evolving world with changing needs" (Floyd, 1987, p. 194). From this perspective, software requirements are never static entities; they change both as the user's needs change and as the software is being developed, through evaluation and feedback from different groups of people. Since a software specification cannot be separated from the people for whom the software is produced, and since communication about the specification is necessary, the essential part of the specification should be written in natural language. Thus Floyd views social processes as crucial in the entire software development process. In this regard she goes beyond DeMillo, Lipton, and Perlis, for whom the existence of social processes was necessary only for arguing against formal verification.

Naur's contribution to this part, 'The Place of Strictly Defined Notation in Human Insight' (Naur, 1989), is not nearly so conciliatory an attempt to analyze the relationship between formal and informal methods of software development. Instead, it is an indictment of what he considers to be an overemphasis on 'minor' issues, as he calls them, in the area of constructed models. One such issue is "the construction of a model M_b, from building elements E_b, when model M_a, built from element E_a, is given. This is often encountered in a form where E_a is a so-called specification language, M_a is a specification, and E_b is a programming language" (Naur, 1989, p. 8). Thus he relegates to minor status the actual building of programs from specifications, and with it the verification, formal or otherwise, of a program in relation to its specification.

Major issues, according to Naur, with regard to the modeling of anything, including the modeling of real world problems in terms of specifications or programs, involve the modelee, or thing being modeled, and thus are not primarily concerned with questions of formalism, or the use of strictly defined notation according to abstract rules. These issues are (1) the *modeling insight*, or "grasping a relation between aspects of the world that have no inherent strictly defined properties and the strictly defined items of the model", and (2) the *model building insight*, or finding a way of "combining the available model building elements so as to arrive at a model having appropriate properties" (Naur, 1989, p. 7). With regard to software, the major issues are therefore the designing and the building of programs, and the fundamental problems here are not solved by adherence to a formalism. The relationship of a perhaps formally stated design *specification* and a program, while having a role, is a minor one, and does not justify the ignoring of *in*formal concerns pervading contemporary work in programming logic, including formal verification. Naur gives many examples of current research in which informal concerns about formal matters are expressed in passing but given scant attention to make way for formal results and techniques. For Naur, this results in a 'deplorable' situation and elicits a scathing assessment of the state of programming logic research:

It is curious to observe how the authors in this field, who in the formal aspects of their work require painstaking demonstration and proof, in the informal aspects are satisfied with subjective claims that have not the slightest support, neither in argument nor in verifiable evidence. Surely common sense will indicate that such a manner is scientifically unacceptable. (Naur, 1989, p. 14)

Philosophical foundations of what it means to be a model are also central to Smith's analysis in 'Limits of Correctness in Computers' (Smith, 1985, reprinted in this volume). While not arguing *against* formal verification *per se*, he points out that what it really accomplishes is the establishment of the relative consistency between a computer program and an abstract specification of the problem it is supposed to solve. This relation leaves utterly unaddressed the question of whether the program solves the problem in the real world, that is, the question of whether the program (or its specification) is an *accurate* model of some aspect of the real world. Thus his primary observation is very similar to Naur's, but he uses it as a launching point for an analysis of program correctness. The question of program correctness, Smith argues, should not be confused with the question of consistency with its specifications. In fact, 'correct' can mean many other things: "Is a program correct when it does what we have instructed it to do? or what we wanted it to do? or what history would dispassionately say it should have done?" (Smith, 1985, p. 12).

Because 'correct' can mean so many things, Smith suggests that 'reliable' is the adjective we should attach as our desideratum for programs. Reliability is a concept which we know how to apply to humans, and we should, claims Smith, apply no stricter reliability criteria to computer programs, for they cannot be expected to do better. If anything, a program is destined to perform worse, since the realm it is dealing with is not the real world, but a model of it. Since this model can never be complete, it makes no sense to ask that a program be 'correct' as its behavior relates to the real world:

Plans to build automated systems capable of making a 'decision', in a matter of seconds, to annihilate Europe, say, should make you uneasy; requiring a person to make the same decision in a matter of the same few seconds should make you uneasy too, and for very similar reasons. The problem is that there is simply no way that reasoning of any sort can do justice to the inevitable complexity of the situation, because of what reasoning is. Reasoning is based on partial models. Which means it cannot be guaranteed to be correct. (Smith, 1985, pp. 18—19)

While correctness, in a narrow technical sense, can be seen to be a relation between a program and a model of a problem in terms of abstract specifications, reliability must be a relation between the program and its actual functioning in the real world.

FOCUS ON FORMAL VERIFICATION

Verificationists are concerned with reasoning about a model M_b, such that it can be shown to be consistent with another model M_a. As we have seen, the debate over program verification began with DeMillo, Lipton, and Perlis' insistence on the role of social processes in mathematical proofs and became inflamed with Fetzer's charge that the very notion of program verification trades on ambiguity. There is usually no question that M_a, a program specification, is taken to be a formal model. But it is not clear whether M_b, a program, is also a formal model, so Fetzer's ambiguity comes into play. The lure of formal methods with respect to verification extends beyond the realm of programs and into hardware and system areas. We can assess the appropriateness of formal methods in these areas by learning from the program verification case.

For a decade or more, the National Security Agency has funded research in the design of 'provably secure' computers, meaning computers that are mathematically guaranteed not to compromise a given multi-level security policy (DoD, 1987). The goal here is not to prove the consistency of a software specification and program, as in program verification, but to prove the consistency of a security model M_a and a computer system design M_b, thus showing that there are no security leaks inherent in the design. Thus the goal is *design* verification *vis à vis* a security model. Since both M_a and M_b are formal models with no causal efficacy, Fetzer's ambiguity charge does not pose a problem here unless one makes the misguided claim that by proving the system design M_b to be secure, any computer built to that design is also secure.

But Fetzer's ambiguity does come into play in another area of verification, namely hardware verification. One of the most successful application areas for automated theorem proving techniques has been that of 'proving' computer hardware circuits. The formal model M_a is taken to be a hardware circuit specification. From a logic description, M_b, of a hardware circuit, M_a and M_b are shown to be consistent, verifying the circuit with respect to its specification. But the notion of a hardware circuit, like a program, is ambiguous. Is the conclusion, that the circuit is verified, referring to the circuit as a causal or a formal model? If the latter, then the conclusion may be an analytic statement and indeed follow with the logical certainty reserved for conclusions in mathematics. But if the former, then the conclusion is a synthetic statement and follows only in conjunction with accompanying descriptions of further facts concerning the circuit's fabrication and physical environment when it is used.

While distinguishing between entities as causal and formal models may seem to be belaboring the obvious, confusing them can lead to disturbing results. For Cohn, a researcher in hardware verification, these manifest themselves in oversold claims made in promotional material concerning a military microprocessor named Viper. According to one such claim, Viper is called "the first commercially available microprocessor with . . . a formal specification and a proof that the chip conforms to it" (Cohn, 1989, p. 135). Cohn, a member of the Viper project team, carefully shows, however, that "Neither an intended behavior nor a physical chip is an object to which the word 'proof' meaningfully applies" (Cohn, 1989, p. 135). While she does not explicitly refer to the formal model/causal model ambiguity, it underlies her caution with respect to the meaning of 'verified': "Since any model is an abstraction of a material device, it is *never* correct to call a system 'verified' without reference to the level of the models used" (Cohn, 1989, p. 136). The relative, rather than absolute, nature of hardware verification claims is as clear for Cohn as the relative, rather than absolute, nature of program verification claims is for Fetzer:

A proof that one specification implements another — despite being completely rigorous, expressed in an explicit and well-understood logic, and even checked by another system — should still be viewed in context of the many other extra-logical factors that affect the correct functioning of hardware systems. In addition to the abstract design, everything from the system operators to the mechanical parts must function correctly — and correctly together — to avoid catastrophe. (Cohn, 1989, p. 136)

There seems, in the end, to be no way to verify either a program's or a hardware device' reliability other than by testing it empirically and observing its behavior. In 'Program Verification, Defeasible Reasoning, and Two Views of Computer Science' (Colburn, 1991, reprinted in this volume), Colburn tries to make this explicit through a delineation of five different ways programs can be run as tests. Only one of these — namely testing a program to determine if its (abstract) behavior conforms to its specifications — can conceivably be replaced by purely deductive, mathematical reasoning. The other ways — for example, testing a program to determine if its specification actually solves the given problem — cannot be so replaced because the activities involved cannot be reduced to proofs of consistency between syntactic or formal objects; they are fundamentally different from that, employing the element of empirical discovery common to natural science, and not to mathematics.

We come, for Colburn, to a view of computer programming as part of an overall process of computer science which is analogous to natural or, even more broadly, experimental science. In this view, objects such as algorithms or program specifications are like hypotheses which can only be tested by writing and running programs. Such tests provide insight into the adequacy of the model/real world relationship which we exploit in the program. As such, the program is only a tool in an overall process of inquiry, like experiment, apparatus and setup is in hypothesis confirmation or rejection. This is not to say that the programming/ mathematics analogy which has dogged the discussion up till now is invalid, but that it should be seen as subordinate to the overall computer science/experimental science analogy which is the primary one. The problem of producing a program with certain specifications must still be solved, and a formal approach may be appropriate, but ordinarily this is a subproblem in an overall problem of, variously, testing an algorithm, or testing the specifications themselves, or explaining a phenomenon through simulation.

It is fitting that the last word in this volume be given by Fetzer, who aroused such passion in the computer science community in 1988. For the two years following, he became enmeshed in a debate of monumental proportions carried out in multiple venues, including the editorial pages of the *Communications of the ACM*, publicly staged counterpoints, and electronic mail discussion groups. Only the most carefully reasoned philosophical position could hold up to this kind of scrutiny (not to mention hysterical diatribes), and in the Epilogue ('Philosophical Aspects of Program Verification', Fetzer, 1991, reprinted in this volume), Fetzer recounts the position, supplemented by the benefits of two years' further reflection. His basic charge is the same: "[I]t should be apparent that the very idea of the mathematical paradigm for computer science trades on ambiguity" (Fetzer, 1991, p. 209). Strong claims of formal verificationists are victim to this ambiguity through the ignoring of several distinctions: between programs running on abstract machines with *no* physical counterpart and programs running on abstract machines *with* a physical counterpart; between 'programs-as-texts' and 'programs-as-causes'; and between pure and applied mathematics. Recognizing these distinctions, for Fetzer, reveals that the claim that it is possible to reason in a purely *a priori* manner about the behavior of a program is true if the behavior is merely abstract; false and dangerously misleading otherwise. Fetzer appears, then, to emerge from the fracas none the worse:

And it should also be obvious that the conception of computer science as a branch of pure mathematics cannot be sustained. The proper conception is that of computer science as a branch of applied mathematics, where even that position may not go far enough in acknowledging the limitations imposed by physical devices. (Fetzer, 1991, p. 212)

For myself, I see a possible replacement of the mathematical paradigm by an experimental science paradigm for computer science, and like many authors I see it in a way which should reconcile heretofore conflicting views. But whether the paradigm be experimental science, or applied mathematics, or engineering, it is interesting to note, and a tribute to the value of philosophizing in any age, that after studying the important contributions of the writers in this book I come finally to an observation by Carnap made well before the Age of Computers:

All of logic including mathematics, considered from the point of view of the total language, is thus no more than an auxiliary calculus for dealing with synthetic statements. *Formal science* has no independent significance, but is an auxiliary component introduced for technical reasons in order to faciliate linguistic transformations in the *factual sciences*. The great importance of the formal sciences, that is, of logic and mathematics, within the total system of science is thereby not in the least denied but instead, through a characterization of this special function, emphasized. (Carnap, 1934, p. 127)

Had Carnap lived to take part in the program verification debate, he might have couched this observation thus:

All of logic including formal verification, considered from the point of view of the total language of computer science, is not more than an auxiliary calculus for dealing with broader issues of computer problem-solving as an experimental discipline. *Formal verification* has no independent significance, but is an auxiliary component introduced for technical reasons in order to facilitate linguistic transformations from formal program-specification models to formal program-execution models. The importance of formal verification within the total system of computer science is thereby not in the least denied but instead, through a characterization of this special function, emphasized.

I couldn't have put it better myself.

Department of Computer Science,
University of Minnesota.

REFERENCES

Blum, B.: 1991, 'Formalism and Prototyping in the Software Process', to be published in *Information and Decision Technologies*.

Carnap, R.: 1953, 'Formal and Factual Science', in Feigl, H. and Brodbeck, M. (eds.), *Readings in the Philosophy of Science*, New York: Appleton-Century-Crofts, pp. 123—128.

Cohn, A.: 1989, 'The Notion of Proof in Hardware Verification', *Journal of Automated Reasoning* **5**(2), 127—139.

Colburn, T.: 1991, 'Program Verification, Defeasible Reasoning, and Two Views of Computer Science', *Minds and Machines* **1**(1), 97—116.

DeMillo, R., Lipton, R., and Perlis, A.: 1979, 'Social Processes and Proofs of Theorems and Programs', *Communications of the ACM* **22** (May 1979), 271—280.

Department of Defense: 1983, 'Trusted Computer Systems Evaluation Criteria', CSC-STD-001-83, August 15, 1983.

Fetzer, J.: 1988, 'Program Verification: The Very Idea', *Communications of the ACM* **31**(9), 1048—1063.

Fetzer, J.: 1991, 'Philosophical Aspects of Program Verification', *Minds and Machines* **1**(2), 197—216.

Floyd, C.: 1987, 'Outline of a Paradigm Change in Software Engineering', in *Computers and Democracy: A Scandinavian Challenge*, England: Gower Publishing Company, pp. 191—210.

Floyd, R.: 1967, 'Assigning Meanings to Programs', *Proceedings of Symposia in Applied Mathematics, Vol. 19*, pp. 19—32.

Hoare, C. A. R.: 1969, 'An Axiomatic Basis for Computer Programming', *Communications of the ACM* **12** (October 1969), 576—583.

Hoare, C. A. R.: 1986, 'Mathematics of Programming', *BYTE* (August 1986), pp. 115—118, 120, 122, 124, and 148—149.

McCarthy, J.: 1962, 'Towards a Mathematical Science of Computation', *Proceedings of the IFIP Congress* **62**(1963), pp. 21—28.

Meyer, B.: 1985, 'On Formalism in Specifications', *IEEE Software* (January 1985), pp. 6—26.

Naur, P.: 1966, 'Proof of Algorithms by General Snapshots', *BIT* **6**, 310—316.

Naur, P.: 1982, 'Formalization in Program Development', *BIT* **22**, 437—453.

Naur, P.: 1989, 'The Place of Strictly Defined Notation in Human Insight', unpublished paper from 1989 Workshop on Programming Logic, Bastad, Sweden.

Scherlis, W. L. and Scott, D. S.: 1983, 'First Steps Towards Inferential Programming', *Information Processing* **83**, 199—212.

Smith, B. C.: 1985, 'Limits of Correctness in Computers', Center for the Study of Language and Information Report No. CSLI-85-36 (October 1985).

PART I
THE MATHEMATICAL PARADIGM

TOWARDS A MATHEMATICAL SCIENCE
OF COMPUTATION

1. INTRODUCTION

In this paper I shall discuss the prospects for a mathematical science of computation. In a mathematical science, it is possible to deduce from the basic assumptions, the important properties of the entities treated by the science. Thus, from Newton's law of gravitation and his laws of motion, one can deduce that the planetary orbits obey Kepler's laws.

What are the entities with which the science of computation deals?

What kinds of facts about these entities would we like to derive?

What are the basic assumptions from which we should start?

What important results have already been obtained?

How can the mathematical science help in the solution of practical problems?

I would like to propose some partial answers to these questions. These partial answers suggest some problems for future work. First I shall give some very sketchy general answers to the questions. Then I shall present some recent results on three specific questions. Finally, I shall try to draw some conclusions about practical applications and problems for future work.

This paper should be considered together with my earlier paper [1]. The main results of the present paper are new but there is some overlap so that this paper would be self-contained. However some of the topics in this paper such as conditional expressions, recursive definition of functions, and proof by recursion induction are considered in much greater detail in the earlier paper.

2. WHAT ARE THE ENTITIES WITH WHICH
COMPUTER SCIENCE DEALS?

These are problems, procedures, data spaces, programs representing procedures in particular programming languages, and computers.

Problems and procedures are often confused. A problem is defined by the criterion which determines whether a proposed solution is

35

Timothy R. Colburn et al. (eds.), Program Verification, 35–56.
© 1993 *Kluwer Academic Publishers.*

accepted. One can understand a problem completely without having any method of solution.

Procedures are usually built up from elementary procedures. What these elementary procedures may be, and how more complex procedures are constructed from them, is one of the first topics in computer science. This subject is not hard to understand since there is a precise notion of a computable function to guide us, and computability relative to a given collection of initial functions is easy to define.

Procedures operate on members of certain data spaces and produce members of other data spaces, using in general still other data spaces as intermediates. A number of operations are known for constructing new data spaces from simpler ones, but there is as yet no general theory of representable data spaces comparable to the theory of computable functions. Some results are given in [1].

Programs are symbolic expressions representing procedures. The same procedure may be represented by different programs in different programming languages. We shall discuss the problem of defining a programming language semantically by stating what procedures the programs represent. As for the syntax of programming languages, the rules which allow us to determine whether an expression belongs to the language have been formalized, but the parts of the syntax which relate closely to the semantics have not been so well studied. The problem of translating procedures from one programming language to another has been much studied, and we shall try to give a definition of the correctness of the translation.

Computers are finite automata. From our point of view, a computer is defined by the effect of executing a program with given input on the state of its memory and on its outputs. Computer science must study the various ways elements of data spaces are represented in the memory of the computer and how procedures are represented by computer programs. From this point of view, most of the current work on automata theory is beside the point.

3. WHAT KINDS OF FACTS ABOUT PROBLEMS, PROCEDURES, DATA SPACES, PROGRAMS, AND COMPUTERS WOULD WE LIKE TO DERIVE?

Primarily, we would like to be able to prove that given procedures solve given problems. However, proving this may involve proving a whole host of other kinds of statement such as:

1. Two procedures are equivalent, i.e. compute the same function.
2. A number of computable functions satisfy a certain relationship, such as an algebraic identity or a formula of the functional calculus.
3. A certain procedure terminates for certain initial data, or for all initial data.
4. A certain translation procedure correctly translates procedures between one programming language and another.
5. One procedure is more efficient than another equivalent procedure in the sense of taking fewer steps or requiring less memory.
6. A certain transformation of programs preserves the function expressed but increases the efficiency.
7. A certain class of problems is unsolvable by any procedure, or requires procedures of a certain type for its solution.

4. WHAT ARE THE AXIOMS AND RULES OF INFERENCE OF A MATHEMATICAL SCIENCE OF COMPUTATION?

Ideally we would like a mathematical theory in which every true statement about procedures would have a proof, and preferably a proof that is easy to find, not too long, and easy to check. In 1931, Gödel proved a result, one of whose immediate consequences is that there is no complete mathematical theory of computation. Given any mathematical theory of computation there are true statements expressible in it which do not have proofs. Nevertheless, we can hope for a theory which is adequate for practical purposes, like proving that compilers work; the unprovable statements tend to be of a rather sweeping character, such as that the system itself is consistent.

It is almost possible to take over one of the systems of elementary number theory such as that given in Mostowski's book *Sentences Undecidable in Formalized Arithmetic* since the content of a theory of computation is quite similar. Unfortunately, this and similar systems were designed to make it easy to prove meta-theorems about the system, rather than to prove theorems in the system. As a result, the integers are given such a special role that the proofs of quite easy statements about simple procedures would be extremely long.

Therefore it is necessary to construct a new, though similar, theory in which neither the integers nor any other domain, (e.g. strings of symbols) are given a special role. Some partial results in this direction are described in this paper. Namely, an integer-free formalism for describing computations has been developed and shown to be adequate in the cases where it can be compared with other formalisms. Some methods of proof have been developed, but there is still a gap when it comes to methods of proving that a procedure terminates. The theory also requires extension in order to treat the properties of data spaces.

5. WHAT IMPORTANT RESULTS HAVE BEEN OBTAINED RELEVANT TO A MATHEMATICAL SCIENCE OF COMPUTATION?

In 1936 the notion of a computable function was clarified by Turing, and he showed the existence of universal computers that, with an appropriate program, could compute anything computed by any other computer. All our stored program computers, when provided with unlimited auxiliary storage, are universal in Turing's sense. In some subconscious sense even the sales departments of computer manufacturers are aware of this, and they do not advertise magic instructions that cannot be simulated on competitors machines, but only that their machines are faster, cheaper, have more memory, or are easier to program.

The second major result was the existence of classes of unsolvable problems. This keeps all but the most ignorant of us out of certain Quixotic enterprises such as trying to invent a debugging procedure that can infallibly tell if a program being examined will get into a loop.

Later in this paper we shall discuss the relevance of the results of mathematical logic on creative sets to the problem of whether it is possible for a machine to be as intelligent as a human. In my opinion it is very important to build a firmer bridge between logic and recursive function theory on the one side, and the practice of computation on the other.

Much of the work on the theory of finite automata has been motivated by the hope of applying it to computation. I think this hope is mostly in vain because the fact of finiteness is used to show that the automaton will eventually repeat a state. However, anyone who waits for an IBM 7090 to repeat a state, solely because it is a finite automaton, is in for a very long wait.

6. HOW CAN A MATHEMATICAL SCIENCE OF COMPUTATION HELP IN THE SOLUTION OF PRACTICAL PROBLEMS?

Naturally, the most important applications of a science cannot be foreseen when it is just beginning. However, the following applications can be foreseen.

1. At present, programming languages are constructed in a very unsystematic way. A number of proposed features are invented, and then we argue about whether each feature is worth its cost. A better understanding of the structure of computations and of data spaces will make it easier to see what features are really desirable.

2. It should be possible almost to eliminate debugging. Debugging is the testing of a program on cases one hopes are typical, until it seems to work. This hope is frequently vain.

Instead of debugging a program, one should prove that it meets its specifications, and this proof should be checked by a computer program. For this to be possible, formal systems are required in which it is easy to write proofs. There is a good prospect of doing this, because we can require the computer to do much more work in checking each step than a human is willing to do. Therefore, the steps can be bigger than with present formal systems. The prospects for this are discussed in [2].

7. USING CONDITIONAL EXPRESSIONS TO DEFINE FUNCTIONS RECURSIVELY

In ordinary mathematical language, there are certain tools for defining new functions in terms of old ones. These tools are composition and the identification of variables. As it happens, these tools are inadequate computable in terms of old ones. It is then customary to define all functions that can reasonably be regarded as to give a verbal definition. For example, the function $|x|$ is usually defined in words.

If we add a single formal tool, namely *conditional expressions* to our mathematical notation, and if we allow conditional expressions to be used recursively in a manner that I shall describe, we can define, in a completely formal way, all functions that can reasonably be regarded as computable in terms of an initial set. We will use the ALGOL 60 notation of conditional expressions in this paper.

We shall use conditional expressions in the simple form

$$\textbf{if } p \textbf{ then } a \textbf{ else } b$$

where p is a propositional expression whose value is **true** or **false**. The value of the conditional expression is a if p has the value **true** and b if p has the value **false**. When conditional expressions are used, the stock of predicates one has available for constructing ps is just as important as the stock of ordinary functions to be used in forming the as and bs by composition. The following are examples of functions defined by conditional expressions:

(i) $|x| = \textbf{if } x < 0 \textbf{ then } -x \textbf{ else } x$

(ii) $n! = \textbf{if } n = 0 \textbf{ then } 1 \textbf{ else } n \times (n-1)!$

(iii) $n! = g(n, 1)$
where
$g(n, s) = \textbf{if } n = 0 \textbf{ then } s \textbf{ else } g(n-1, n \times s)$

(iv) $n! = f(n, 0, 1)$
where
$f(n, m, p) = \textbf{if } m = n \textbf{ then } p \textbf{ else } f(n, m+1, (m+1) \times p)$

(v) $n^- = pred(n, 0)$
where
$pred(n, m) = \textbf{if } m' = n \textbf{ then } m \textbf{ else } pred(n, m')$

(vi) $m + n = \textbf{if } n = 0 \textbf{ then } m \textbf{ else } m' + n^-$

The first of these is the only non-recursive definition in the group. Next, we have three different procedures for computing $n!$; they can be shown to be equivalent by the methods to be described in this paper. Then we define the predecessor function n^- for positive integers ($3^- = 2$) in terms of the successor function n' ($2' = 3$). Finally, we define addition in terms of the successor, the predecessor and equality. In all of the definitions, except for the first, the domain of the variables is taken to be the set of non-negative integers.

As an example of the use of these definitions, we shall show how to compute 2! by the second definition of $n!$. We have

$$2! = g(2, 1)$$
$$= \textbf{if } 2 = 0 \textbf{ then } 1 \textbf{ else } g(2-1, 2 \times 1)$$
$$= g(1, 2)$$
$$= \textbf{if } 1 = 0 \textbf{ then } 2 \textbf{ else } g(1-1, 1 \times 2)$$

$$= g(0, 2)$$
$$= \text{if } 0 = 0 \text{ then } 2 \text{ else } g(0 - 1, 0 \times 2)$$
$$= 2$$

Note that if we attempted to compute $n!$ for any n but a non-negative integer the sucessive reductions would not terminate. In such cases we say that the computation does not converge. Some recursive functions converge for all values of their arguments, some converge for some values of the arguments, and some never converge. Functions that always converge are called *total* functions, and the others are called *partial* functions. One of the earliest major results of recursive function theory is that there is no formalism that gives all the computable total functions and no partial functions.

We have proved [1] that the above method of defining computable functions, starting from just the successor function n' and equality, yields all the computable functions of integers. This leads us to assert that we have the complete set of tools for defining functions which are computable in terms of given base functions.

If we examine the next to last line of our computation of 2! we see that we cannot simply regard the conditional expression

if p **then** a **else** b

as a function of the three quantities p, a, and b. If we did so, we would be obliged to compute $g(-1, 0)$ before evaluating the conditional expression, and this computation would not converge. What must happen is that when p is true we take a as the value of the conditional expression without even looking at b.

Any reference to recursive functions in the rest of this paper refers to functions defined by the above methods.

8. PROVING STATEMENTS ABOUT RECURSIVE FUNCTIONS

In the previous section we presented a formalism for describing functions which are computable in terms of given base functions. We would like to have a mathematical theory strong enough to admit proofs of those facts about computing procedures that one ordinarily needs to show that computer programs meet their specifications. In [1] we showed that our formalism for expressing computable functions, was strong enough so that all the partial recursive functions of integers could be obtained from the successor function and equality. Now we would like a theory strong enough so that the addition of some form of Peano's axioms would allow the proof of all the theorems of one of the forms of elementary number theory.

We do not yet have such a theory. The difficulty is to keep the axioms and rules of inference of the theory free of the integers or other special domain, just as we have succeeded in doing with the formalism for constructing functions.

We shall now list the tools we have so far for proving relations between recursive functions. They are discussed in more detail in [1].

1. Substitution of expressions for free variables in equations and other relations.

2. Replacement of a sub-expression in a relation by an expression which has been proved equal to the sub-expression. (This is known in elementary mathematics as substitution of equals for equals.) When we are dealing with conditional expressions, a stronger form of this rule than the usual one is valid. Suppose we have an expression of the form

if p **then** a **else** (**if** q **then** b **else** c)

and we wish to replace b by d. Under these circumstances we do not have to prove the equation $b = d$ in general, but only the weaker statement

$$\sim p \wedge q \supset b = d$$

This is because b affects the value of the conditional expression only in case $\sim p \wedge q$ is true.

3. Conditional expressions satisfy a number of identities, and a complete theory of conditional expressions, very similar to propositional calculus, is thoroughly treated in [1]. We shall list a few of the identities taken as axioms here.

(i)   (**if true then** a **else** b) $= a$

(ii)   (**if false then** a **else** b) $= b$

(iii)   **if** p **then** (**if** q **then** a **else** b) **else** (**if** q **then** c **else** d)
         $=$ **if** q **then** (**if** p **then** a **else** c) **else** (**if** p **then** b **else** d)

(iv) f (**if** p **then** a **else** b) $=$ **if** p **then** $f(a)$ **else** $f(b)$

4. Finally, we have a rule called *recursion induction* for making

arguments that in the usual formulations are made by mathematical induction. This rule may be regarded as taking one of the theorems of recursive function theory and giving it the status of a rule of inference. Recursion induction may be described as follows.

Suppose we have a recursion equation defining a function f, namely

$$(1) \qquad f(x_1, \ldots, x_n) = \varepsilon \{f, x_1, \ldots, x_n\}$$

where the right-hand side will be a conditional expression in all non-trivial cases, and suppose that

(i) the calculation of $f(x_1, \ldots, x_n)$ according to this rule converges for a certain set A of n-tuples,
(ii) the functions $g(x_1, \ldots, x_n)$ and $h(x_1, \ldots, x_n)$ each satisfy Equation (1) when g or h is substituted for f.

Then we may conclude that $g(x_1, \ldots, x_n) = h(x_1, \ldots, x_n)$ for all n-tuples $(x_1, \ldots, x_n)$ in the set A.

As an example of the use of recursion induction we shall prove that the function g defined by

$$g(n, s) = \textbf{if } n = 0 \textbf{ then } s \textbf{ else } g(n - 1, n \times s)$$

satisfies $g(n, s) = n! \times s$ given the usual facts about $n!$. We shall take for granted the fact that $g(n, s)$ converges for all non-negative integral n and s. Then we need only show that $n! \times s$ satisfies the equation for $g(n, s)$. We have

$$n! \times s = \textbf{if } n = 0 \textbf{ then } n! \times s \textbf{ else } n! \times s$$
$$\textbf{if } n = 0 \textbf{ then } s \textbf{ else } (n - 1)! \times (n \times s),$$

and this has the same form as the equation satisfied by $g(n, s)$. The steps in this derivation are fairly obvious but also follow from the abovementioned axioms and rules of inference. In particular, the extended rule of replacement is used. A number of additional examples of proof by recursion induction are given in [1], and it has been used to prove some fairly complicated relations between computable functions.

9. RECURSIVE FUNCTIONS, FLOW CHARTS, AND ALGOLIC PROGRAMS

In this section I want to establish a relation between the use of recursive functions to define computations, the flow chart notation, and programs expressed as sequences of ALGOL-type assignment statements, together

with conditional **go to**s. The latter we shall call *Algolic* programs with
the idea of later extending the notation and methods of proof to cover
more of the language of ALGOL. Remember that our purpose here is
not to create yet another programming language, but to give tools for
proving facts about programs.

In order to regard programs as recursive functions, we shall define
the *state vector* ξ of a program at a given time, to be the set of current
assignments of values to the variables of the program. In the case of a
machine language program, the state vector is the set of current
contents of those registers whose contents change during the course of
execution of the program.

When a block of program having a single entrance and exit is
executed, the effect of this execution on the state vector may be
described by a function $\xi' = r(\xi)$ giving the new state vector in terms
of the old. Consider a program described by the flow chart of Figure 1.

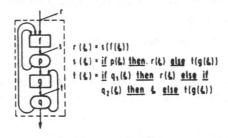

Fig. 1. Flow chart for equations shown.

The two inner computation blocks are described by functions $f(\xi)$
and $g(\xi)$, telling how these blocks affect the state vector. The decision
ovals are described by predicates $p(\xi)$, $q_1(\xi)$, and $q_2(\xi)$ that give the
conditions for taking the various paths. We wish to describe the
function $r(\xi)$ that gives the effect of the whole block, in terms of the
functions f and g and the predicates p, q_1, and q_2. This is done with the
help of two auxiliary functions s and t. $s(\xi)$ Describes the change in ξ
between the point labelled s in the chart and the final exit from the
block; t is defined similarly with respect to the point labelled t. We can
read the following equations directly from the flow chart:

$$r(\xi) = s(f(\xi))$$
$$s(\xi) = \text{if } p(\xi) \text{ then } r(\xi) \text{ else } t(g(\xi))$$
$$t(\xi) = \text{if } q_1(\xi) \text{ then } r(\xi) \text{ else if } q_2(\xi) \text{ then } \xi \text{ else } t(g(\xi))$$

In fact, these equations contain exactly the same information as does the flow chart.

Consider the function mentioned earlier, given by

$$g(n, s) = \text{if } n = 0 \text{ then } s \text{ else } g(n - 1, n \times s)$$

It corresponds, as we leave the reader to convince himself, to the Algolic program

$$a: \quad \text{if } n = 0 \text{ then go to } b;$$
$$s := n \times s;$$
$$n := n - 1;$$
$$\textbf{go to } a;$$
$$b:$$

Its flow chart is given in Figure 2.

This flow chart is described by the function $fac(\xi)$ and the equation

$$fac(\xi) = \text{if } p(\xi) \text{ then } \xi \text{ else } fac(g(f(\xi)))$$

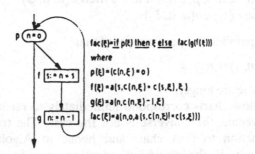

Fig. 2. Flow chart for $fac(\xi)$.

where $p(\xi)$ corresponds to the condition $n = 0$, $f(\xi)$ to the statement $n := n - 1$, and $g(\xi)$ to the statement $s := n \times s$.

We shall regard the state vector as having many components, but the only components affected by the present program are the s-component and the n-component. In order to compute explicitly with the state vectors, we introduce the function $c(var, \xi)$ which denotes the value assigned to the variable var in the state vector ξ, and we also introduce the function $a(var, val, \xi)$ which denotes the state vector that results when the value assigned to var in ξ is changed to val, and the values of the other variables are left unchanged.

The predicates and functions of state vectors mentioned above then become:

$$p(\xi) = (c(n, \xi) = 0)$$
$$g(\xi) = a(n, c(n, \xi) - 1, \xi)$$
$$f(\xi) = a(s, c(n, \xi) \times c(s, \xi), \xi)$$

We can prove by recursion induction that

$$fac(\xi) = a(n, 0, a(s, c(n, \xi)! \times c(s, \xi), \xi)),$$

but in order to do so we must use the following facts about the basic state vector functions:

(i) $c(u, a(v, \alpha, \xi)) = \text{if } u = v \text{ then } \alpha \text{ else } c(u, \xi)$

(ii) $a(v, c(v, \xi), \xi) = \xi$

(iii) $a(u, \alpha, a(v, \beta, \xi)) = \text{if } u = v \text{ then } a(u, \alpha, \xi)$
         $\text{else } a(v, \beta, a(u, \alpha, \xi))$

The proof parallels the previous proof that

$$g(n, s) = n! \times s,$$

but the formulae are longer.

While all flow charts correspond immediately to recursive functions of the state vector, the converse is not the case. The translation from recursive function to flow chart and hence to Algolic program is immediate, only if the recursion equations are in *iterative* form. Suppose we have a recursion equation

$$r(x_1, \ldots, x_n) = \mathscr{E}\{r, x_1, \ldots, x_n, f_1, \ldots, f_m\}$$

where $\mathscr{E}\{r, x_1, \ldots, x_n, f_1, \ldots, f_m\}$ is a conditional expression defining the function r in terms of the functions $f_1, \ldots, f_m$. $\mathscr{E}$ is said to be in iterative form if r never occurs as the argument of a function but only in terms of the main conditional expression in the form

$$\ldots \text{then } r(\ldots).$$

In the examples, all the recursive definitions except

$$n! = \textbf{if } n = 0 \textbf{ then } 1 \textbf{ else } n \times (n - 1)!$$

are in iterative form. In that one, $(n - 1)!$ occurs as a term of a product. Non-iterative recursive functions translate into Algolic programs that use themselves recursively as procedures. Numerical calculations can usually be transformed into iterative form, but computations with symbolic expressions usually cannot, unless quantities playing the role of push-down lists are introduced explicitly.

10. RECURSION INDUCTION ON ALGOLIC PROGRAMS

In this section we extend the principle of recursion induction so that it can be applied directly to Algolic programs without first transforming them to recursive functions. Consider the Algolic program

a: **if** $n = 0$ **then go to** b;
    $s := n \times s$;
    $n := n - 1$;
    go to a;
b:

We want to prove it equivalent to the program

a: $s := n! \times s$
    $n := 0$
b:

Before giving this proof we shall describe the principle of recursion induction as it applies to a simple class of flow charts. Suppose we have a block of program represented by Figure 3a. Suppose we have proved that this block is equivalent to the flow chart of Figure 3b where the shaded block is the original block again. This corresponds to the idea of a function satisfying a functional equation. We may now conclude that the block of Figure 3a is equivalent to the flow chart of Figure 3c for those state vectors for which the program does not get stuck in a loop in Figure 3c.

This can be seen as follows. Suppose that, for a particular input, the program of Figure 3c goes around the loop n times before it comes out. Consider the flow chart that results when the whole of Figure 3b is substituted for the shaded block, and then substituted into this figure

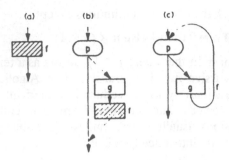

Fig. 3. Recursion induction applied to flow charts.

again, etc. for a total of n substitutions. The state vector in going through this flow chart will undergo exactly the same tests and changes as it does in going through Figure 3c, and therefore will come out in n steps without ever going through the shaded block. Therefore, it must come out with the same value as in Figure 3c. Therefore, for any vectors for which the computation of Figure 3c converges, Figure 3a is equivalent to Figure 3c.

Thus, in order to use this principle to prove the equivalence of two blocks we prove that they satisfy the same relation, of type 3a—3b, and that the corresponding 3c converges.

We shall apply this method to showing that the two programs mentioned above are equivalent. The first program may be transformed as follows:

from

a: **if** $n = 0$ **then go to** b; $s := n \times s$; $n := n - 1$;
   go to a; b:

to

a: **if** $n = 0$ **then go to** b; $s := n \times s$; $n := n - 1$;
   a_1: **if** $n = 0$ **then go to** b; $s := n \times s$; $n := n - 1$;
   go to a_1; b:

The operation used to transform the program is simply to write separately the first execution of a loop.

In the case of the second program we go:

from

$$a: s: = n! \times s; n: = 0; b:$$

to

$$a: \textbf{if } n = 0 \textbf{ then go to } c; s: = n! \times s; n: = 0;$$
$$\textbf{go to } b; c: s: = n! \times s; n: = 0; b:$$

to

$$a: \textbf{if } n = 0 \textbf{ then go to } c; s: = (n - 1)! \times (n \times s); n: = 0;$$
$$\textbf{go to } b; c:, b:$$

to

$$a: \textbf{if } n = 0 \textbf{ then go to } b; s: = n \times s; n: = n - 1; s:$$
$$= n! \times s; n: = 0; b:$$

The operations used here are: first, the introduction of a spurious branch, after which the same action is performed regardless of which way the branch goes; second, a simplification based on the fact that if the branch goes one way then $n = 0$, together with rewriting one of the right-hand sides of an assignment statement; and third, elimination of a label corresponding to a null statement, and what might be called an anti-simplification of a sequence of assignment statements. We have not yet worked out a set of formal rules justifying these steps, but the rules can be obtained from the rules given above for substitution, replacement and manipulation of conditional expressions.

The result of these transformations is to show that the two programs each satisfy a relation of the form: *program* is equivalent to

$$a: \textbf{if } n = 0 \textbf{ then go to } b; s: = n \times s; n: = 0; program.$$

The program corresponding to Figure 3c in this case, is precisely the first program which we shall assume always converges. This completes the proof.

11. THE DESCRIPTION OF PROGRAMMING LANGUAGES

Programming languages must be described syntactically and semantically. Here is the distinction. The syntactic description includes:

1. A description of the *morphology*, namely what symbolic expressions represent grammatical programs.
2. The rules for the analysis of a program into parts and its synthesis from the parts. Thus we must be able to recognize a term representing a sum and to extract from it the terms representing the summands.

The semantic description of the language must tell what the programs mean. The meaning of a program is its effect on the state vector in the case of a machine independent language, and its effect on the contents of memory in the case of a machine language program.

12. ABSTRACT SYNTAX OF PROGRAMMING LANGUAGES

The Backus normal form that is used in the ALGOL report, describes the morphology of ALGOL programs in a *synthetic* manner. Namely, it describes how the various kinds of program are built up from their parts. This would be better for translating into ALGOL than it is for the more usual problem of translating from ALGOL. The form of syntax we shall now describe differs from the Backus normal form in two ways. First, it is analytic rather than synthetic; it tells how to take a program apart, rather than how to put it together. Second, it is abstract in that it is independent of the notation used to represent, say sums, but only affirms that they can be recognized and taken apart.

Consider a fragment of the ALGOL arithmetic terms, including constants and variables and sums and products. As a further simplification we shall consider sums and products as having exactly two arguments, although the general case is not much more difficult. Suppose we have a term t. From the point of view of translating it we must first determine whether it represents a constant, a variable, a sum, or a product. Therefore we postulate the existence of four predicates: $isconst(t)$, $isvar(t)$, $issum(t)$ and $isprod(t)$. We shall suppose that each term satisfies exactly one of these predicates.

Consider the terms that are variables. A translator will have to be able to tell whether two symbols are occurrences of the same variable, so we need a predicate of equality on the space of variables. If the variables will have to be sorted an ordering relation is also required.

Consider the sums. Our translator must be able to pick out the summands, and therefore we postulate the existence of functions $addend(t)$ and $augend(t)$ defined on those terms that satisfy $issum(t)$.

Similarly, we have the functions $mplier(t)$ and $mpcand(t)$ defined on the products. Since the analysis of a term must terminate, the recursively defined predicate

$$term(t) = isvar(t) \lor isconst(t) \lor issum(t) \land term$$
$$(addend(t)) \land term(augend(t)) \lor isprod(t) \land$$
$$term(mplier(t)) \land term(mpcand(t))$$

must converge and have the value **true** for all terms.

The predicates and functions whose existence and relations define the syntax, are precisely those needed to translate from the language, or to define the semantics. That is why we need not care whether sums are represented by $a + b$, or $+ab$, or (PLUS A B), or even by Gödel numbers $7^a 11^b$.

It is useful to consider languages which have both an analytic and a synthetic syntax satisfying certain relations. Continuing our example of terms, the synthetic syntax is defined by two functions $mksum(t, u)$ and $mkprod(t, u)$ which, when given two terms t and u, yield their sum and product respectively. We shall call the syntax *regular* if the analytic and the synthetic syntax are related by the plausible relations:

(i) $issum(mksum(t, u))$ and $isprod(mkprod(t, u))$

(ii) $addend(mksum(t, u)) = t$; $mplier(mkprod(t, u)) = t$
       $augend(mksum(t, u)) = u$; $mpcand(mkprod(t, u)) = u$

(iii) $issum(t) \supset mksum(addend(t), augend(t)) = t$ and
       $isprod(t) \supset mkprod(mplier(t), mpcand(t)) = t$

Once the abstract syntax of a language has been decided, then one can choose the domain of symbolic expressions to be used. Then one can define the syntactic functions explicitly and satisfy oneself, preferably by proving it, that they satisfy the proper relations. If both the analytic and synthetic functions are given, we do not have the difficult and sometimes unsolvable analysis problems that arise when languages are described synthetically only.

In ALGOL the relations between the analytic and the synthetic functions are not quite regular, namely the relations hold only up to an equivalence. Thus, redundant parentheses are allowed, etc.

13. SEMANTICS

The analytic syntactic functions can be used to define the semantics of a programming language. In order to define the meaning of the arithmetic terms described above, we need two more functions of a semantic nature, one analytic and one synthetic. If the term t is a constant then $val(t)$ is the number which t denotes. If α is a number $mkconst(\alpha)$ is the symbolic expression denoting this number. We have the obvious relations

(i) $val(mkconst(\alpha)) = \alpha$

(ii) $isconst(mkconst(\alpha))$

(iii) $isconst(t) \supset mkconst(val(t)) = t$

Now we can define the meaning of terms by saying that the value of a term for a state vector ξ is given by

$$value(t, \xi) = \text{if } isvar(t) \text{ then } c(t, \xi) \text{ else if } isconst(t) \text{ then } val(t) \text{ else if } issum(t) \text{ then } value(addend(t), \xi) + value(augend(t), \xi) \text{ else if } isprod(t) \text{ then } value(mplier(t), \xi) \times value(mpcand(t), \xi)$$

We can go farther and describe the meaning of a program in a programming language as follows:

The meaning of a program is defined by its effect on the state vector.

Admittedly, this definition will have to be elaborated to take input and output into account.

In the case of ALGOL we should have a function $\xi' = algol(\pi, \xi)$ which gives the value ξ' of the state vector after the ALGOL program π has stopped, given that it was started as its beginning and that the state vector was initially ξ. We expect to publish elsewhere, a recursive description of the meaning of a small subset of ALGOL.

Translation rules can also be described formally. Namely,
1. A machine is described by a function

$$x' = machine(p, x)$$

giving the effect of operating the machine program p on a machine vector x.

2. An invertible representation $x = rep(\xi)$ of a state vector as a machine vector is given.

3. A function $p = trans(\pi)$ translating source programs into machine programs is given.

The correctness of the translation is defined by the equation

$$rep(algol\,(\pi,\,\xi)) = machine(trans(\pi),\,rep(\xi)).$$

It should be noted that $trans(\pi)$ is an abstract function and not a machine program. In order to describe compilers, we need another abstract invertible function, $u = repp(\pi)$, which gives a representation of the source program in the machine memory ready to be translated. A *compiler* is then a machine program such that $trans(\pi) = machine(compiler,\,repp(\pi))$.

14. THE LIMITATIONS OF MACHINES AND MEN AS PROBLEM-SOLVERS

Can a man make a computer program that is as intelligent as he is? The question has two parts. First, we ask whether it is possible in principle, in view of the mathematical results on undecidability and incompleteness. The second part is a question of the state of the programming art as it concerns artificial intelligence. Even if the answer to the first question is 'no', one can still try to go as far as possible in solving problems by machine.

My guess is that there is no such limitation in principle. However, a complete discussion involves the deeper parts of recursive function theory, and the necessary mathematics has not all been developed.

Consider the problem of deciding whether a sentence of the lower predicate calculus is a theorem. Many problems of actual mathematical or scientific interest can be formulated in this form, and this problem has been proved equivalent to many other problems including problem of determining whether a program on a computer with unlimited memory will ever stop. According to Post, this is equivalent to deciding whether an integer is a member of what Post calls a *creative* set. It was proved by Myhill that all creative sets are equivalent in a quite strong sense, so that there is really only one class of problems at this level of unsolvability.

Concerning this problem the following acts are known.

(1) There is a procedure which will do the following: If the number is in the creative set, the procedure will say 'yes', and if the number is not in the creative set the procedure will not say 'yes', it may say 'no' or it may run on indefinitely.

(2) There is no procedure which will always say 'yes' when the answer is 'yes', and will always say 'no' when the answer is 'no'. If a procedure has property (1) it must sometimes run on indefinitely. Thus there is no procedure which can always decide definitely whether a number is a member of a creative set, or equivalently, whether a sentence of the lower predicate calculus is a theorem, or whether an ALGOL program with given input will ever stop. This is the sense in which these problems are unsolvable.

(3) Now we come to Post's surprising result. We might try to do as well as possible. Namely, we can try to find a procedure that always says 'yes' when the answer is 'yes', and never says 'yes' when the answer is 'no', and which says 'no' for as large a class of the negative cases as possible, thus narrowing down the set of cases where the procedure goes on indefinitely as much as we can. *Post showed that one cannot even do as well as possible.* Namely, Post gave a procedure which, when given any other partial decision procedure, would give a better one. The better procedure decides all the cases the first procedure decided, and an infinite class of new ones. At first sight this seems to suggest that Emil Post was more intelligent than any possible machine, although Post himself modestly refrained from drawing this conclusion. Whatever program you propose, Post can do better. However, this is unfair to the programs. Namely, Post's improvement procedure is itself mechanical and can be carried out by machine, so that the machine can improve its own procedure or can even improve Post, if given a description of Post's own methods of trying to decide sentences of the lower predicate calculus. It is like a contest to see who can name the largest number. The fact that I can add one to any number you give, does not prove that I am better than you are.

(4) However, the situation is more complicated than this. Obviously, the improvement process may be carried out any finite number of times and a little thought shows that the process can be carried out an infinite

number of times. Namely, let p be the original procedure, and let Post's improvement procedure be called P, then $P^n p$ represents the original procedure improved n times. We can define $P^\omega p$ as a procedure that applies p for a while, then Pp for a while, then p again, then Pp, then $P^2 p$, then p, then Pp, then $P^2 p$, then $P^3 p$, etc. This procedure will decide any problem that any P^n will decide, and so may justifiably be called $P^\omega p$. However, $P^\omega p$ is itself subject to Post's original improvement process and the result may be called $P^{\omega+1} p$. How far can this go? The answer is technical. Namely, given any recursive transfinite ordinal α, one can define $P^\alpha p$. A recursive ordinal is a recursive ordering of the integers that is a well-ordering in the sense that any subset of the integers has a least member in the sense of the ordering. Thus, we have a contest in trying to name the largest recursive ordinal. Here we seem to be stuck, because the limit of the recursive ordinals is not recursive. However, this does not exclude the possibility that there is a different improvement process Q, such that Qp is better than $P^\alpha p$ for any recursive ordinal α.

(5) There is yet another complication. Suppose someone names what he claims is a large recursive ordinal. We, or our program, can name a larger one by adding one, but how do we know that the procedure that he claims is a recursive well-ordering, really is? He says he has proved it in some system of arithmetic. In the first place we may not be confident that his system of arithmetic is correct or even consistent. But even if the system is correct, by Gödel's theorem it is incomplete. In particular, there are recursive ordinals that cannot be proved to be so in that system. Rosenbloom, in his book *Elements of Mathematical Logic*, drew the conclusion that man was in principle superior to machine because, given any formal system of arithmetic, he could give a stronger system containing true and provable sentences that could not be proved in the original system. What Rosenbloom missed, presumably for sentimental reasons, is that the improvement procedure is itself mechanical, and subject to iteration through the recursive ordinals, so that we are back in the large ordinal race. Again we face the complication of proving that our large ordinals are really ordinals.

(6) These considerations have little practical significance in their present form. Namely, the original improvement process is such that the improved process is likely to be too slow to get results in a

reasonable time, whether carried out by man or by machine. It may be possible, however, to put this hierarchy in some usable form.

In conclusion, let me re-assert that the question of whether there are limitations in principle of what problems man can make machines solve for him as compared to his own ability to solve problems, really is a technical question in recursive function theory.

Computation Center,
Stanford University, California, U.S.A.

REFERENCES

1. McCarthy, J.: 1963, 'A Basis for a Mathematical Theory of Computation', in: Braffort, P. and Hirshberg, D. (eds.), *Computer Programming and Formal Systems*, North-Holland, Amsterdam.
2. McCarthy, J.: 1962, 'Checking Mathematical Proofs by Computer', *Proc. Symp. on Recursive Function Theory* (1961). Amer. Math. Soc.

PETER NAUR

PROOF OF ALGORITHMS BY GENERAL SNAPSHOTS

INTRODUCTION

It is a deplorable consequence of the lack of influence of mathematical thinking on the way in which computer programming is currently being pursued, that the regular use of systematic proof procedures, or even the realization that such proof procedures exist, is unknown to the large majority of programmers. Undoubtedly, this fact accounts for at least a large share of the unreliability and the attendant lack of over-all effectiveness of programs as they are used to-day.

Historically this state of affairs is easily explained. Large scale computer programming started so recently that all of its practitioners are, in fact, amateurs. At the same time the modern computers are so effective that they offer advantages in use even when their powers are largely wasted. The stress has been on always larger, and, allegedly, more powerful systems, in spite of the fact that the available programmer competence often is unable to cope with their complexities.

However, a reaction is bound to come. We cannot indefinitely continue to build on sand. When this is realized there will be an increased interest in the less glamorous, but more solid, basic principles. This will go in parallel with the introduction of these principles in the elementary school curricula. One subject which will then come up for attention is that of proving the correctness of algorithms. The purpose of the present article is to show in an elementary way that this subject not only exists, but is ripe to be used in practise. The illustrations are phrased in ALGOL 60, but the technique may be used with any programming language.

THE MEANING OF PROOF

The concept of proof originated in mathematics where we prove that a theorem, i.e. some statement about the properties of certain concepts, is logically contained in, or follows from, certain given axioms. A proof

Timothy R. Colburn et al. (eds.), Program Verification, 57—64.
© 1993 *Kluwer Academic Publishers*.

always involves two things, e.g. a theorem and a set of axioms, which in consequence of the proof are shown to be consistent.

In data processing we need proofs to relate the transformation defined by an algorithm to a description of the transformation in some other terms, usually a description of the static properties of the result of the transformation. The need for a proof arises because we want to convince ourselves that the algorithm we have written is correct. Ideally our way of proceeding is as follows. We first have the description of the desired result in terms of static properties. We then proceed to construct an algorithm for calculating that result, using examples and intuition to guide us. Having constructed the algorithm we want to prove that it does indeed produce a result having the desired properties.

As a very simple example, let the problem be to find the greatest of N given numbers. The numbers may be given as an **array** $A[1 : N]$. The result, R, may be described statistically as being the value of such a member of the array A, $A[r]$ say, that it holds for any i between 1 and N that $A[i] \leq A[r]$. Note that this description is static. It describes the desired result as being greater than or equal to any other element of the array, not by way of a process showing how to find it. Note also that there is no guarantee implied that such a result exists at all. Proving in general that objects having certain properties exist is an important mathematical pursuit. One way of expressing such a proof is simply to provide an algorithm which finds the object. For our purposes this is often fully sufficient, so normally we do not have to worry about any other existence proof.

We now enter into the construction of an algorithm for finding the greatest number. We may be more or less clever in doing this. A good idea, which may or may not occur to us, is to run through the numbers once, keeping at all stages track of the greatest number found so far. We thus arrive at something like the following algorithm, which incidentally still can be made more effective:

PROGRAM 1. Greatest number.

```
r: = 1;
for i: = 2 step 1 until N do
    if A[i] > A[r] then r: = i;
R: = A[r];
```

This looks pretty good to me. However, can I prove that it is correct?

Many people, including many experienced programmers, will tend to dismiss the question, claiming that it is obvious that the solution is correct. I wish to oppose this view. Although the algorithm is not very complicated, it does contain some 7 operators and 12 operands. This is more than what can be grasped immediately by anyone, and I am sure that even the experienced programmer, in studying the algorithm, makes use of mental images and a certain decomposition of the process before he accepts it. I wish to claim that he goes through a proof, and I want this to be brought into the open, using a technique which can be used also in more complicated cases.

SNAPSHOTS

Our proof problem is one of relating a static description of a result to a dynamic description of a way to obtain the result. Basically there are two ways of bringing the two descriptions closer together, either we may try to make the static description more dynamic, with a hope of getting to the given algorithm, or we may try to make the dynamic description more static. Of these the second is clearly preferable because we have far more experience in manipulating static descriptions, through practise in dealing with mathematical formulae. Therefore, if only we can derive some static description from the dynamic one, there is good hope that we may manipulate it so as to show that it is identical with the given static description.

	r	i	R	N	$A[1]$	$A[2]$	$A[3]$	$A[4]$
Initial				4	2	1	5	2
Following $r:=1$	1							
— for $i:=2$		2						
Second time in loop		3						
Following $r:=i$	3							
Third time in loop		4						
Following $R:=A[r]$			5					

We are thus led to the use of snapshots of the dynamic process for purposes of proof, because a snapshot is an instantaneous, and therefore static, picture of the development of the process. In its most primitive form a snapshot refers to one particular moment in the development of the process applied to one particular set of data. In a slightly more developed use we give a series of snapshots, referring to successive moments in the process, but still as applied to one particular set of data. To illustrate this technique, every detail of an example of the use of Program 1 is given above. Successive snapshots are given in successive lines, where for clarity a value given in a column holds unchanged in following lines unless another value is given.

This snapshot technique is quite useful as an aid to understanding a given algorithm. However, it is not a proof technique because it depends entirely on the choice of the data set. In order to achieve a proof we shall need more general snapshots.

GENERAL SNAPSHOTS

By a General Snapshot I shall mean a snapshot of a dynamic process which is associated with one particular point in the actual program text, and which is valid every time that point is reached in the execution of the process.

From this definition it immediately follows that the values of variables given in a General Snapshot normally at best can be expressed as general, mathematical expressions or by equivalent formulations. I have to say "at best" because in many cases we can only give certain limits on the value, and I have to admit "equivalent formulations" because we do not always have suitable mathematical notation available.

In order to illustrate this notion, here is a version of the above algorithm expanded with General Snapshots at six different points:

PROGRAM 2. Greatest number, with snapshots.

```
comment General Snapshot 1: 1 ≤ N;
r: = 1;
comment General Snapshot 2: 1 ≤ N, r = 1;
for i: = 2 step 1 until N do
    begin comment General Snapshot 3: 2 ≤ i ≤ N,
    1 ≤ r ≤ i − 1,
    A[r] is the greatest among the elements A[1], A[2], . . . ,
    A[i − 1];
    if A[i] > A[r] then r: = i;
    comment General Snapshot 4: 2 ≤ i ≤ N, 1 ≤ r ≤ i,
    A[r] is the greatest among the elements A[1], A[2], . . . ,
    A[i];
    end;
```

> **comment** *General Snapshot* 5: $1 \leqq r \leqq N$, $A[r]$ *is the greatest among the elements* $A[1], A[2], \ldots, A[N]$;
> $R := A[r]$;
> **comment** *General Snapshot* 6: R *is the greatest value of any element,* $A[1], A[2], \ldots, A[N]$;

In the first instance the reader may feel that these snapshots are as unsystematic as the original algorithm. The reason for this is that I have only given the final result, not the steps leading to it. The best way to see how they have arisen is to start with number 3. At this stage we are primarily interested in the values of i and r. The condition on i, $2 \leqq i \leqq N$, seems a reasonable first guess in view of the preceding for clause. The guess on r, $1 \leqq r \leqq i - 1$, expresses that we have so far scanned the elements up to $A[i - 1]$, and the explanation of $A[r]$ hopefully will seem a sensible claim. With snapshot 3 as our initial hypothesis we can now execute one step of the algorithm, to snapshot 4. The step is conditional, so we have to consider the two cases separately.

First, suppose that $A[i] > A[r]$ is true. Then clearly $A[i]$ is the greatest among the elements up to $A[i]$. Changing r to i as done in the assignment then makes it again true to say that $A[r]$ is the greatest. In this case the previous condition on r becomes false and has to be changed to $1 \leqq r \leqq i$.

Second, suppose that $A[i] > A[r]$ is false. Then $A[r]$ remains the greatest element so far found. The extended range for r is also valid. Altogether we have seen that if snapshot 3 is correct, then snapshot 4 follows.

Continuing now from snapshot 4 we must be quite clear on the meaning of the for clause governing the execution, from which it follows that the situations described in snapshot 4 must be divided into two cases:

(1) $2 \leqq i \leqq N - 1$, leading to stepping on of i and a repetition at snapshot 3, and (2) $i = N$, continuing the execution beyond the following **end**. In case (1) the stepping on of i leads to a situation which may be derived from snapshot 4, case (1), by replacing all occurrences of i by $i - 1$. The point is that what is now denoted i is unity larger than that value of i for which the snapshot is valid. By this substitution we arrive at a situation in which we have $2 \leqq i - 1 \leqq N - 1$, or equivalently $3 \leqq i \leqq N$, and otherwise precisely the situation of snapshot 3. We have thus proved that the hypothetical snapshot 3 is

consistent with the action taking place in the for loop. To complete the proof of the correctness of snapshot 3 we still have to check the first entry into the loop, i.e. the transition from snapshot 2 to 3. Again remembering the definition of the for clause, we only have to consider the subcases of snapshot 2 for which $2 \leq N$. These subcases clearly lead to a situation consistent with snapshot 3 and moreover show that the cases $2 = i$ and $1 = r$ are realized, a fact which does not follow from the transition from snapshot 4.

The situation of snapshot 5 can be entered either from snapshot 2 in case $1 = N$, or from snapshot 4 in case $i = N$ with $2 \leq N$. Both of these entries combine to form the situation described in snapshot 5, which is thereby proved. It should be carefully noted that snapshot 5 requires $1 \leq N$ to make sense. This condition is therefore necessary both in snapshots 2 and 1. Appearing before the complete algorithm it becomes a condition on the proper use of it. This is emphasized in the final transition from snapshot 5 to 6 where the variable $A[r]$ occurs. If the array is declared as **array** $A[1 : N]$, this variable is defined only if $1 \leq N$.

The proof of snapshot 6 establishes the correctness of the algorithm.

EXHAUSTIVE ENUMERATION OF ENTRY AND EXIT CONDITIONS

One of the most common mistakes in programming is that special cases, needing special treatment in the program, are overlooked. The General Snapshots offer obvious help in this matter. As an illustration, consider the situation of General Snapshot 4. The situation is entered only by one route, and we have

Entry condition: $2 \leq i \leq N$

However, the effect of the for clause is to divide the further process into two branches:

Exist condition: (1) $\quad 2 \leq i \leq N - 1,$ (2) $\quad i = N$

To check that no cases have been overlooked we have to see that the union of the exist conditions exhaust the entry condition.

More generally a situation may be entered by many different routes. General Snapshot 3 shows a situation having two entries. Similarly a situation may continue along many different routes. To check for the completeness we must in any case establish an *exhaustive enumeration* of the entry conditions and see that it is matched perfectly in the exit.

EXTERNAL PARAMETERS OF A PROCESS

When describing a General Snapshot it is important to realize that all variables have their current values, which are generally not the same from one snapshot to another. This may throw doubt on the real significance of a result derived by means of a proof. As an example, we may be concerned whether in General Snapshot 6 above the values of the elements $A[1]$, $A[2]$, ..., $A[N]$ are in fact the ones provided when the algorithm was entered. In order to remove such doubts we may add further information to the snapshots, relating the values of variables in different snapshots to one another. Thus we may add in each of the snapshots 2 to 6 in Program 3.2 the claim:

> **comment** *The values of N, $A[1]$, $A[2]$, ..., $A[N]$, are the same as when General Snapshot 1 was last encountered*;

A statement of this kind must be proved along with all other claims made in a General Snapshot.

This technique particularly helps in describing parameters of subprocesses designed to be joined together to more complicated programs.

CONSTRUCTION OF ALGORITHMS

The principal difficulty in the construction of algorithms lies with the loops. A process written as part of a loop must so to speak bite its own tail properly. At the same time most of the execution time of normal programs is spent in a loop, often the innermost one of a nest of loops. When constructing algorithms we should therefore start by establishing the loop structure and then attack each nest of loops by starting with the innermost one. While doing this we must decide on the most convenient representation of the data processed.

When writing the actual statements of loops, General Snapshots may be used as aids to construction, as follows: Consider the method of algorithmic solution which you intend to use. Imagine that you are right in the middle of using this method, in other words that the algorithm is going, but that nothing has been completed yet. Now describe the situation, in terms of a General Snapshot. Introduce any variables you need to make this description specific. Write down the limits on your variables and other descriptions of their current values. When you have done this, write down the essential, central part of your process and convince yourself that the picture you have made is in fact adequate as a basis for doing the central process. Then write the proper preparation of the process. In particular, for every single variable used make sure that it is (1) declared, (2) initialized, (3) changed, and (4) used.

A/S Regnecentralen, Copenhagen, Denmark

BIBLIOGRAPHY

The basic question of proof has so far been ignored in data processing to an incredible degree. A review of recent work is given in:

McCarthy, J.: 1965, 'Problems in the Theory of Computation', *Proc. IFIP Congress 65*, Vol. 1, pp. 219—222.

Most of the work described there is heavily oriented towards the basic theoretical problems of computation. The General Snapshots of the present chapter are related to the state vectors of J. McCarthy:

McCarthy, J.: 1964, 'A Formal Description of a Subset of ALGOL', *Proc. of a Conference on Formal Language Description Languages*, Vienna.

However, the present approach, which is directly applicable in proving and constructing practical programs, is believed to be new. Similar concepts have been developed independently by Robert W. Floyd (unpublished paper, communicated privately).

ROBERT W. FLOYD

ASSIGNING MEANINGS TO PROGRAMS[1]

INTRODUCTION

This paper attempts to provide an adequate basis for formal definitions
of the meanings of programs in appropriately defined programming
languages, in such a way that a rigorous standard is established for
proofs about computer programs, including proofs of correctness,
equivalence, and termination. The basis of our approach is the notion
of an interpretation of a program: that is, an association of a proposition with each connection in the flow of control through a program,
where the proposition is asserted to hold whenever that connection is
taken. To prevent an interpretation from being chosen arbitrarily, a
condition is imposed on each command of the program. This condition
guarantees that whenever a command is reached by way of a connection whose associated proposition is then true, it will be left (if at all) by
a connection whose associated proposition will be true at that time.
Then by induction on the number of commands executed, one sees that
if a program is entered by a connection whose associated proposition is
then true, it will be left (if at all) by a connection whose associated
proposition will be true at that time. By this means, we may prove
certain properties of programs, particularly properties of the form: 'If
the initial values of the program variables satisfy the relation R_1, the
final values on completion will satisfy the relation R_2'. Proofs of
termination are dealt with by showing that each step of a program
decreases some entity which cannot decrease indefinitely.

These modes of proof of correctness and termination are not
original; they are based on ideas of Perlis and Gorn, and may have
made their earliest appearance in an unpublished paper by Gorn. The
establishment of formal standards for proofs about programs in languages which admit assignments, transfer of control, etc., and the
proposal that the semantics of a programming language may be defined
independently of all processors for that language, by establishing
standards of rigor for proofs about programs in the language, appear to

65

Timothy R. Colburn et al. (eds.), Program Verification, 65–81.
© 1993 *Kluwer Academic Publishers.*

be novel, although McCarthy [1, 2] has done similar work for programming languages based on evaluation of recursive functions.

A semantic definition of a programming language, in our approach, is founded on a syntactic definition. It must specify which of the phrases in a syntactically correct program represent commands, and what conditions must be imposed on an interpretation in the neighborhood of each command.

We will demonstrate these notions, first on a flowchart language, then on fragments on ALGOL.

DEFINITIONS. A *flowchart* will be loosely defined as a directed graph with a command at each vertex, connected by *edges* (arrows) representing the possible passages of control between the commands. An edge is said to be an *entrance* to (or an *exit* from) the command c at vertex v if its destination (or origin) is v. An *interpretation I* of a flowchart is a mapping of its edges on propositions. Some, but not necessarily all, of the free variables of these propositions may be variables manipulated by the program. Figure 1 gives an example of an interpretation. For any edge e, the associated proposition $I(e)$ will be called the *tag* of e. If e is an entrance (or an exit) of a command c, $I(e)$ is said to be an *antecedent* (or a *consequent*) of c.

For any command c with k entrances and l exits, we will designate the entrances to c by $a_1, a_2, \ldots, a_k$, and the exits by $b_1, b_2, \ldots, b_l$. We will designate the tag of a_i by P_i ($1 \leqq i \leqq k$), and that of b_i by Q_i ($1 \leqq i \leqq l$). Boldface letters will designate vectors formed in the natural way from the entities designated by the corresponding nonboldface letters: for cxample, $\mathbf{P}$ represents $(P_1, P_2, \ldots, P_k)$.

A *verification* of an interpretation of a flowchart is a proof that for every command c of the flowchart, if control should enter the command by an entrance a_i with P_i true, then control must leave the command, if at all, by an exit b_j with Q_j true. A *semantic definition* of a particular set of command types, then, is a rule for constructing, for any command c of one of these types, a *verification condition* $V_c(\mathbf{P}; \mathbf{Q})$ on the antecedents and consequents of c. This verification condition must be so constructed that a proof that the verification condition is satisfied for the antecedents and consequents of each command in a flowchart is a verification of the interpreted flowchart. That is, if the verification condition is satisfied, and if the tag of the entrance is true when the

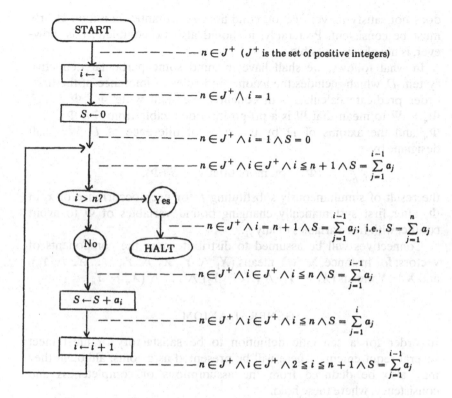

$$n \in J^+ \quad (J^+ \text{ is the set of positive integers})$$

$$n \in J^+ \wedge i = 1$$

$$n \in J^+ \wedge i = 1 \wedge S = 0$$

$$n \in J^+ \wedge i \in J^+ \wedge i \leq n + 1 \wedge S = \sum_{j=1}^{i-1} a_j$$

$$n \in J^+ \wedge i = n + 1 \wedge S = \sum_{j=1}^{i-1} a_j; \text{ i.e., } S = \sum_{j=1}^{n} a_j$$

$$n \in J^+ \wedge i \in J^+ \wedge i \leq n \wedge S = \sum_{j=1}^{i-1} a_j$$

$$n \in J^+ \wedge i \in J^+ \wedge i \leq n \wedge S = \sum_{j=1}^{i} a_j$$

$$n \in J^+ \wedge i \in J^+ \wedge 2 \leq i \leq n + 1 \wedge S = \sum_{j=1}^{i-1} a_j$$

Fig. 1. Flowchart of program to compute $S = \sum_{j=1}^{n} a_j$ $(n \geq 0)$.

statement is entered, the tag of the exit selected will be true after execution of the statement.

A *counterexample* to a particular interpretation of a single command is an assignment of values (e.g., numbers in most programming languages) to the free variables of the interpretation, and a choice of entrance, such that on entry to the command, the tag of the entrance is true, but on exit, the tag of the exit is false for the (possibly altered) values of the free variables. A semantic definition is *consistent* if there is no counterexample to any interpretation of any command which satisfies its verification condition. A semantic definition is *complete* if there is a counterexample to any interpretation of any command which does not satisfy its verification condition. A semantic definition clearly must be consistent. Preferably, it should also be complete; this, however, is not always possible.

In what follows, we shall have in mind some particular deductive system D, which includes the axioms and rules of inference of the first-order predicate calculus, with equality. We shall write $\Phi_1, \Phi_2, \ldots,$ $\Phi_n \vdash \Psi$ to mean that Ψ is a proposition deducible from $\Phi_1, \Phi_2, \ldots,$ Φ_n and the axioms of D by the rules of inference of D. We shall designate by

$$S_{f_1, f_2, \ldots, f_n}^{x_1, x_2, \ldots, x_n}(\Phi) \quad \text{or, more briefly,} \quad S_f^x(\Phi),$$

the result of simultaneously substituting f_i for each occurrence of x_i in Φ, after first systematically changing bound variables of Φ to avoid conflict with free variables of any f_i.

Connectives will be assumed to distribute over the components of vectors; for instance, $\mathbf{X} \wedge \mathbf{Y}$ means $(X_1 \wedge Y_1, X_2 \wedge Y_2, \ldots, X_n \wedge Y_n)$, and $\mathbf{X} \vdash \mathbf{Y}$ means $(X_1 \vdash Y_1) \wedge (X_2 \vdash Y_2) \wedge \ldots \wedge (X_n \vdash Y_n)$.

GENERAL AXIOMS

In order for a semantic definition to be satisfactory, it must meet several requirements. These will be presented as axioms, although they may also be deduced from the assumptions of completeness and consistency, where these hold.

If $V_c(\mathbf{P}; \mathbf{Q})$ and $V_c(\mathbf{P}'; \mathbf{Q}')$, then:

AXIOM 1. $V_c(\mathbf{P} \wedge \mathbf{P}'; \mathbf{Q} \wedge \mathbf{Q}')$;

AXIOM 2. $V_c(\mathbf{P} \vee \mathbf{P}'; \mathbf{Q} \vee \mathbf{Q}')$;

AXIOM 3. $V_c((\exists x)(\mathbf{P}); (\exists x)(\mathbf{Q}))$.

Also,

AXIOM 4. If $V_c(\mathbf{P}; \mathbf{Q})$ and $\mathbf{R} \vdash \mathbf{P}, \mathbf{Q} \vdash \mathbf{S}$, then $V_c(\mathbf{R}; \mathbf{S})$.

COROLLARY 1. *If $V_c(\mathbf{P}; \mathbf{Q})$ and* $\vdash (\mathbf{P} \equiv \mathbf{R})$, $\vdash (\mathbf{Q} \equiv \mathbf{S})$, *then $V_c(\mathbf{R}; \mathbf{S})$.*

Axiom 1, for example, essentially asserts that if whenever P is true on entering command c, Q is true on exit, and whenever P' is true on entry, Q' is true on exit, then whenever both P and P' are true on entrance, both Q and Q' are true on exit. Thus Axiom 1 shows that if separate proofs exist that a program has certain properties, then these proofs may be combined into one by forming the conjunction of the several tags for each edge. Axiom 2 is useful for combining the results of a case analysis, for instance, treating several ranges of initial values of certain variables. Axiom 3 asserts that if knowing that the value of the variable x has property P before executing a command assures that the (possibly altered) value will have property Q after executing the command, then knowing that a value exists having property P before execution assures that a value exists having property Q after execution. Axiom 4 asserts that if P and Q are verifiable as antecedent and consequent for a command, then so are any stronger antecedent and weaker consequent.

To indicate how these axioms are deducible from the hypotheses of completeness and consistency for V_c, consider Axiom 1 as an example. Suppose $V_c(\mathbf{P}; \mathbf{Q})$ and $V_c(\mathbf{P}'; \mathbf{Q}')$. Consider any assignment of initial values $\mathbf{V}$ to the free variables $\mathbf{X}$ of the interpretation such that P_i is true (that is, $\vdash S_{\mathbf{V}}^{\mathbf{x}}(P_i)$) and P_i' is true. Then, if the statement is entered by a_i, the exit chosen will be some b_j such that Q_j is true at that time (that is, $\vdash S_{\mathbf{w}}^{\mathbf{x}}(Q_j)$, where $\mathbf{W}$ is the vector of final values of $\mathbf{X}$ after execution of c), and Q_j' is also true, by the assumption of consistency. Thus, there can be no counterexample to the interpretation $I(\mathbf{a}) = (\mathbf{P} \wedge \mathbf{P}')$, $I(\mathbf{b}) = (\mathbf{Q} \wedge \mathbf{Q}')$, and by the assumption of completeness, $V_c(\mathbf{P} \wedge \mathbf{P}'; \mathbf{Q} \wedge \mathbf{Q}')$.

A FLOWCHART LANGUAGE

To make these notions more specific, consider a particular flowchart language with five statement types, represented pictorially as in Figure 2, having the usual interpretations as an assignment operation, a conditional branch, a join of control, a starting point for the program, and a halt for the program.

Take specifically the assignment operator $x \leftarrow f(x, \mathbf{y})$, where x is a variable and f is an expression which may contain occurrences of x and of the vector $\mathbf{y}$ of other program variables. Considering the effect of the command, it is clearly desirable that if P_1 is $(x = x_0 \wedge R)$, and Q_1 is $(x = f(x_0, \mathbf{y}) \wedge R)$, where R contains no free occurrences of x, then $V_c(P_1; Q_1)$. Applying the axioms, we shall establish a definition of $V_{x \leftarrow f(x, \mathbf{y})}$ which is complete and consistent if the underlying deductive

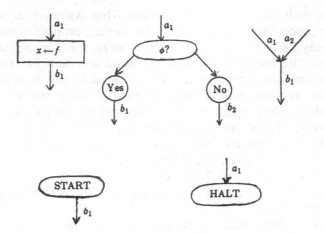

Fig. 2.

system is, and which is, in that sense, the most general semantic definition of the assignment operator.

Designating the command $x \leftarrow f(x, y)$ by c, we apply Axiom 3 to $V_c(P_1, Q_1)$, to obtain

$$V_c((\exists x_0)P_1; (\exists x_0)Q_1).$$

Because $[(\exists x) (x = e \wedge P(x))] \equiv P(e)$, provided x does not occur free in e, we apply Corollary 1, to get $V_c(R(x, y); (\exists x_0) (x = f(x_0, y) \wedge R(x_0, y)))$.

Finally, by Corollary 1, we have

The Verification Condition for Assignment Operators

(1)   If P_1 has the form $R(x, y)$ and if $(\exists x_0) (x = f(x_0, y) \wedge R(x_0, y)) \vdash Q_1$, then $V_{x \leftarrow f(x, y)}(P_1, Q_1)$.

Taking this as the semantic definition of $x \leftarrow f(x, y)$, and assuming the completeness and consistency of the deductive system D, we show that the semantic definition is complete and consistent.

To show consistency, assume that x_1 and y_1 are initial values of x and y such that $\vdash R(x_1, y_1)$. Then after execution of $x \leftarrow f(x, y)$, the values x_2 and y_2 are such that $x = x_2 = f(x_1, y_1)$, $y = y_2 = y_1$; thus $x_2 = $

$f(x_1, y_2) \wedge R(x_1, y_2)$, or $(\exists x_0)\ (x_2 = f(x_0, y_2) \wedge R(x_0, y_2))$. Designating $(\exists x_0)\ (x = f(x_0, y) \wedge R(x_0, y))$ as $T_c(R(x, y))$, we have shown that upon exit from c, $S_{x_2 y_2}^{xy}(T_c(R(x, y)))$ is true. Now since $T_c(R(x, y)) \vdash Q$, we find $\vdash S_{x_2 y_2}^{xy}(Q)$, by the assumption of the consistency of D, so that V_c is consistent.

To show completeness, assume it false that $T_c(R(x, y)) \vdash Q$. Then, by the completeness of D, there is a set of values x_2 and y_2 for x and y such that $S_{x_2 y_2}^{xy}(T_c(R(x, y)))$ is true, but $S_{x_2 y_2}^{xy}(Q)$ is false. Thus, $(\exists x_0)\ (x_2 = f(x_0, y_2) \wedge R(x_0, y_2))$. Let x_1 be a particular value of x_0 for which $x_2 = f(x_1, y_2) \wedge R(x_1, y_2)$. Now using x_1 and y_2 as initial values for x and y, we may generate a counterexample to the interpretation $I(a_1) = R(x, y), I(b_1) = Q$.

Thus we have shown that V_c is complete (consistent) if D is complete (consistent). By consideration of vacuous statements such as $x \leftarrow x$, we could change each 'if' to 'if and only if'. Thus, the semantic definition (1) we have given is the natural generalization of the original sufficient condition for verification; V_c is both necessary and sufficient.

The other command types of Figure 2 are more easily dealt with. For the branch command, $V_c(P_1; Q_1, Q_2)$ is $(P_1 \wedge \Phi \vdash Q_1) \wedge (P_1 \wedge \neg\Phi \vdash Q_2)$. For the join command, $V_c(P_1, P_2; Q_1)$ is $(P_1 \vee P_2 \vdash Q_1)$. For the start command the condition $V_c(Q_1)$, and for the halt command the condition $V_c(P_1)$ are identically true. All of these semantic definitions accord with the usual understanding of the meanings of these commands, and in each case V_c is complete and consistent if D is.

Using these semantic definitions, it is not hard to show that Figure 1 is a verifiable interpretation of its flowchart provided D contains a suitable set of axioms for the real numbers, summation, the integers, inequalities, and so forth. Thus, if the flowchart is entered with n a positive integer, the value of i on completion will be $n + 1$ (assuming that the program terminates) and the value of S will be $\Sigma_{j=1}^{n} a_j$. Presumably, the final value of i is of no interest, but the value of S is the desired result of the program, and the verification proves that the program does in fact compute the desired result if it terminates at all. Another section of this paper deals with proofs of termination.

Each of the given semantic definitions of the flowchart commands takes the form that $V_c(\mathbf{P}, \mathbf{Q})$ if and only if $(T_1(\mathbf{P}) \vdash Q_1) \wedge \ldots \wedge (T_l(\mathbf{P}) \vdash Q_l)$, where T_j is of the form $T_{j_1}(P_1) \vee T_{j_2}(P_2) \vee \ldots \vee T_{j_k}(P_k)$. In particular there is the following:

(1) For an assignment operator $x \leftarrow f$

$$T_1(P_1) \quad \text{is} \quad (\exists x_0)\,(x = S^x_{x_0}(f) \wedge S^x_{x_0}(P_1)).$$

(2) For a branch command

$$T_1(P_1) \quad \text{is} \quad P_1 \wedge \Phi,$$
$$T_2(P_1) \quad \text{is} \quad P_1 \wedge \neg \Phi.$$

(3) For a join command

$$T_1(P_1, P_2) \quad \text{is} \quad P_1 \vee P_2; \quad \text{that is,}$$
$$T_{11}(P_1) \quad \text{is} \quad P_1, \qquad T_{12}(P_2) \quad \text{is} \quad P_2.$$

(4) For a start command, $T_1(\ )$ is **false**.
    Thus, $V_c(Q_1)$ is identically true.
(5) For a halt command, the set of T_js and Q_js is empty.
    Thus $V_c(P_1)$ is identically true.
For any set of semantic definitions such that

$$V_c(\mathbf{P}, \mathbf{Q}) \equiv (T_1(\mathbf{P}) \vdash Q_1) \wedge \ldots \wedge T_l(\mathbf{P}) \vdash Q_l),$$

in any verifiable interpretation, it is possible to substitute $T_j(\mathbf{P})$ for Q_j as a tag for any particular exit of a command without loss of verifiability. It is obvious that this substitution satisfies the semantic definition of the command whose exit is b_j; since $\vdash (T_j(\mathbf{P}) \supset Q_j)$, by Axiom 4 the substitution satisfies the semantic definition of the command whose entrance is b_j, and there are no other commands whose verification condition involves $I(b_j)$.

It is, therefore, possible to extend a partially specified interpretation to a complete interpretation, without loss of verifiability, provided that initially there is no closed loop in the flowchart all of whose edges are not tagged and that there is no entrance which is not tagged. This fact offers the possibility of automatic verification of programs, the programmer merely tagging entrances and one edge in each innermost loop; the verifying program would extend the interpretation and verify it, if possible, by mechanical theorem-proving techniques.

We shall refer to $\mathbf{T}_c(\mathbf{P})$ as the *strongest verifiable consequent* of the command c, given an antecedent $\mathbf{P}$. It seems likely that most semantic definitions in programming languages can be cast into the form $V_c(\mathbf{P}, \mathbf{Q}) \equiv (\mathbf{T}_c(\mathbf{P}) \vdash \mathbf{Q})$, where $\mathbf{T}_c$ has several obvious properties:
(1) If $\mathbf{P} \supset \mathbf{P}_1, \mathbf{T}_c(\mathbf{P}) \supset \mathbf{T}_c(\mathbf{P}_1)$.

(2) If upon entry by entrance a_i with initial values $\mathbf{V}$, a command is executed and left by exit b_j with final values $\mathbf{W}$, then $\mathbf{T}_c(\mathbf{P}) \equiv \mathbf{Q}$, where P_α is defined as **false** for $\alpha \neq i$, $\mathbf{X} = \mathbf{V}$ for $\alpha = i$, and Q_β is defined as **false** for $\beta \neq j$, $\mathbf{X} = \mathbf{W}$ if $\beta = j$.

(3) If $\mathbf{P} = \mathbf{P}_1 \wedge \mathbf{P}_2, \mathbf{T}_c(\mathbf{P}) \equiv \mathbf{T}_c(\mathbf{P}_1) \wedge \mathbf{T}_c(\mathbf{P}_2)$.

If $\mathbf{P} = \mathbf{P}_1 \vee \mathbf{P}_2, \mathbf{T}_c(\mathbf{P}) \equiv \mathbf{T}_c(\mathbf{P}_1) \vee \mathbf{T}_c(\mathbf{P}_2)$.

If $\mathbf{P} = (\exists x)(\mathbf{P}_1), \mathbf{T}_c(\mathbf{P}) \equiv (\exists x)(\mathbf{T}_c(\mathbf{P}_1))$.

That is, the transformation $\mathbf{T}_c$ distributes over conjunction, disjunction, and existential quantification. A semantic definition having these properties satisfies Axioms 1—4.

AN ALGOL SUBSET

To apply the same notions to a conventional programming language on the order of ALGOL, one might adopt a formal syntax for the language, such as the existing syntactic definition of ALGOL; designate certain phrase types as semantic units, such as the statements in ALGOL; and provide semantic definitions for these semantic units. Let us say that each statement Σ in an ALGOLic language is tagged with an antecedent and a consequent proposition (P_Σ and Q_Σ respectively), said to hold whenever control enters and leaves the statement in the normal sequential mode of control.

Now we may readily set up a verification condition for each common statement type.

(1) If Σ is an assignment statement, $x: = f$, then

$$V_\Sigma(P_\Sigma; Q_\Sigma) \quad \text{is} \quad (\exists x_0)(S_{x_0}^x(P_\Sigma) \wedge x = S_{x_0}^x(f)) \vdash Q_\Sigma.$$

This assumes for simplicity that f is a true function of its free variables and has no side effects.

(2) If Σ is a conditional statement of the form **if** Φ **then** Σ_1 **else** Σ_2,

$$V_\Sigma(P_\Sigma, Q_{\Sigma_1}, Q_{\Sigma_2}; P_{\Sigma_1}, P_{\Sigma_2}, Q_\Sigma) \quad \text{is} \quad (P_\Sigma \wedge \Phi \vdash P_{\Sigma_1})$$
$$\wedge (P_\Sigma \wedge \neg \Phi \vdash P_{\Sigma_2}) \wedge (Q_{\Sigma_1} \vee Q_{\Sigma_2} \vdash Q_\Sigma).$$

Observe that here the exits of Σ_1 and Σ_2 become entrances to Σ, and so on. Consideration of the equivalent flowchart (Figure 3) indicates why this is true.

(3) If Σ_1 is a go-to statement of the form **go to** l, then $V_\Sigma(P_\Sigma; Q_\Sigma)$ is the identically true condition (**false** $\vdash Q_\Sigma$), because the sequential exit is never taken.

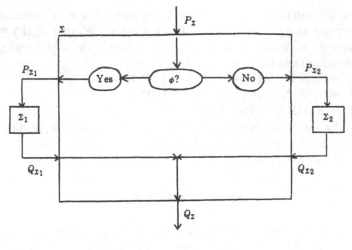

Fig. 3.

(4) If Σ is a labeled statement of the form $l: \Sigma_1$), then

$$V_\Sigma(P_\Sigma, P_l, Q_{\Sigma_1}; Q_\Sigma, P_{\Sigma_1})$$

is $(P_\Sigma \vee P_l \vdash P_{\Sigma_1}) \wedge (Q_{\Sigma_1} \vdash Q_\Sigma)$, where P_l is the disjunction of the antecedents of all statements of the form **go to** l.

(5) If Σ is a for-statement of the form **for** $x: = a$ **step** b **until** c **do** Σ_1, where x is a variable and a, b, c are expressions, the verification rule is most easily seen by constructing the equivalent flowchart (Figure 4).

The strongest verifiable proposition P_α on edge α is

$$(\exists x_0)\,(S_{x_0}^x(P_\Sigma) \wedge x = S_{x_0}^x(a)).$$

The strongest verifiable proposition P_β on edge β is

$$(\exists x_0)\,(S_{x_0}^x(Q_{\Sigma_1}) \wedge x = x_0 + (S_{x_0}^x(b)),$$

which, if b contains no free occurrences of x, can be simplified to $S_{x-b}^x(Q_{\Sigma_1})$. The strongest verifiable proposition P_γ on edge γ is $P_\alpha \vee P_\beta$. Now the condition of verification is

$$(P_\gamma \wedge (x - c) \times \text{sign}(b) > 0 \vdash Q_\Sigma) \wedge (P_\gamma \wedge (x - c) \times \\ \text{sign}\,(b) \leq 0 \vdash P_{\Sigma_1}).$$

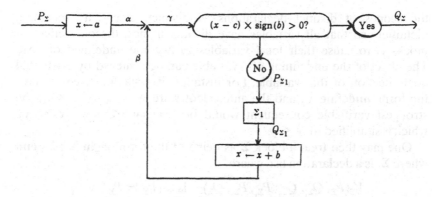

Fig. 4.

More precisely, since the definition of ALGOL 60 states that x is undefined after exhaustion of the **for** statement, the first half of the verification condition should be $(\exists x) (P_\gamma \wedge (x - c) \times \mathrm{sign}(b) > 0) \vdash Q_\Sigma$. In typical cases, these conditions may be greatly simplified, since normally a, b, and c do not contain x, Σ_1 does not alter x, $\mathrm{sign}(b)$ is a constant, etc.

(6) Compound statements. A compound statement Σ is of the form **begin** Σ_1 **end**, where Σ_1 is a statement list. Then $V_\Sigma(P_\Sigma, Q_{\Sigma_1}; P_{\Sigma_1}, Q_\Sigma)$ $\equiv (P_\Sigma \vdash P_{\Sigma_1}) \wedge (Q_{\Sigma_1} \vdash Q_\Sigma)$. Alternatively, one might identify P_Σ with P_{Σ_1} and Q_Σ with Q_{Σ_1}. A statement list Σ is either a statement, or is of the form $\Sigma_1; \Sigma_2$ where Σ_1 is a statement list and Σ_2 is a statement. In the latter case,

$$V_\Sigma(P_\Sigma, Q_{\Sigma_1}, Q_{\Sigma_2}; P_{\Sigma_1}, P_{\Sigma_2}, Q_\Sigma) \quad \text{is} \quad (P_\Sigma \vdash P_{\Sigma_1})$$
$$\wedge (Q_{\Sigma_1} \vdash P_{\Sigma_2}) \wedge (Q_{\Sigma_2} \vdash Q_\Sigma).$$

Alternately, identify P_Σ with P_{Σ_1}, Q_{Σ_1} with P_{Σ_2}, and Q_{Σ_2} with Q_Σ.

(7) A null statement Σ is represented by the empty string. $V_\Sigma(P_\Sigma; Q_\Sigma)$ is $P_\Sigma \vdash Q_\Sigma$. Verification conditions are very similar to relations of deducibility, and in this case the verification condition reduces to precisely a relation of deducibility. One might say facetiously that the subject matter of formal logic is the study of the verifiable interpretations of the program consisting of the null statement.

Blocks (compound statements with bound local variables) cause some difficulties. If we treat occurrences of the same identifier within the scopes of distinct declarations as distinct, essentially renaming identifiers so that all variables have distinct names, the only effect of blocks is to cause their local variables to become undefined on exit. The effect of the undefining of a variable can be achieved by existential quantification of that variable. For instance, if a statement could have the form 'undefine x', and the antecedent were $w < x \wedge x < y$, the strongest verifiable consequent would be $(\exists x)\, (w < x \wedge x < y)$, which is simplified to $w < y$.

One may then treat a block Σ as being of the form begin $\Sigma_1; \Sigma_2$ end where Σ_1 is a declaration list, where

$$V_\Sigma(P_\Sigma, Q_{\Sigma_1}, Q_{\Sigma_2}; P_{\Sigma_1}, P_{\Sigma_2}, Q_\Sigma) \quad \text{is} \quad (P_\Sigma \vdash P_{\Sigma_2} \wedge (Q_{\Sigma_2} \vdash P_{\Sigma_1}) \wedge (Q_{\Sigma_1} \vdash Q_\Sigma).$$

A declaration Σ of the form (say) real x is treated as undefining x when executed at the end of the execution of a block. Thus $V_\Sigma(P_\Sigma; Q_\Sigma)$ is $(\exists X)P_\Sigma \vdash Q_\Sigma$.

A declaration list Σ is either a declaration, or is of the form $\Sigma_1; \Sigma_2$ where Σ_1 is a declaration list and Σ_2 is a declaration;

$$V_\Sigma(P_\Sigma, Q_{\Sigma_1}, Q_{\Sigma_2}; P_{\Sigma_1}, P_{\Sigma_2}, Q_\Sigma) \quad \text{is} \quad (P_\Sigma \vdash P_{\Sigma_1}) \wedge (Q_{\Sigma_1} \vdash P_{\Sigma_2}) \wedge (Q_{\Sigma_2} \vdash Q_\Sigma).$$

The above is a poor approximation to the actual complexities of ALGOL block structure; for example, it does not reflect the fact that transfers out of a block by go-to statements cause local variables of the block to become undefined. It may serve, however, to indicate how a complete treatment could be carried out. Note that it does not say that local variables lose their values upon leaving a block, but that preservation of their values may not be assumed in proofs of programs.

The ALGOL procedure statement offers even more complexities, with its several types of parameters, the dynamic-own feature, the possibility of recursive call, side effects, etc. We will not consider procedure statements in detail, but will illustrate the treatment of side effects by analyzing extended assignment statements allowing embedded assignments as sub-expressions. For example, consider the statement $a := c + (c := c + 1) + c$, which has the effect of assigning $3c_0 + 2$ to a, where c_0 is the initial value of c, and assigning $c_0 + 1$ to c. Such a treatment requires saving the value of the leftmost c before executing the embedded assignment. Let us reluctantly postulate a processor, with a pushdown accumulator stack S. Introducing appropriate stacking and unstacking operators, we say that S_h (the *head* of S) is the contents of the top cell of S; that S_t (the *tail* of S) is the remainder of S, the value S would have if the stack were popped; and $x : S$ is the value S would have if x were stacked on S. These three operators are recognizable as the CAR, CDR, and CONS operators of LISP. The axioms governing them are $(x : S)_h = x$ and $(x : S)_t = S$. Now we may say that if an assignment statement has the form $x: = f$, the processor should perform **begin** STACK(f); UNSTACK(x) **end**. If f is of the forming $g + h$, STACK(f) is **begin** STACK(g); STACK(h); ADD **end**, where ADD pops the two top stack cells, adds their contents, and stacks the result; ADD is $S: = (S_h + (S_t)_h) : ((S_t)_t)$. If x is a variable, STACK(x) is $S: = x : S$. If f is of the form $x: = g$, STACK(f) is **begin** STACK(g); STORE(x) **end**, where STORE(x) is $x: = S_h$. UNSTACK(x) is **begin** $x: = S_h$; $S: = S_t$ **end**.

On this basis, any assignment statement is equivalent to a sequence of simple assignments without side effects; for instance,

$$a: = c + (c: = c + 1) + c$$

is equivalent to

begin $S: = c : S$; $S: = c : S$; $S: = 1 : S$;
$S: = ((S_t)_h + S_h) : ((S_t)_t)$; $c: = S_h$;
$S: = ((S_t)_h + S_h) : ((S_t)_t)$; $S: = c : S$;
$S: = ((S_t)_h + S_h) : ((S_t)_t)$; $a: = S_h$; $S: = S_t$ **end**.

If the antecedent of the original statement is $P(a, c, S)$, the strongest verifiable consequents of the successive statements in the equivalent compound statement are:

(1) $(S: = c : S) : (\exists S')(S = c : S' \wedge P(a, c, S'))$.

(2) $(S: = c : S) : (\exists S'')(\exists S')(S = c : S'' \wedge S'' = c : S'$
$\wedge P(a, c, S'))$,   or
$(\exists S')(S = c : (c : S') \wedge P(a, c, S'))$.

(3) $(S: = 1 : S) : (\exists S')(S = 1 : (c : (c : S')) \wedge P(a, c, S'))$.

(4) $(S: = ((S_t)_h + S_h) : ((S_t)_t)) : (\exists S'')(\exists S')(S = ((S_t'')_h + S_h''):$
$((S_t'')_t) \wedge S'' = 1 : (c: (c : S')) \wedge P(a, c, S'))$

which simplifies, by application of the equation $S'' = 1 : (c: (c : S'))$, to $(\exists S')(S = (c + 1) : (c : S') \wedge P(a, c, S'))$.

(5) $(c := S_h) : (\exists c')\,(\exists S')\,(c = S_h \wedge S = (c' + 1) :$
$(c' : S') \wedge P(a, c', S')).$

Noting that $S_h = c' + 1$, or $c' = S_h - 1 = c - 1$, this becomes $(\exists S')$
$(S = c : (c - 1 : S') \wedge P(a, c - 1, S')).$

(6) $(S := ((S_t)_h + S_h) : ((S_t)_t)) : (\exists S')\,(S = 2c - 1 : S'$
$\wedge\ P(a, c - 1, S')).$

(7) $(S := c : S) : (\exists S')\,(S = c : (2c - 1 : S') \wedge P(a, c - 1, S')).$

(8) $(S := ((S_t)_h + S_h) : ((S_t)_t)) : (\exists S')\,(S = 3c - 1 : S'$
$\wedge\ P(a, c - 1, S')).$

(9) $(a := S_h) : (\exists a')\,(\exists S')\,(a = S_h \wedge S = 3c - 1 : S'$
$\wedge\ P(a', c - 1, S')),\ \text{or}$
$(\exists a')\,(\exists S')\,(a = 3c - 1 \wedge S = 3c - 1 : S'$
$\wedge\ P(a', c - 1, S'))$

(10) $(S \leftarrow S_t) : (\exists S'')\,(\exists a')\,(\exists S')\,(S = S_t'' \wedge a = 3c - 1$
$\wedge\ S'' = 3c - 1 : S' \wedge P(a', c - 1, S')),\ \text{or}$
$(\exists a')\,(\exists S')\,(S = S' \wedge a = 3c - 1$
$\wedge\ P(a', c - 1, S')),\ \text{or}$
$(\exists a')\,(a = 3c - 1 \wedge P(a, c - 1, S)).$

For this statement, then, the condition of verification $V_\Sigma(P(a,\ c,\ S);$ $Q)$ is $((\exists a')\,(a = 3c - 1 \wedge P(a', c - 1, S))) \vdash Q$, which is exactly the verification condition for either of

Begin $c := c + 1;\qquad a := 3c - 1$ **end**

and

Begin $a := 3c + 2;\qquad c := c + 1$ **end**.

Thus, the three statements are shown to be precisely equivalent, at least under the axioms (of exact arithmetic, etc.) used in the proof.

PROOFS OF TERMINATION

If a verified program is entered by a path whose tag is then true, then at every subsequent time that a path in the program is traversed, the corresponding proposition will be true, and in particular if the program ever halts, the proposition on the path leading to the selected exit will be true. Thus, we have a basis for proofs of relations between input and output in a program. The attentive reader, however, will have observed that we have not proved that an exit will ever be reached; the methods so far described offer no security against nonterminating loops. To some extent, this is intrinsic; such a program as, for example, a mechanical proof procedure, designed to recognize the elements of a recursively enumerable but not recursive set, cannot be guaranteed to terminate without a fundamental loss of power. Most correct programs, however, can be proved to terminate. The most general method appears to use the properties of well-ordered sets. A well-ordered set W is an ordered set in which each nonempty subset has a least member; equivalently, in which there are no infinite decreasing sequences.

Suppose, for example, that an interpretation of a flowchart is supplemented by associating with each edge in the flowchart an expression for a function, which we shall call a W-function, of the free variables of the interpretation, taking its values in a well-ordered set W. If we can show that after each execution of a command the current value of the W-function associated with the exit is less than the prior value of the W-function associated with the entrance, the value of the function must steadily decrease. Because no infinite decreasing sequence is possible in a well-ordered set, the program must sooner or later terminate. Thus, we prove termination, a global property of a flowchart, by local arguments, just as we prove the correctness of an algorithm.

To set more precisely the standard for proofs of termination, let us introduce a new variable δ, not used otherwise in an interpreted program. Letting W designate the well-ordered set in which the W-functions are to be shown decreasing, and letting $\otimes$ be the ordering relation of W, it is necessary to prove for a command c whose entrance is tagged with proposition P and W-function ϕ, and whose exit is tagged with proposition Q and W-function ψ that

$$V_c(P \wedge \delta = \phi \wedge \phi \in W; Q \wedge \psi \otimes \delta \wedge \psi \in W).$$

Carrying out this proof for each command in the program, with obvious generalizations for commands having multiple entrances and exits, suffices not only to verify the interpretation, but also to show that the program must terminate, if entered with initial values satisfying the tag of the entrance.

The best-known well-ordered set is the set of positive integers, and the most obvious application of well-orderings to proofs of termination is to use as the W-function on each edge a formula for the number of program steps until termination, or some well-chosen upper bound on this number. Experience suggests, however, that it is sometimes much more convenient to use other well-orderings, and it may even be necessary in some cases. Frequently, an appropriate well-ordered set is the set of n-tuples of positive (or nonnegative) integers, for some fixed n, ordered by the assumption that $(i_1, i_2, \ldots, i_n) \odot (j_1, j_2, \ldots, j_n)$ if, for some k, $i_1 = j_1, i_2 = j_2, \ldots, i_{k-1} = j_{k-1}, i_k < j_k, 1 \leq k \leq n$. The flowchart of Figure 5 shows an interpretation using this well-ordering, for $n = 2$, to prove termination. It is assumed in the interpretation that the variables range over the integers; that is, the deductive system used in verifying the interpretation must include a set of axioms for the integers.

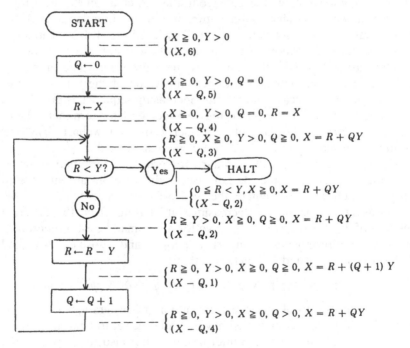

Fig. 5. Algorithm to compute quotient Q and remainder R of $X \div Y$, for integers $X \geq 0, Y > 0$.

Carnegie Institute of Technology,
Pittsburgh, Pennsylvania, U.S.A.

NOTE

[1] This work was supported by the Advanced Research Projects Agency of the Office of the Secretary of Defense (SD-146).

REFERENCES

1. McCarthy, J.: 1963, 'A Basis for a Mathematical Theory of Computation', in *Computer Programming and Formal Systems*, North-Holland, Amsterdam, pp. 33—70.
2. McCarthy, J.: 1962, 'Towards a Mathematical Science of Computation', *Proc. IFIP Congr. 62*, North-Holland, Amsterdam, pp. 21—28.

C. A. R. HOARE

AN AXIOMATIC BASIS FOR
COMPUTER PROGRAMMING

1. INTRODUCTION

Computer programming is an exact science in that all the properties of
a program and all the consequences of executing it in any given
environment can, in principle, be found out from the text of the
program itself by means of purely deductive reasoning. Deductive
reasoning involves the application of valid rules of inference to sets of
valid axioms. It is therefore desirable and interesting to elucidate the
axioms and rules of inference which underlie our reasoning about
computer programs. The exact choice of axioms will to some extent
depend on the choice of programming language. For illustrative purposes, this paper is confined to a very simple language, which is
effectively a subset of all current procedure-oriented languages.

2. COMPUTER ARITHMETIC

The first requirement in valid reasoning about a program is to know the
properties of the elementary operations which it invokes, for example,
addition and multiplication of integers. Unfortunately, in several respects
computer arithmetic is not the same as the arithmetic familiar to
mathematicians, and it is necessary to exercise some care in selecting an
appropriate set of axioms. For example, the axioms displayed in Table I
are rather a small selection of axioms relevant to integers. From this
incomplete set of axioms it is possible to deduce such simple theorems

$$x = x + y \times 0$$
$$y \leqslant r \supset r + y \times q = (r - y) + y \times (1 + q)$$

The proof of the second of these is:

A5	$(r - y) + y \times (1 + q) = (r - y) + (y \times 1 + y \times q)$
A9	$= (r - y) + (y + y \times q)$
A3	$= ((r - y) + y) + y \times q$
A6	$= r + y \times q$ provided $y \leqslant q$

83

Timothy R. Colburn et al. (eds.), Program Verification, 83–96.
© 1993 *Kluwer Academic Publishers.*

TABLE I

A1	$x + y = y + x$	addition is commutative
A2	$x \times y = y \times x$	multiplication is commutative
A3	$(x + y) + z = x + (y + z)$	addition is associative
A4	$(x \times y) \times z = x \times (y \times z)$	multiplication is associative
A5	$x \times (y + z) = x \times y + x \times z$	multiplication distributes through addition
A6	$y \leqslant x \supset (x - y) + y = x$	addition cancels subtraction
A7	$x + 0 = x$	
A8	$x \times 0 = 0$	
A9	$x \times 1 = x$	

The axioms A1 to A9 are, of course, true of the traditional infinite set of integers in mathematics. However they are also true of the finite sets of 'integers' which are manipulated by computers provided that they are confined to *nonnegative* numbers. Their truth is independent of the size of the set; furthermore, it is largely independent of the choice of technique applied in the event of 'overflow'; for example:

(1) Strict interpretation: the result of an overflowing operation does not exist; when overflow occurs, the offending program never completes its operation. Note that in this case, the equalities of A1 to A9 are strict, in the sense that both sides exist or fail to exist together.

(2) Firm boundary: the result of an overflowing operation is taken as the maximum value represented.

(3) Modulo arithmetic: the result of an overflowing operation is computed modulo the size of the set of integers represented.

These three techniques are illustrated in Table II by addition and multiplication tables for a trivially small model in which 0, 1, 2, and 3 are the only integers represented.

It is interesting to note that the different systems satisfying axioms A1 to A9 may be rigorously distinguished from each other by choosing a particular one of a set of mutually exclusive supplementary axioms. For example, infinite arithmetic satisfies the axiom:

$$A10_I \quad \neg \exists x \forall y \quad (y \leqslant x),$$

where all finite arithmetics satisfy:

$$A10_F \quad \forall x \quad (x \leqslant \text{max})$$

TABLE II

1. Strict Interpretation

+	0	1	2	3		×	0	1	2	3
0	0	1	2	3		0	0	0	0	0
1	1	2	3	*		1	0	1	2	3
2	2	3	*	*		2	0	2	*	*
3	3	*	*	*		3	0	3	*	*

* nonexistent

2. Firm Boundary

+	0	1	2	3		×	0	1	2	3
0	0	1	2	3		0	0	0	0	0
1	1	2	3	3		1	0	1	2	3
2	2	3	3	3		2	0	2	3	3
3	3	3	3	3		3	0	3	3	3

3. Modulo Arithmetic

+	0	1	2	3		×	0	1	2	3
0	0	1	2	3		0	0	0	0	0
1	1	2	3	0		1	0	1	2	3
2	2	3	0	1		2	0	2	0	2
3	3	0	1	2		3	0	3	2	1

where 'max' denotes the largest integer represented.

Similarly, the three treatments of overflow may be distinguished by a choice of one of the following axioms relating to the value of max $+ 1$:

$$A11_S \quad \neg \exists x \quad (x = \max + 1) \qquad \text{(strict interpretation)}$$
$$A11_B \quad \max + 1 = \max \qquad\qquad \text{(firm boundary)}$$
$$A11_M \quad \max + 1 = 0 \qquad\qquad\quad \text{(modulo arithmetic)}$$

Having selected one of these axioms, it is possible to use it in deducing the properties of programs; however, these properties will not necessarily obtain, unless the program is executed on an implementation which satisfies the chosen axiom.

3. PROGRAM EXECUTION

As mentioned above, the purpose of this study is to provide a logical basis for proofs of the properties of a program. One of the most important properties of a program is whether or not it carries out its intended function. The intended function of a program, or part of a program, can be specified by making general assertions about the values which the relevant variables will take *after* execution of the program. These assertions will usually not ascribe particular values to each variable, but will rather specify certain general properties of the values and the relationships holding between them. We use the normal notations of mathematical logic to express these assertions, and the familiar rules of operator precedence have been used wherever possible to improve legibility.

In many cases, the validity of the results of a program (or part of a program) will depend on the values taken by the variables before that program is initiated. These initial preconditions of successful use can be specified by the same type of general assertion as is used to describe the results obtained on termination. To state the required connection between a precondition (P), a program (Q) and a description of the result of its execution (R), we introduce a new notation:

$$P\{Q\}R.$$

This may be interpreted 'If the assertion P is true before initiation of a program Q, then the assertion R will be true on its completion'. If there are no preconditions imposed, we write **true** $\{Q\}R.$[1]

The treatment given below is essentially due to Floyd [8] but is applied to texts rather than flowcharts.

3.1. *Axiom of Assignment*

Assignment is undoubtedly the most characteristic feature of programming a digital computer, and one that most clearly distinguishes it from other branches of mathematics. It is surprising therefore that the axiom governing our reasoning about assignment is quite as simple as any to be found in elementary logic.

Consider the assignment statement:

$$x: = f$$

where: x is an identifier for a simple variable; f is an expression of a programming language without side effects, but possibly containing x.

Now any assertion $P(x)$ which is to be true of (the value of) x *after* the assignment is made must also have been true of (the value of) the expression f, taken *before* the assignment is made, i.e. with the old value of x. Thus if $P(x)$ is to be true after the assignment, then $P(f)$ must be true before the assignment. This fact may be expressed more formally:

D0   Axiom of Assignment

$$\vdash P_0\{x:=f\}P$$

where: x is a variable identifier; f is an expression; P_0 is obtained from P by substituting f for all occurrences of x.

It may be noticed that D0 is not really an axiom at all, but rather an axiom schema, describing an infinite set of axioms which share a common pattern. This pattern is described in purely syntactic terms, and it is easy to check whether any finite text conforms to the pattern, thereby qualifying as an axiom, which may validly appear in any line of a proof.

3.2. *Rules of Consequence*

In addition to axioms, a deductive science requires at least one rule of inference, which permits the deduction of new theorems from one or more axioms or theorems already proved. A rule of inference takes the form 'If $\vdash X$ and $\vdash Y$ then $\vdash Z$', i.e. if assertions of the form X and Y have been proved as theorems, then Z also is thereby proved as a theorem. The simplest example of an inference rule states that if the execution of a program Q ensures the truth of the assertion R, then it also ensures the truth of every assertion logically implied by R. Also, if P is known to be a precondition for a program Q to produce result R, then so is any other assertion which logically implies P. These rules may be expressed more formally:

D1   Rules of Consequence

$$\text{If } \vdash P\{Q\}R \text{ and } \vdash R \supset S \text{ then } \vdash P\{Q\}S$$
$$\text{If } \vdash P\{Q\}R \text{ and } \vdash S \supset P \text{ then } \vdash S\{Q\}R$$

3.3. *Rule of Composition*

A program generally consists of a sequence of statements which are executed one after another. The statements may be separated by a semicolon or equivalent symbol denoting procedural composition: $(Q_1; Q_2; \ldots; Q_n)$. In order to avoid the awkwardness of dots, it is possible to deal initially with only two statements $(Q_1; Q_2)$, since longer sequences can be reconstructed by nesting, thus $(Q_1; (Q_2; (\cdots (Q_{n-1}; Q_n) \ldots)))$. The removal of the brackets of this nest may be regarded as convention based on the associativity of the ';-operator', in the same way as brackets are removed from an arithmetic expression $(t_1 + (t_2 + (\ldots (t_{n-1} + t_n) \ldots)))$.

The inference rule associated with composition states that if the proven result of the first part of a program is identical with the precondition under which the second part of the program produces its intended result, then the whole program will produce the intended result, provided that the precondition of the first part is satisfied.

In more formal terms:

D2   Rule of Composition

$$\text{If } \vdash P\{Q_1\}R_1 \text{ and } \vdash R_1\{Q_2\}R \text{ then } \vdash P\{(Q_1; Q_2)\}R$$

3.4. *Rule of Iteration*

The essential feature of a stored program computer is the ability to execute some portion of program (S) repeatedly until a condition (B) goes false. A simple way of expressing such an iteration is to adapt the ALGOL 60 **while** notation:

while B **do** S

In executing this statement, a computer first tests the condition B. If this is false, S is omitted, and execution of the loop is complete. Otherwise, S is executed and B is tested again. This action is repeated until B is found to be false. The reasoning which leads to a formulation of an inference rule for iteration is as follows. Suppose P to be an assertion which is always true on completion of S, provided that it is also true on initiation. Then obviously P will still be true after any number of iterations of the statement S (even no iterations). Furthermore, it is known that the controlling condition B is false when the iteration finally terminates. A slightly more powerful formulation is possible in light of the fact that B may be assumed to be true on initiation of S:

D3   Rule of Iteration

$$\text{If } \vdash P \wedge B\{S\}P \text{ then } \vdash P\{\textbf{while } B \textbf{ do } S\}\neg B \wedge P$$

3.5. *Example*

The axioms quoted above are sufficient to construct the proof of properties of simple programs, for example, a routine intended to find the quotient q and remainder r obtained on dividing x by y. All variables are assumed to range over a set of nonnegative integers conforming to the axioms listed in Table I. For simplicity we use the trivial but inefficient method of successive subtraction. The proposed program is:

$$((r:=x; q:=0); \textbf{while}$$
$$y \leqslant r \textbf{ do } (r:=r-y; q:=1+q))$$

An important property of this program is that when it terminates, we can recover the numerator x by adding to the remainder r the product of the divisor y and the quotient q (i.e. $x = r + y \times q$). Furthermore, the remainder is less than the divisor. These properties may be expressed formally:

$$\textbf{true}\{Q\}\neg y \leqslant r \wedge x = r + y \times q$$

where Q stands for the program displayed above. This expresses a necessary (but not sufficient) condition for the 'correctness' of the program.

A formal proof of this theorem is given in Table III. Like all formal proofs, it is excessively tedious, and it would be fairly easy to introduce notational conventions which would significantly shorten it. An even more powerful method of reducing the tedium of formal proofs is to derive general rules for proof construction out of the simple rules accepted as postulates. These general rules would be shown to be valid by demonstrating how every theorem proved with their assistance could equally well (if more tediously) have been proved without. Once a powerful set of supplementary rules has been developed, a 'formal proof' reduces to little more than an informal indication of how a formal proof could be constructed.

TABLE III

Line number	Formal proof	Justification
1	$\mathbf{true} \supset x = x + y \times 0$	Lemma 1
2	$x = x + y \times 0 \{r := x\} x = r + y \times 0$	D0
3	$x = r + y \times 0 \{q := 0\} x = r + y \times q$	D0
4	$\mathbf{true} \{r := x\} x = r + y \times 0$	D1 (1, 2)
5	$\mathbf{true} \{r := x; q := 0\} x = r + y \times q$	D2 (4, 3)
6	$x = r + y \times q \wedge y \leqslant r \supset x = (r - y) + y \times (1 + q)$	Lemma 2
7	$x = (r - y) + y \times (1 + q) \{r := r - y\} x = r + y \times (1 + q)$	D0
8	$x = r + y \times (1 + q) \{q := 1 + q\} x = r + y \times q$	D0
9	$x = (r - y) + y \times (1 + q) \{r := r - y; q := 1 + q\} x = r + y \times q$	D2 (7, 8)
10	$x = r + y \times q \wedge y \leqslant r \{r := r - y; q := 1 + q\} x = r + y \times q$	D1 (6, 9)
11	$x = r + y \times q \{\mathbf{while}\ y \leqslant r\ \mathbf{do}\ (r := r - y; q := 1 + q)\}$ $\neg y \leqslant r \wedge x = r + y \times q$	D3 (10)
12	$\mathbf{true} \{((r := x; q := 0); \mathbf{while}\ y \leqslant r\ \mathbf{do}\ (r := r - y; q := 1 + q))\}$ $\neg y \leqslant r \wedge x = r + y \times q$	D2 (5, 11)

Notes
1. The left hand column is used to number the lines, and the right hand column to justify each line, by appealing to an axiom, a lemma or a rule of inference applied to one or two previous lines, indicated in brackets. Neither of these columns is part of the formal proof. For example, line 2 is an instance of the axiom of assignment (D0); line 12 is obtained from lines 5 and 11 by application of the rule of composition (D2).
2. Lemma 1 may be proved from axioms A7 and A8.
3. Lemma 2 follows directly from the theorem proved in Section 2.

4. GENERAL RESERVATIONS

The axioms and rules of inference quoted in this paper have implicitly assumed the absence of side effects of the evaluation of expressions and conditions. In proving properties of programs expressed in a language permitting side effects, it would be necessary to prove their absence in each case before applying the appropriate proof technique. If the main purpose of a high level programming language is to assist in the construction and verification of correct programs, it is doubtful whether the use of functional notation to call procedures with side effects is a genuine advantage.

Another deficiency in the axioms and rules quoted above is that they give no basis for a proof that a program successfully terminates. Failure to terminate may be due to an infinite loop; or it may be due to violation of an implementation-defined limit, for example, the range of numeric operands, the size of storage, or an operating system time limit. Thus the notation '$P\{Q\}R$' should be interpreted 'provided that the program successfully terminates, the properties of its results are described by R'. It is fairly easy to adapt the axioms so that they cannot be used to predict the 'results' of nonterminating programs; but the actual use of the axioms would now depend on knowledge of many implementation-dependent features, for example, the size and speed of the computer, the range of numbers, and the choice of overflow technique. Apart from proofs of the avoidance of infinite loops, it is probably better to prove the "conditional" correctness of a program and rely on an implementation to give a warning if it has had to abandon execution of the program as a result of violation of an implementation limit.

Finally it is necessary to list some of the areas which have not been covered: for example, real arithmetic, bit and character manipulation, complex arithmetic, fractional arithmetic, arrays, records, overlay definition, files, input/output, declarations, subroutines, parameters, recursion, and parallel execution. Even the characterization of integer arithmetic is far from complete. There does not appear to be any great difficulty in dealing with these points, provided that the programming language is kept simple. Areas which do present real difficulty are labels and jumps, pointers, and name parameters. Proofs of programs which made use of these features are likely to be elaborate, and it is not surprising that this should be reflected in the complexity of the underlying axioms.

5. PROOFS OF PROGRAM CORRECTNESS

The most important property of a program is whether it accomplishes the intentions of its user. If these intentions can be described rigorously by making assertions about the values of variables at the end (or at intermediate points) of the execution of the program, then the techniques described in this paper may be used to prove the correctness of the program, provided that the implementation of the programming language conforms to the axioms and rules which have been used in the proof. This fact itself might also be established by deductive reasoning, using an axiom set which describes the logical properties of the hardware circuits. When the correctness of a program, its compiler, and the hardware of the computer have all been established with mathematical certainty, it will be possible to place great reliance on the results of the program, and predict their properties with a confidence limited only by the reliability of the electronics.

The practice of supplying proofs for nontrivial programs will not become widespread until considerably more powerful proof techniques become available, and even then will not be easy. But the practical advantages of program proving will eventually outweigh the difficulties, in view of the increasing costs of programming error. At present, the method which a programmer uses to convince himself of the correctness of his program is to try it out in particular cases and to modify it if the results produced do not correspond to his intentions. After he has found a reasonably wide variety of example cases on which the program seems to work, he believes that it will always work. The time spent in this program testing is often more than half the time spent on the entire programming project; and with a realistic costing of machine time, two thirds (or more) of the cost of the project is involved in removing errors during this phase.

The cost of removing errors discovered after a program has gone into use is often greater, particularly in the case of items of computer manufacturer's software for which a large part of the expense is borne by the user. And finally, the cost of error in certain types of program may be almost incalculable — a lost spacecraft, a collapsed building, a crashed aeroplane, or a world war. Thus the practice of program proving is not only a theoretical pursuit, followed in the interests of academic respectability, but a serious recommendation for the reduction of the costs associated with programming error.

The practice of proving programs is likely to alleviate some of the other problems which afflict the computing world. For example, there is the problem of program documentation, which is essential, firstly, to inform a potential user of a subroutine how to use it and what it accomplishes, and secondly, to assist in further development when it becomes necessary to update a program to meet changing circumstances or to improve it in the light of increased knowledge. The most rigorous method of formulating the purpose of a subroutine, as well as the conditions of its proper use, is to make assertions about the values of variables before and after its execution. The proof of the correctness of these assertions can then be used as a lemma in the proof of any program which calls the subroutine. Thus, in a large program, the structure of the whole can be clearly mirrored in the structure of its proof. Furthermore, when it becomes necessary to modify a program, it will always be valid to replace any subroutine by another which satisfies the same criterion of correctness. Finally, when examining the detail of the algorithm, it seems probable that the proof will be helpful in explaining not only *what* is happening but *why*.

Another problem which can be solved, insofar as it is soluble, by the practice of program proofs is that of transferring programs from one design of computer to another. Even when written in a so-called machine-independent programming language, many large programs inadvertently take advantage of some machine-dependent property of a particular implementation, and unpleasant and expensive surprises can result when attempting to transfer it to another machine. However, presence of a machine-dependent feature will always be revealed in advance by the failure of an attempt to prove the program from machine-independent axioms. The programmer will then have the choice of formulating his algorithm in a machine-independent fashion, possibly with the help of environment enquiries; or if this involves too much effort or inefficiency, he can deliberately construct a machine-dependent program, and rely for his proof on some machine-dependent axiom, for example, one of the versions of A11 (Section 2). In the latter case, the axiom must be explicitly quoted as one of the preconditions of successful use of the program. The program can still, with complete confidence, be transferred to any other machine which happens to satisfy the same machine-dependent axiom; but if it becomes necessary to transfer it to an implementation which does not, then all the places where changes are required will be clearly annotated by the fact that the proof at that point appeals to the truth of the offending machine-dependent axiom.

Thus the practice of proving programs would seem to lead to solution of three of the most pressing problems in software and programming, namely, reliability, documentation, and compatibility. However, program proving, certainly at present, will be difficult even for programmers of high caliber; and may be applicable only to quite simple program designs. As in other areas, reliability can be purchased only at the price of simplicity.

6. FORMAL LANGUAGE DEFINITION

A high level programming language, such as ALGOL, FORTRAN, or COBOL, is usually intended to be implemented on a variety of computers of differing size, configuration and design. It has been found a serious problem to define these languages with sufficient rigour to ensure compatibility among all implementors. Since the purpose of compatibility is to facilitate interchange of programs expressed in the language, one way to achieve this would be to insist that all implementations of the language shall 'satisfy' the axioms and rules of inference which underlie proofs of the properties of programs expressed in the language, so that all predictions based on these proofs will be fulfilled, except in the event of hardware failure. In effect, this is equivalent to accepting the axioms and rules of inference as the ultimately definitive specification of the meaning of the language.

Apart from giving an immediate and possibly even provable criterion for the correctness of an implementation, the axiomatic technique for the definition of programming language semantics appears to be like the formal syntax of the ALGOL 60 report, in that it is sufficiently simple to be understood both by the implementor and by the reasonably sophisticated user of the language. It is only by bridging this widening communication gap in a single document (perhaps even provably consistent) that the maximum advantage can be obtained from a formal language definition.

Another of the great advantages of using an axiomatic approach is that axioms offer a simple and flexible technique for leaving certain aspects of a language *undefined*, for example, range of integers, accuracy of floating point, and choice of overflow technique. This is absolutely essential for standardization purposes, since otherwise the language will be impossible to implement efficiently on differing hardware designs. Thus a programming language standard should consist of a set of axioms of universal applicability, together with a choice from a set of supplementary axioms describing the range of choices facing an implementator. An example of the use of axioms for this purpose was given in Section 2.

Another of the objectives of formal language definition is to assist in the design of better programming languages. The regularity, clarity, and ease of implementation of the ALGOL 60 syntax may at least in part be due to the use of an elegant formal technique for its definition. The use of axioms may lead to similar advantages in the area of 'semantics', since it seems likely that a language which can be described by a few 'self-evident' axioms from which proofs will be relatively easy to construct will be preferable to a language with many obscure axioms which are difficult to apply in proofs. Furthermore, axioms enable the language designer to express his general *intentions* quite simply and directly, without the mass of detail which usually accompanies algorithmic descriptions. Finally, axioms can be formulated in a manner largely independent of each other, so that the designer can work freely on one axiom or group of axioms without fear of unexpected interaction effects with other parts of the language.

ACKNOWLEDGMENTS

Many axiomatic treatments of computer programming [1—3] tackle the problem of proving the equivalence, rather than the correctness, of algorithms. Other approaches [4, 5] take recursive functions rather than programs as a starting point for the theory. The suggestion to use axioms for defining the primitive operations of a computer appears in [6, 7]. The importance of program proofs is clearly emphasized in [9], and an informal technique for providing them is described. The suggestion that the specification of proof techniques provides an adequate formal definition of a programming language first appears in [8]. The formal treatment of program execution presented in this paper is clearly derived from Floyd. The main contributions of the author appear to be: (1) a suggestion that axioms may provide a simple solution to the problem of leaving certain aspects of a language undefined; (2) a comprehensive evaluation of the possible benefits to be gained by adopting this approach both for program proving and for formal language definition.

However, the formal material presented here has only an expository status and represents only a minute proportion of what remains to be done. It is hoped that many of the fascinating problems involved will be taken up by others.

Department of Computer Science,
Queen's University of Belfast, Northern Ireland.

NOTE

[1] If this can be proved in our formal system, we use the familiar logical symbol for theoremhood: $\vdash P\{Q\}R$.

REFERENCES

1. Yanov, Yu I.: 1958, 'Logical Operator Schemes', *Kybernetika 1*.
2. Igarashi, S.: 1968, 'An Axiomatic Approach to Equivalence Problems of Algorithms with Applications', Ph.D. Thesis, 1964. Rep. Compt. Centre, U. Tokyo, pp. 1–101.
3. De Bakker, J. W.: 1968, 'Axiomatics of Simple Assignment Statements'. M.R. 94, Mathematisch Centrum, Amsterdam, June.
4. McCarthy, J.: 1963, 'Towards a Mathematical Theory of Computation', *Proc. IFIP Cong. 1962*, North Holland Pub. Co., Amsterdam.
5. Burstall, R.: 1968, 'Proving Properties of Programs by Structural Induction', *Experimental Programming Reports*: No. 17 DMIP, Edinburgh, Feb.
6. Van Wijngaarden, A.: 1966, 'Numerical Analysis as an Independent Science', *BIT* **6**, 66–81.
7. Laski, J.: 1968, 'Sets and other Types', *ALGOL Bull.* **27**.
8. Floyd, R. W.: 'Assigning Meanings to Programs', *Proc. Amer. Math. Soc. Symposia in Applied Mathematics*, Vol. 19, pp. 19–31.
9. Naur, P.: 1966, 'Proof of Algorithms by General Snapshots', *BIT* **6**, 310–316.

PART II
ELABORATING THE PARADIGM

WILLIAM L. SCHERLIS AND DANA S. SCOTT

FIRST STEPS TOWARDS INFERENTIAL PROGRAMMING

Our basic premise is that the ability to construct and modify programs will not improve without a new and comprehensive look at the entire programming process. Past theoretical research, say, in the logics of programs, has tended to focus on methods for reasoning about individual programs; little has been done, it seems to us, to develop a sound understanding of the *process* of programming — the process by which programs evolve in concept and in practice. At present, we lack the means to describe the techniques of program construction and improvement in ways that properly link verification, documentation and adaptability.

The attitude that takes these factors and their dynamics into account we propose to call 'inferential programming'. The problem of developing this attitude and the tools required is far from easy and needs extensive investigation, and in this paper we are only going to be able to discuss it in rather broad terms. We wish to suggest, in particular, two goals for research that can build on past experience but that enter at what we feel is the right level of generality. This is the topic of Section 1. In Section 2 we set forth our views on the conceptualization of the notion of program derivation and the reasons why we think it is needed. The discussion inevitably brings up the question of the rôle of formalization, which is the subject of Section 3. In Section 4 we try to lay out the difficulties of the inferential programming problem; while in Section 5 we speculate briefly on the future of programming.

1. A RESEARCH PROGRAMME

The initial goal of research must be to discern the basic structural elements of the process of programming and to cast them into a precise framework for expressing them and reasoning about them. This understanding must be embodied in a *logic of programming* — a mathematical system that allows us not only to reason about individual programs, but also to carry out *operations* (or transformations) on programs and, most importantly, to reason about the operations. It can be argued —

99

Timothy R. Colburn et al. (eds.), Program Verification, 99—133.
© 1993 *Kluwer Academic Publishers.*

not without controversy — that logicans did this service for mathematical proof, and we will discuss the pros and cons presently when we argue that programming is even more in need to this kind of attention. The development of such a logic of programming will of course require a practical understanding of the semantics of programming concepts. There is no reason to suppose that the formalization of this logic will look like standard existing systems, since the principles must be adapted to the motivating problem and the appropriate concepts.

The second goal for research is to build the experimental environment. The logic of programming is intended to be a 'systematized' one, codifying our competence about programming in a way that can be used as the basis for an implemented computer system. It will be necessary to construct prototype *interactive* systems to facilitate the exploration of logics of programming and, eventually, to lead us to the natural development of practical semantically-based programming tools. In particular, it is important that such systems will permit study of the *dynamics* of programming: the relation between the way in which derivations of programs from specifications are structured conceptually and the process, over time, by which they actually evolve and are maintained and modified. This dynamical aspect we find lacking in current proposals.

This programme of research is aimed at discovering the principles that we feel can be embodied in the programming tools of the future. Such principles must be independent of individual programs — or even programming languages — if they are going to have reasonably universal significance. By developing conceptual and mechanical tools for expressing these principles and for reasoning with them, we hope to demonstrate that programming can be made a more straightforward and exact process and that the power of the programmer to think directly about the problems he has to solve can be significantly increased.

In thinking about the problem, we have found that it is important to distinguish program derivation — the *conceptual* history of a program — from what we have called *inferential programming*, by which we mean the collection of methods (and associated tools) for constructing, modifying, and reasoning about such derivations over time. Let us discuss this distinction in more detail.

Program Derivation

Contemporary programming languages and methodologies, we claim, encourage programmers to try to capture an entire process of program development in the *text* of a single program; programmers find they are attempting to write programs that — in themselves — can be easily understood and modified and yet have acceptable performance properties. Inevitably, there must be sacrifices in order to obtain the right balance between clarity and efficiency; often, perhaps more often than not, the greater sacrifice is from clarity, and the resulting programs become so complex and interconnected that eventual modification becomes prohibitively costly.

Many programmers feel it is more natural to describe a program in terms of its *derivation* or *evolution* — the sequence of insights that is required to derive the implementation from straightforward specifications [18, 38, 46]. By representing the process of program development as a *sequence* of programs, arranged as if the final implementation were developed from an initial specification by a series of refinement steps, we can maintain a structure in which clarity and efficiency coexist. Separations between program abstractions (such as abstract data types or generic procedures) and their representations do not exist *within* individual programs in derivations, but rather are spread over a sequence of program derivation steps. Abstractions introduced in early derivation steps are replaced, in later steps, by their intended representations, allowing more specialized and, hence, faster code to be ultimately obtained. The programmer need never confront the possibility of having to maintain the abstraction *and* an efficient implementation simultaneously in a single program.

Programming, even more so than mathematics, is a highly stylized affair, in which certain patterns of activity are shared in large numbers of applications. This, indeed, is an argument that programming is largely a *skill*; the good programmers are not only smarter, but they have command of a larger collection of standard programming patterns. Although attempts have been made to describe these patterns in terms of the program *text* through which they are made manifest, we believe that the patterns are really patterns of *derivation*, and the textual similarities are only superficial.

Inferential Programming

Inferential programming, on the other hand, is like the *process* of building mathematical proofs: Mathematicians do not develop proofs by starting at line one and considering their possible moves from there. Rather, they formulate a strategy and fill in gaps until they have enough detail to make a convincing argument. The proof text that emerges is a highly structured justification for a mathematical fact — even if it is written in ordinary language. The *process* of building the proof, on the other hand, is a somewhat undisciplined and exploratory activity, in which insights are gained and expressed and finally woven together to form a mathematical argument. By analogy, then, we prefer to separate program derivations — highly structured justifications for programs — from inferential programming — the process of building, manipulating, and reasoning about program derivations.

It follows that in proposing structure for program derivations, we are in no way attempting to coerce programmers to follow a specific temporal discipline in building their programs. We intend, rather, that they be provided with inferential programming *tools* — conceptual and mechanical aids — to facilitate the expression of the program and its justification and to help in the process of program development. In an inferential programming framework programmers can focus their thoughts less on expressing actual programs, and more on expressing *how* the programs come about. As a consequence, we claim, programming language design ceases to be the critical issue in programming methodology. It is likely that ultimately conventional programming languages will be required only for the very last refinement steps — and only because these steps precede a conventional compilation step.

As regards verification, inferential programming will offer a more natural method for proving correctness of complicated programs than do conventional techniques. The common approach to program proof has been to develop program and specification first, and then prove correctness as a separate step. The tediousness and difficulty of these proofs has prompted much of the software engineering community to abandon hope for the economic development of provably correct programs. Indeed, this style of proof requires programmers to rediscover the insights that went into the original development of the program and express them in a formal logical language. Unlike conven-

tional approaches to correctness, inferential programming techniques —
particularly when embodied in mechanical tools — will effectively allow
programs and proofs to be developed simultaneously. By representing
programs as sequences of derivation steps and using systematic techniques to move from one step to the next, correctness of the final
program follows directly from the correctness of the derivation techniques. With cleverly constructed mechanical tools, the business of
proof could be so effectively hidden from users that the development of
correctness proofs seems to be automatic. Of course, those steps in a
derivation whose correctness has not been proven can be isolated to
facilitate debugging and testing.

We must emphasize that the idea of controlling program derivation
is by no means new, and there has already been considerable activity
(take [1, 6, 11, 12, 20, 25, 37, 45], just to name a few). These groups
have found that, in the process of trying to build advanced program
development tools and heuristic systems for reasoning about programs,
it was very difficult to reason about the structure and meaning of
programs within a purely program-oriented framework and, instead,
some sort of evolutionary transformational paradigm must be used. A
variety of systems have emerged, but they all seem to have this aspect
in common. The experience has taught us much about the structure of
programs and the methods by which they are developed, though much
of this work has been misinterpreted, we feel. There is still confusion on
certain important points, and it is our aim here to correct some miscon-
ceptions we have perceived about where this research is going and to
suggest some directions for future work.

2. THE NATURE OF PROGRAM DERIVATIONS

The traditional 'correctness' proof — that a program is consistent with
its specifications — does not constitute a derivation of the program.
Conventional proofs, as currently presented in the literature, do little to
justify the *structure* of the program being proved, and they do even less
to aid in the development of new programs that may be similar to the
program that was proved. That is, they neither *explicate* existing
programs nor *aid* in their modification and adaptation.

We intend that program derivations serve as conceptual or idealized
histories of the development of programs. That is, a program derivation
can be considered an idealized record of the sequence of design

decisions that led to a particular realization of a specification. It is not a true history of the discovery of the program in that it does *not* include the usual blind alleys and false starts, and it does not reveal how the initial specifications were actually arrived at. But it does, nevertheless, show how the shape of the final implementation is determined by a combination of design decisions and commitments. The importance of choosing the right level of abstraction is substantiated by consideration of the necessary *changes* that have to be made in programs when new needs are imposed.

Modification and Adaptability

The constructive quality of program derivations is exactly what makes them particularly useful in environments in which programs eventually must be adapted to other uses. Indeed, a very important advantage that we see coming from the inferential programming technique will concern program modification — an activity which reportedly demands the largest proportion of available programmer time in industry and government today.

It is common wisdom that in many circumstances it is better to modify an old program than to develop an entirely new program. This is clearly appropriate when the developments of the old and new programs would have much in common — in our terms, when there would be significant sharing in the program derivations. This can be the case even if the resulting target programs have little in common. Modification is difficult in a conventional framework because, like *a posteriori* verification, it requires rediscovery of concepts used during development of an implementation. Simple conceptual changes to specification often require complicated and extensive changes to code. In an inferential programming system, not only can the conceptual changes be made directly at the appropriate places in program derivations, but the supporting system can be used to help propagate these changes correctly into implementations.

This kind of adaptability is important not only in the broad 'software engineering' context, but in the development of localized fragments of code as well. Much of current programming practice consists of adapting general algorithms and techniques from textbooks or other programs to particular applications. Abstraction mechanisms in languages can alleviate much of this problem, by permitting a single generalized template of an algorithm or other programming abstraction to serve in many contexts. There are many cases, however, in which the application required does not match a true instance of the template developed. In these cases, the connection between original algorithm and intended use is one of *analogy*, and a much more sophisticated mechanism than simple instantiation is required to establish the connection. Because the program derivation reveals so much more about the structure of programs, we believe that patterns of analogy are more likely to be apprehended and expressed when derivations are made objects of study. An example is mentioned below.

It is clear, in fact, that a vast portion of the development of *new* programs is carried out by programmers on the basis of their prior experience in similar situations. Programming consists largely of choosing appropriate known techniques and adapting them to the problem at hand. That ordinary programming language is insufficient to *express* these techniques has been widely suggested by researchers interested in automating programming and program understanding. Our hypothesis (shared by others) is that *derivation structure* is the appropriate vehicle for expression; unlike programming language, derivation structure provides a way of making explicit the *rationale* for program structure.

Programming language designers have long sought to provide language constructs that reflect as closely as possible our thinking about the structure of algorithms. For example, some years ago Dijkstra and Knuth showed that nearly all uses of **gotos** in programs were actually parts of higher control abstractions such as **while** loops, case analysis, and so on [17, 28]. The number of distinct control constructs turned out to be small enough that they could be — and were — included as primitive in programming languages, even though they brought no additional real expressive power. The point here is that *program derivations* allow us to express our thinking about the correctness and modification of programs in a much more natural — and useful — way than do conventional proofs.

In this regard we mention the 'Programmer's Apprentice' automatic programming system designed by Rich, Shrobe, and Waters at MIT [34, 44]. One of the key notions in this system is the programming *cliché*, which is a programming-language syntactic manifestation of a program derivation pattern. It was found that it is not adequate to describe clichés purely in terms of program text; some external structuring must also be specified. For this purpose, the notion of 'plan' was introduced.

A plan is an abstraction based on program structure; it provides a much richer way of describing relationships in programs than ordinary program text. The primary limitations on this enterprise derive from the fact that plans are basically abstractions from program structure; they do not express evolution or *rationale* in any direct way. The connections with the progressive commitments to implementation are also not easy to formulate in this way. Therefore, we feel we have to re-examine completely the idea of derivation in order to have a notion that captures the right features of the programming process.

Conceptualizing Program Derivation

Specifications differ from programs in that they describe aspects or restrictions on the functionality of a desired algorithm without imposing constraints on *how* that functionality is to be achieved. That is, from the point of view of specification (by our definition), the means by which the desired functionality is obtained is not relevant. In this sense, specifications for programs are *static*; they constrain implementations by constraining the *relationship* between input and output parameters. But even this distinction between input and output can be regarded as a temporal, implementation-based notion (as is seen in the example of *PROLOG*), so specifications that appear fully abstract often are not [14]. Implementations, similarly, are not usually fully constrained either; programmers frequently level many crucial representational decisions to their compilers, involving themselves in sticky details only when performance is exceptionally poor.

Thus, following [5, 36], we can be led to view this difference between specification and implementation simply as one of *degree*. But let us note here that the 'wide-spectrum' languages that have been proposed are only a partial solution; it is sometimes — perhaps usually — necessary that semantic meanings of programming language constructs change as derivation proceeds. This, as is remarked below, is a form of *commitment*, and we feel it is a sensitive issue.

In *achieving* a specification by programming, then, many decisions have to be made about how abstractions used in a specification are to be realized in a limited language of actual or virtual machine operations. This involves representing abstract objects in the form of data or control structures or by a combination of both. For example, a *function* could become manifest as an array (in which indices are mapped to cell contents), or as a procedure calculating output values from input values, or as a list of input/output pairs that must be searched. The range of possibilities is vast, and a great variety are used in practice.

Indeed, programmers are so familiar with the many techniques for representation that they often jump directly from informal specification to realization without ever being too conscious of the act of choice made for the abstraction being realized. The choice of representation, however, can depend on many factors, and in practice trial and error is required to obtain the right structure, which makes the programmer more conscious of what he is doing. Of course, much backtracking and revision is required to obtain a workable *specification* in the first place, but let us remember that at this point in the discussion we are concerned only with the structural relationship between specification and implementation.

We thus move, in a program derivation, along the near continuum from specification to implementation by means of a sequence of representational decisions, or *commitments* — whether entirely consciously or not. We use the term 'commitment' to refer to the process of introduction of structure that goes along with realizing or representing an abstraction. In this sense, commitment and abstraction, as processes, are inverse notions.

The commitments that allow passage from specification to implementation are linked together and, indeed, advantage is realized from them, by means of the *simplification* steps they permit us to make. That is, commitments introduce structure, which in turn facilitates simplification, which then suggests further commitments, and so on. In this way, the problem of implementation becomes one of selecting an appropriate sequence of commitments.

What is simplification? Any such notion must necessarily be based on some idea of *cost* or *complexity*, since otherwise the term 'simplification' is meaningless! At specification time, the only cost is conceptual cost, and simplifications made to 'improve' specifications are intended to lower conceptual cost. But if we are to move towards implementation, this notion of conceptual cost must give way to the more usual (and better understood) notion of computational cost. If we could minimize conceptual cost and computational cost simultaneously, then there would be no need for this notion of program derivation at all. Practice, unfortunately, has shown this to be impossible, so we must develop structure in which movement can be made along the cost axis depending on the needs of the 'reader' (human or mechanical). Program derivations are thus *directional* in this sense, but their *orientation* depends on whether conceptual cost or computational cost is being minimized. (Actually, we should be speaking of a cost *space*, whose axes include not only computational and conceptual costs, but also costs such as numerical accuracy.)

Thus, representational commitments increase conceptual cost, but they are necessary if computational cost is to be decreased and because *execution* of programs requires us to realize the abstractions of the specification in the limited, concrete terms that computers can accept and manipulate. Thus, our program derivation structure emerges: a progression of increasingly committed representations of programs leading from specification to implementation.

Let us recall, however, that a program derivation describes a structural relationship only; we can use it as a setting for discussion not only of the traditional problem of obtaining useful implementations from specifications, but of the inverse problem as well. We obtain a useful *specification* for a program by selectively ignoring implementation details — by a sequence of *abstraction* steps. In this scenario, the cost being minimized in simplification steps becomes conceptual cost. Although non-effective specifications are often the most natural ones, it is usually inappropriate (as we remarked above) to find *maximally* abstract specifications. In fact, for certain applications such as editors and operating systems the most useful specifications tend to be less abstract (and much larger) than for other applications. There is, of course, nothing inherent in program derivation structure that forces a certain level of abstraction in specifications. But as we understand more about how to write useful specifications, the distance between specification and implementation in program derivations will increase and the derivations will become correspondingly more useful.

Although it is not structurally necessary, it will simplify our discussion if we consider separately the business of deducing facts in derivations — facts about the situation at a particular point in a derivation or simply imported from outside bodies of knowledge. Of course, these *observation* steps don't affect *cost* in the same way that simplification or commitment/abstraction steps do, but they do provide a mechanism by which the business of *establishing* a precondition for some other step can be separated from the step itself.

We remark here that an essential goal in this development is to find a repertoire of program derivation steps that fairly closely reflect the way we think about program development informally [21]. For this reason, we have justified our division of program derivation steps into the categories of commitment/abstraction and simplification on purely philosophical grounds. This is the same kind of necessarily informal argument that some logicians use to argue that, say, Gentzen-style formal proofs are more *natural* than Hilbert-style formal proofs, and are therefore more suited to informal application. In this case, however, it is still a bit early to tell if the proposed structure is exactly the right one. (Comparison with a number of informal and formal derivations in the literature suggests that while in some cases the structure we suggest is indeed directly apparent, in most it becomes apparent only when *aggregates* of steps are considered. For example derivations, see [9, 13, 24, 32, 33].)

Examples from Practice

We have just argued that commitments and simplifications, along with observations, are the fundamental steps of program derivations. We wish now to show that, while this categorization is still perhaps a bit speculative, many conventional programming techniques fall very nicely into these classes of steps. We mention several specific forms that these steps can take on.

Perhaps the most immediate form of commitment is the representation of abstract structures such as functions (as discussed above), or sets, or graphs. Graphs, for example, can be represented as adjacency matrices, as linked physical structure, as relation subprograms, and so on. There are, however, other kinds of commitments that are made when passing from specification to implementation. Specifications describe relations, while executions of programs are inherently sequential activities. In order to develop implementations, then, we must determine an *order of computation* — not necessarily total, as for concurrent or nondeterministic computations. Once ordering commitments are made, then simplification can be made that allow sections of code to take advantage of results computed earlier. For example, traversing the nodes of a graph after making a commitment to an order of computation that is consistent with a depth-first traversal allows for a very efficient implementation. It might be the case, however, that for certain applications this order of enumeration of nodes is not appropriate, and less efficient methods must be used.

A much simpler example concerning order of computation arises, say, in the usual iterative integer square root program. In this trivial program, the integer square root i of a positive number n is found by initializing i to 0 and incrementing its value until $i^2 \leqslant n \leqslant (i + 1)^2$. By making this commitment to testing values of i in ascending order, the value of $(i + 1)^2$ from a previous iteration can be used directly as the value of i^2 on the next. Further, the *next* value of i^2 can be computed directly from the preceding value by a simple addition. But note these algebraic simplifications are possible *only* because of the commitment to incrementing the value of i.

A less obvious form of commitment is commitment to *order of presentation*. To illustrate this kind of commitment, we mention briefly an example of a programming language that permits *avoidance* of commitment to order of presentation. In PROLOG, one defines *relations* and is permitted to access the relations in many ways. If, for example, a relation $R(x, y, z)$ is defined (possibly recursively), then the definition could be used in a straightforward way to test if a given triple is in the relation. But, in certain circumstances, it could also be used when given, say, particular values for x and z only, leaving the system to search for values of y for which the relation holds. It is a specific language design decision to discourage explicit commitments on how defined relations are used. For many applications, this avoidance of commitment had a distinct cognitive advantage and can be realized fairly efficiently using a parameter binding mechanism based on unification. This notion, on the other hand, is so foreign to users of conventional languages that it rarely appears in published specifications. The lesson, then, is that many commitments are made in specification concerning *input* variables and *output* variables that often unnecessarily constrain the range of possible implementations (as a program derivation setting).

On the other hand, some languages such as POP-2 provide a limited explicit mechanism for *introducing* such commitments. Partial application, called by Ershov *mixed computation*, is an example of such mechanism [7, 19]. Indeed, as Ershov points out, compilation is itself just the simplification that results naturally from an order-of-presentation commitment. In this case, procedure text and perhaps certain actual arguments are provided before others (i.e., at *compile* time), and then the natural simplifications are made. Note that these 'natural' simplifications may depend on deep outside theorems introduced as *observations*.

Language design that permits explicit commitments to be made or avoided certainly broadens the range of applicability of that language. In the program derivation framework, however, we seek language features for specifications that permit avoidance of the various kinds of commitments whenever possible. Machine language programs, in which very few such decisions are left unresolved, are at the other extreme of the spectrum.

Commitments are made not only to eliminate from specifications those abstractions not directly realized by computers, but also to permit *simplifications*. Let us consider, by way of example, the use of the *divide-and-conquer* paradigm to obtain the usual binary array search algorithm. The problem is to determine if a key k appears in an array segment $A[l \ldots u]$. The essential assumption is that the array elements are given in increasing order. We can then take advantage of this assumption after a commitment is made to testing a particular array element, say $A[m]$. (This is the 'divide' step.) If the test fails, we want to test recursively the remainder of the array. A simplification allows us to use additional information about the outcome of the test to exclude more than just that single element $A[m]$ from consideration; indeed, the entire initial portion of the array segment from position m to the beginning can be excluded if $A[m] < k$. We are then left with a considerably smaller segment for the recursive call (which means much less to 'conquer').

Another form of simplification is the elimination of simple *redundancy*, and this is the basis for the so-called *dynamic programming* paradigm for algorithm design. The Cocke—Kasami—Younger parsing algorithm, for example, results from a simple recursive definition of the *derives* relation for Chomsky—Normal—Form context-free grammars. By making an appropriate order-of-computation commitment — which in this case is really only a partial commitment — and eliminating redundancy in the resulting definitions, an exponential-time algorithm is transformed into a polynomial-time algorithm.

3. THE RÔLE OF FORMALIZATION

If we are to succeed at building semantically based tools that will have any significance for programming methodology, our conceptualization of program derivation, which we have just discussed in general terms, must lead to some of the same kinds of understanding that logic brought to the perception of the structure of mathematical proof. The parallel must not be regarded as perfect, however. Formalization in logic is the means by which *proofs* are introduced as legitimate objects of mathematical attention; that is, a formal proof itself becomes a mathematical structure that can be reasoned about. Our conceptualization of program derivations, similarly, must lead us to a point where program derivations are considered as *formal* structure representing the evolutions of programs. As we remarked, we are still far from being able to propose the exact details of the required formalization; but even the suggestion that it is *possible* to find such a definition raises many questions — and generally raises blood pressure as well! — since the significance of formalization in mathematics remains hotly debated.

Who Deals with Formal Structures?

We do not mean to suggest that programmers must deal directly with the minute formal structure of a derivation — say, the filling in of the names of the principles employed at each step. Indeed, it will be asked: do we need to formalize program derivations at all? Mathematicians do not really build formal proofs in practice; why should programmers? Real programming, like the proving of real theorems, is a process unhampered by the observance of niggling little rules, a process requiring, rather, insight and creativity. How can formalization help? Perhaps formal logic improved our understanding of the *structure* of mathematical proof, but it hasn't helped us prove *new theorems*, has it? Well then, neither will formalization help us find new programs.

The tenor of this objection is common, and it sounds very reasonable at face value. We find a basic fallacy here, however. In a way, formalization plays an even more important rôle in computer science than in mathematics because — as is obvious to all programmers — programming languages are by necessity formal languages. The programming process is a process of building and reasoning about formal program structures; people can consume informal proofs (indeed more easily than their formal counterparts), but computers do not run 'informal' programs. To run, programs must be syntactically absolutely correct. Much of the good advice of program methodology is aimed at keeping the complexity of the formal programs under control, and much systems building is devoted to constructing aids for providing a programming environment allowing the programmer to rise far above the counting of parentheses and the filling in of semicolons. That is not all there is to formalization, of course.

To be fair to classical mathematics, its level of precision has improved by leaps and bounds since the turn of the century. The field of algebraic geometry is an outstanding example, and this improvement has almost nothing to do with the history of mathematical logic and logical formalization. To cite a trivial example of precision (that has been around for a very long time but required clear thinking when it was originated), consider proving, say, the convergence of some infinite series. Many manipulations must be done, and every student has experienced the problem of getting some signs wrong. Going back over the formulae until the missing minus sign is located is just exactly the same as debugging a program. Much of the language of algebra, calculus, and more advanced analysis *is* a formal language, and people prove things by learning the rules of the formalization. Today, the MACSYMA system is able to deal with these manipulations in a pretty direct way. And much, much more of mathematics has been formalized in this sense: take large portions of algebraic topology, for example, which require heavy formal machinery now often expressed in category theory. Mathematicians are able to prove important theorems with the aid of these formalizations that were unthinkable 100 years ago. The point is that the difficulty of the abstractions has forced *real* mathematicians to introduce what can only be called formal methods. The argument that mathematicians have with logicians, on the other hand, is over the further question of whether there is any sense in looking at a *complete* formalization of a whole proof. Often there is not.

Returning to the domain of programming, we think that there is a difference in scale and a difference in kind between programming and mathematics. It will be agreed that much of programming deals with systems: languages, compilers, interfaces. All these topics are formal by nature, as structured messages have to be interpreted and either executed or turned into something else. That is, computation is a symbol manipulation activity, and so formalization lies at the basis of all automated systems. This is the difference in kind.

But, even relatively elementary programs tend to be more complicated than elementary theorems. Why? Because in a certain sense

more of their structure has to be made explicit. In mathematical literature, it is often sufficient to do one precise calculation, and for the other
cases say 'similarly'. A proof is often more a genteel tourist guide than
instructions for a treasure hunt. Programs, on the other hand, not only
operate on very highly structured data, but they must do so in unobvi-
ous ways in order to be efficient. All this pressure requires a high
degree of formal treatment. But, just as before, we ask: even if programming is more concerned with formalization than mathematics,
must the *whole* process be formalized? This leads us directly to the
next question.

How to Go: Formal or Informal?

Like proofs, program derivations can certainly be presented both
formally and informally. Formal derivations, like any formal proof or
other structure, are not intended for everyday consumption. Much
misunderstanding has resulted when this important distinction has not
been recognized. For example, David Gries has this pessimistic quotation on program transformation in his new book [26] (p. 235).

It is extremely difficult to understand a program by studying a long sequence of
transformations; one quickly becomes lost in details or bored with the process.

We do not intend here to ascribe this misunderstanding to Gries. We
suspect, rather, that his remark is a natural response to the way in
which transformation sequences have sometimes been *presented* in the
literature, not to their *structure*. It is just as inappropriate — to understanding — to use a sequence of *formal* transformation steps to describe a program as it is to justify a theorem by giving a full formal
proof. We are quite comfortable with informal mathematical proofs; we
must learn better, however, how to present transformation sequences
(which are really program derivations) in an informal way so that they
will be a useful tool for explaining programs.

It should not be forgotten that we have *never* had useful informal
ways of justifying the correctness *and* structure of programs until
recently! Indeed, most of our experience in this area is with *formal*
proofs; the historical development of programming methodology has
been completely unlike that of ordinary mathematics in this regard. The
point of most of the early program-transformation papers, for example,
was to expose new program-transformation techniques, not to give

accounts of algorithms [8, 9, 30, 32, 43]. More recently, papers have appeared in which the transformational method is used to explicate existing algorithms — and occasionally even *new* algorithms — but we have still not yet developed concise informal language for describing the transformation sequences [13, 24, 33].

Gries went so far as to suggest that programs subject to transformation should be proved afresh at every stage of development. We suggest, contrariwise, that the transformation process, properly presented, constitutes a much more useful proof than the usual sort of static program proof. Of course, transformations can transform proof along with program, so this requirement could always be satisfied trivially — but this is not the point.

Here is the key design problem: we must conceptualize program derivation in such a way that both informal and formal presentations will be (appropriately) useful! That is, formal program derivations will be useful only if the structure they make manifest is in some way a reflection of the way we think (intuitively) about the evolution of programs. We must build formal structures that will permit not only the *presentation* of arguments (in the form of program derivations) but also the *development* of new arguments and the adaptation and modification of old ones — by means of inferential programming techniques.

An alternative approach to formalizing our reasoning about the evolution of programs is based, rather directly, on the evolution of formal proofs in mathematics, combined with rules for generating an algorithm from the proof [3, 4, 22, 29, 31]. Improvements to the algorithm are made either by restructuring the formal proof or through the use of automatic optimization tools. Commitment steps, in this approach, correspond to *proof-development* steps. The inferential programming approach, on the other hand, is based on the thesis that the passage from informal to formal can bring us directly into a program-oriented language, even if the 'programs' are nothing more than abstract — even noneffective — specifications.

When we think again about who deals with formal structures, the answer that we hope will emerge is that the programmer deals with them on one level (just as mathematicians have to) while manipulating them in informal ways in his head or in his documentation, and the system deals with them on quite another level interacting with the programmer(s). The kind of system we would like to see built will keep track of all the little syntactical niceties as a matter of course, but at a different stage of interaction will dovetail steps of the *derivation* under direction of the programmer. This view gives us a chance to suggest a new answer to the next question.

At what Point do We Find Verification?

It is commonly argued that formalization inhibits the social process of acceptance of proofs. Indeed, why should we *believe* formal proofs — even if they are produced by a machine — since they are almost immune to the usual kind of social process of having friends, enemies, referees, students, and others check the details in their heads or with pencil and paper. In objecting specifically to the suggestions for methods of (automatic) program verification, DeMillo, Lipton, and Perlis say at the start of their well-known paper [15] (p. 271):

We believe that, in the end, it is a social process that determines whether mathematicians feel confident about a theorem — and we believe that, because no comparable social process can take place among program verifiers, program verification is bound to fail. We can't see how it's going to be able to affect anyone's confidence about programs.

Further on they are even more outspoken and say (p. 275):

The proof by itself is nothing; only when it has been subjected to the social processes of the mathematical community does it become believable.

In this regard — though they mention the early muddy history of the 'solutions' to the Four-Color Conjecture — De Millo, Lipton and Perlis do not discuss the subsequent machine-aided proof of this world-famous conjecture. (Their paper was received by the editors in October 1978 and probably written much earlier. The first announcement by Haken, Appel, and Koch of the solution of the conjecture was made in September 1976, and the paper was published in September 1977, see [27].) As this example fits exactly into the area of methods of proof that they are criticizing, we would like to go more into the circumstances of the proof and its believability.

Fortunately, the computer proof — a checking of a large number of cases — of the Four-Color Theorem (hereafter, 4CT) has caused considerable comment, so we shall not have to be too explicit here about mathematical details and can refer to the published literature. Everyone has surely heard the statement of the problem: Does every finite planar map require only four colors to color all regions so that no two adjacent regions have the same color? The conjecture received many inadequate proofs over the years. In February 1979, in an article '*The Four-Color Problem and Its Philosophical Significance*', Thomas Tymoczko published a long broadside [42] against the claimed solution in which he asserted (p. 58):

What reason is there for saying that the 4CT is not really a theorem or that mathematicians have not really produced a proof of it? Just this: no mathematician has seen a proof of the 4CT, nor has any seen a proof that it has a proof. Moreover, it is very unlikely that any mathematician will ever see a proof of the 4CT.

What reason is there then to accept the 4CT as proved? Mathematicians know that it has a proof according to the most rigorous standards of formal proof — a computer told them! Modern high-speed computers were used to verify some crucial steps in an otherwise mathematically acceptable argument for the 4CT, and other computers were used to verify the work of the first.

Thus, the answer to whether the 4CT has been proved turns on an account of the role of computers in mathematics. Even the most natural account leads to serious philosophical problems. According to that account, such use of computers in mathematics, as in the 4CT, introduced empirical experiments into mathematics. Whether or not we choose to regard the 4CT as proved, we must admit that the current proof is no traditional proof, no *a priori* deduction of a statement from premises. It is a traditional proof with a lacuna, or gap, which is filled by the results of a well-thought-out experiment. This makes the 4CT the first mathematical proposition to be known *a posteriori* and raises again for philosophy the problem of distinguishing mathematics from natural sciences.

Subsequent commentators are far from agreeing with Tymoczko that "we are committed to changing the sense of 'theorem', or more to the point, to changing the sense of the underlying concept of 'proof'." In the same journal in December 1980 in a reply, '*Computer Proof*', Paul Teller argues cogently against all Tymoczko's conclusions [41]. In the very same number of the journal, Michael Detlefsen and Mark Luker in another long reply, '*The Four-Color Theorem and Mathematical Proof*', point out that computer checking of certain proof steps is hardly a novelty (they give several telling references and quotations) and they discuss rather fully the question of 'empiricism' [16]. They say (p. 804):

We do not disagree with Tymoczko's claim that evidence of an empirical sort is utilized in the proof of the 4CT. What we find unacceptable is the claim that this is in any sense novel.

It is best for the reader to read the original paper and the replies himself to judge the issues. They make interesting reading.

Another author, however, who touches on the points closest to our present discussion is E. R. Swart in yet another reply to Tymoczko in November 1980 in an article '*The Philosophical Implications of the Four-Color Theorem*' [39]. Swart — owing to his detailed and professional knowledge of the field — points out that in fact Tymoczko fails to recognize the actual weak point of the Haken—Appel proof. Moreover, he suggests a cure. He also points out that other machine-aided proofs of the 4CT are now available, referring to extensive work of F. Allaire. He says (p. 698):

... Allaire's proof also involves a discharging/reducibility approach but only requires some 50 hours of computer time. It is moreover, based on an entirely different discharging procedure and a completely independently developed reducibility testing program. At the very least Allaire's proof must rank as an independent corroboration of the truth of the four-color conjecture, and there can be little doubt that even if the Haken/Appel proof is flawed the theorem is nevertheless true.

Swart goes on convincingly to support the thesis that computers are more reliable than humans in checking details and says (p. 700):

Human beings get tired, and their attention wanders, and they are all too prone to slips of various kinds: a hand-checked proof may justifiably be said to involve a "complex set of empirical factors." Computers do not get tired and almost never introduce errors into a valid implementation of a logically impeccable algorithm.

He understands, of course, that the original algorithm design is the important point here.

In our view what Swart demonstrates in a fully documented way is just how computer-aided proof *can* be subjected to the 'social process' — which is just what De Millo, Lipton and Perlis denied was going to happen. In the case of the 4CT, Swart criticizes the original method, he points out how it can be strengthened, he discusses an alternative approach, and he comments on the general reliability of algorithms. This is more attention to particulars than many results get, and in any case he has gone into the mathematical details elsewhere and clearly knows what he has been talking about. A particularly encouraging statement about the 'social' nature of the activity occurs on p. 704 of his paper:

It is perhaps appropriate to begin to draw this section to a close with some discussion of the obvious first requirement of mathematical proofs — namely, that they should be convincing. At this juncture in history the 4CT has not been properly integrated into graph theory as a whole and stands to some extent as a monument on its own, but there is little doubt that this is not its permanent lot.

Indeed, it already has strong connections with at least some branches of graph theory that have no direct reliance on computer programs. Several mathematicians, such as Walter Stromquist, Frank Bernhart, and Frank Allaire, who did research on the question of reducibility also developed a coherent theory of irreducibility that is in complete agreement with the reducibility results that have been obtained thus far on the computer. Moreover, in the light of such irreducibility theory, it became possible to determine anti-configurations for all planar configurations that are not freely reducible. And Frank Allaire was able to make excellent use of such anti-configurations in finding reducers for intractable reducible configurations. Such developments can surely only serve to strengthen the confidence that mathematicians have in the truth of the 4CT.

And in the years to come, when the theorem is even more inextricably intertwined with graph theory as a whole, it will seem not a little quaint to even suggest that it is not an *a priori* theorem with a surveyable proof. The four-color conjecture served as an excellent stimulus to graph-theoretic research, and the 4CT may continue to exert a benign influence on graph theory until such time as it has been brought into "the body of the kirk."

We think this answers De Millo, Lipton, and Perlis pretty fully. To be fair to them, they did not have very good paradigms in mind: at the time that they wrote their paper the suggestions for computer-aided verification were low-level. But Swart describes very well what happens to this kind of work when there is actual content involved. He shows that, since people are involved in formulating the problems and in evaluating the results, *thoughts* tend to occur to them. No one is going to be satisfied with the answer 'VERIFIED' — they will want to know how and why and what the context is. Though the 4CT is a highly specialized problem, the algorithms developed have wider significance beyond the proof of one theorem. This is the typical direction of problem solving into generalization, as everyone knows.

Our contention is that the difficulty with the question of program verification — which has almost become a dirty word — is that the question was asked at the wrong level of abstraction. Our approach is to replace this question by the aim of correct program *development* with correctness being checked at each stage. This puts the emphasis on verification in a different light. De Millo, Lipton, and Perlis say (p. 279):

The concept of verifiable software has been with us too long to be easily displaced. For the practice of programming, however, verifiability must not be allowed to overshadow reliability. Scientists should not confuse mathematical models with reality — and verification is nothing but a model of believability. Verifiability is not and cannot be a dominating concern in software design. Economics, deadlines, cost—benefit ratios, personal and group style, the limits of acceptable error — all these carry immensely much more weight in design than verifiability or nonverifiability.

So far, there has been little philosophical discussion of making software reliable rather than verifiable. If verification adherents could redefine their efforts and reorient themselves to this goal, or if another view of software could arise that would draw on social processes of mathematics and the modest expectations of engineering, the interests of real-life programming and theoretical computer science might both be better served.

We agree with them that reliability is the key driving force and that verifiability at some time *after* a giant program has been completely written by different hands is virtually impossible. But we cannot let this be an excuse for not producing *provably* correct software, even if 'correct' only means that documentation is generated concerning the *extent* to which a program is reliable. We ourselves would not be satisfied with such a weak notion of correctness, but informative documentation would be a step forward. The emphasis in our minds is on 'proof' — meaning that the right *observations* from allowable ones are made at the right points in the derivation.

Once we accept the proposition that individual program derivation steps *preserve* correctness, then it is implicit in derivation structure that implementations are consistent with specifications. The confidence of users arises from their knowledge of *how* the implementation is linked to the specification, and no formal proofs need ever change hands. That is, confidence is based on the assured *existence* of the proof and not on its content or structure. It is exactly this kind of reasoning that justifies our confidence that the object code produced by programming language compilers is faithful to the source code. Once we accept the correctness of a compiler — an acceptance which is almost always effected by a social process — we necessarily accept the correctness of the results; this is what compiler correctness *means*. In practice, we rarely bother even to inspect the results of compilations. Our treatment of transformations is going to be like our treatment of compilers — and indeed like arbitrary programs — for, once we accept their correctness, then we necessarily accept the correctness of their results.

What are the Proper Analogies?

De Millo, Lipton, and Perlis contrast two analogies in their paper (p. 275):

The Verifiers' Original Analogy
Mathematics   *Programming*
theorem ... program
proof ... verification

The De Millo—Lipton—Perlis Analogy
Mathematics   *Programming*
theorem ... specification
proof ... program
imaginary formal
demonstration ... verification

We just do not agree with this picture of the activity. We would like to revise the analogy in the light of all the foregoing discussion, since we feel that, if left in the above form, it does a disservice to the concept of formalization and its correct rôle.

The Revised Analogy
Mathematics   *Programming*
problem ... specification
theorem ... program
proof ... program derivation

All of this has to be taken with a big grain of salt, since all these words can mean many things. We prefer to put 'problem' in parallel with 'specification', because the statement of a (mathematical) problem formulates a goal without saying how to solve the problem. Neither does a specification determine an algorithm — nor need a specification be satisfiable at all. In both cases the answer must be found, and it need not be unique.

Now theorems come in many flavors. If De Millo, Lipton, and Perlis understand by 'theorem' a statement like 'The problem has a solution', then we agree there is not much to choose between problem and theorem — and their analogy can stand. But many more theorems are stated 'The solution to the problem is given by this formula: ...', and this is much more like a program. There is no hint in the statement of the theorem *why* the formula gives the answer; one must look to the proof. As we have remarked before, programs tend to be pretty explicit, so let us understand by 'theorem' something more like 'theorem *cum* construction' which is parallel to 'program for a given specification'. But if we agree that the lines of the analogy table are not meant to be read in isolation, then we can take it that each line follows on from the previous one and that our suggestions can remain in the more abbreviated form.

The word 'proof', admittedly, is much harder to pin down; proofs, too, come in all flavors, and they are all too often half baked. A good proof should contain some discussion to let the poor reader know how the solution to the problem was arrived at. There are lots of tedious steps that have to be checked, but the author of the proof should have supplied some organization to the way the steps have been assembled. We are back at our conflict between formal and informal methods. De Millo, Lipton, and Perlis say (p. 275):

There is a fundamental logical objection to verification, an objection on its own ground of formalistic rigor. Since the requirement for a program is informal and the program is formal, there must be a transition, and the transition itself must be necessarily informal. We have been distressed to learn that this proposition, which seems self-evident to us, is controversial. So we should emphasize that as antiformalists, we would not object to verification on these grounds; we only wonder how this inherently informal step fits into the formalist view. Have the adherents of verification lost sight of the informal origins of the formal objects they deal with? Is it their assertion that their formalizations are somehow incontrovertible? We must confess our confusion and dismay.

Confession is always regarded as good for the soul, but somehow we cannot accept that the authors are too sincere in this passage. They must have felt that they had dealt a death blow to the project of verification, and they did not want to gloat too much. But our whole argument in this paper is that the place for verification is *within a program derivation*. Formalization is much concerned with the *way* the steps of the derivation fit together; while the informal understanding of how the solution to the original problem is emerging is in the *choice* of the sequence of steps. (Granted, an informal reader of a program derivation may need some comments to help him remember where he is and which subgoal is in need of attention, but a mathematical proof needs some commentary also — if one is ever going to learn anything beyond an ability to recite the proof verbatim.) There is plenty of room here for the interplay between formal and informal strategy without relegating formal methods to the madhouse.

It is no argument that some mathematicians give vague proofs — so vague that one often wonders how they solve their problems at all. There are excellent authors that craft proofs that can be checked even by undergraduates. There are also proofs that have thorny steps that require machine-aided checks. As we learn to take advantage of the power of the computer, there will be many more of these proofs. Students will soon learn to use the necessary machines (perhaps before their teachers), and proofs will become more complex. There is no conflict between the formal and the informal in this way of telling the story, however, as each has its place.

In making these analogies, one must take care to assess goals. There are problems that must be faced by the computer scientist that mathematicians might regard as unutterably boring. As far as the latter is concerned the problem is solved, and the choice of a particular computation method or a particular representation of the data is of no concern. The programmer will have to slog through many thickets in which the mathematician can see no game. The subjects are different, the interests are different, and the aims are different, but the two hunters after solutions of problems can learn from each other. Just as logic has something to tell us about proofs, we feel inferential programming can provide a framework for presenting and justifying the structure of algorithms and programs.

4. THE INFERENTIAL PROGRAMMING PROBLEM

The success of the inferential programming approach to program development will depend on the design and implementation of the supporting mechanical systems. There is no denying that constructing an adequate design that can be made to work in practice is a truly challenging problem! A major difficulty in building such a system is that there are many, many ways in which we can deal with derivations: extending them in one direction or another, reasoning about them, using them as a basis for constructing new derivations, and so on.

It is not our intention here to champion any particular philosophy of practical program development, nor do we intend for inferential programming systems to embody any preconceived approach to programming methodology or management. We believe, in fact, that programming, like other areas of creative endeavor, should not be too heavily shackled by form or method. *Tools* or *vehicles* for programming, therefore, must be constructed in such a way that programmers will still feel the freedom to explore in unfettered fashion. They must be given, however, the added confidence that the paths they explore are all safe ones — leading to correct implementations and faithful specifications. Awareness, in addition, has to be maintained of the various possible directions still open, and advice must be forthcoming when 'navigation' decisions are to be made. The journeys may still involve experimentation and backtracking and, rarely, unprotected forays, but the experiences ultimately gained have to be captured in the form of 'maps' that others can use to stay safely on course when in similar situations. Thus, as experience accumulates, programmers will find themselves more often in known — or at least easily charted — territory in which they can defer decisions to their guides, interfering only when necessity (or curiosity) demands.

It must be emphasized here that we do not regard programming as an intellectually shallow activity that can be automated simply by finding the right set of 'tricks'. No one in the near future will succeed in fully automating the programming process, and we must not waste our efforts in such an attempt. Instead, we feel we should focus on building powerful *interactive tools*. As our understanding of the process of programming improves, it is true that more aspects of it will be subject to complete systematization and automation — but the completely mechanical programmer is a will-o'-the-wisp. For this reason, we are concentrating our investigations on finding the sorts of components a powerful interactive inferential programming system should have, and on understanding how they could aid in both the formal and heuristic aspects of programming.

Integrating Deduction

There must be, first of all, a mechanism for carrying out the fundamental program-derivation steps. In its most primitive form, this mechanism will apply syntactic transformations in order to make or remove commitments, to carry out simplifications, and to make straightforward observations. More advanced capabilities would include tools for manipulating and reasoning about entire derivations. All of these symbol-manipulation activities can be regarded as forms of deduction, and a powerful single general-purpose deductive mechanism such as that already used in LCF [23] could go a long way towards realizing them.

Whatever the *deductive facility* is, it must at the very least have sufficient capability that users are not bothered with having to make trivial algebraic simplifications and transformations. It goes without saying that the more difficult simplification and observation steps may require considerable activity in formal reasoning with many heuristic decisions — and perhaps with interaction from the user. Although we feel able to address the heuristic issues in a meaningful way, much thought is still required before the right style of interaction between user and system can be arrived at. More will be said on these points below, but it should be clear that the heart of an inferential programming system will be its deductive mechanism.

It seems to be a valid point to make that an effective approach to theorem proving is first to start with a suitably implemented proof-checker technology and then to add heuristic features. The success of LCF is largely due to this correct philosophical attitude. In LCF a clear distinction is made between metalanguage and object language, permitting users to focus separately on facts and the strategies that control the process of inference. A really general programming language (for LCF it is called ML) for controlling inference was also used in the AI languages PLANNER and CONNIVER. The novelty of LCF is the use of the ML type mechanism to maintain important distinctions in the object language, such as between theorems and other formulas. This permits users to experiment with proof strategies and be confident of not disturbing the underlying logic.

As the formal reasoning mechanism will be operating primarily on *program-derivation structure*, we have to ask what the actual shape of this structure will be. Since our project is not yet at the stage of implementation, we can only anticipate now the kinds of problems that will probably arise when we set out to formalize the informal understanding of derivations discussed earlier. At a first approximation, a program derivation is likely to be a directed graph in which nodes are programs and arcs are fundamental program derivation steps (i.e., commitment and simplification). The simplest sort of program derivation yields a linear graph; more complex structure emerges when alternative commitments are pursued and different implementations of the same specification (or multiple specifications for the same implementation) are obtained.

Program derivations are themselves objects, and it is often easiest to obtain a *new* program derivation by a transformation on an existing derivation. A mechanism of this sort (and a language for expressing *relationships* among derivations) will have to be developed if, for example, users are to be able to create new derivations by analogy with existing ones. This question also leads us to thinking about derivation strategies and heuristics and their realizations as higher-level derivations. Much experimentation remains to be done, especially in obtaining a feeling of how deduction is to be combined with the more prosaic steps of program construction.

Adapting Conventional Tools

Many groups are currently at work building and using program-development tools. We cannot survey the whole vast field here, but we do wish to discuss some useful aspects of present efforts in order to be able to explain how the additional features we hope to add to a system will qualify it for use in what we call inferential programming. The existing programming aids help programmers operate on the *syntax* of their programs, but they generally do little to help when the *effects* of programs must be considered. It is an essential part of our thesis that powerful *semantically based* programming tools will utterly change the way in which programs are created and modified. It is not simply a matter of adding these new features to our standard tools but, rather, of creating an entirely new attitude towards the programming process.

Consider *structured editors*, for example. With these tools, programmers are allowed to explore alternative syntactic constructions as they manipulate the text of their programs. They are thus freed from concerns relating to the *syntactical* correctness of their programs. We assert, similarly, that programming tools that operate on derivation structure will help programmers explore a range of implementations while essentially freeing them from *semantical* correctness concerns. The change in attitude here makes programming move closer to problem solving.

The first 'automatic programming' tools were of course the *compilers*. Incremental compilers and associated programming environments — as seen, for example, in modern LISP systems — are among our most powerful contemporary programming tools. Perhaps the principal reason the LISP environment is so attractive is that users have tremendous freedom to modify the text of programs *and* sample their executions without having to follow the rigid discipline of edit/compile/ test associated with traditional compilers. In such an enviroment, users can respond to problems by investigating the execution context of the problem, then, perhaps, by making a local change (while still in the context), and finally by continuing execution. Immediate response and adaptation to small problems is possible, and radical context switches are not required except in unusual cases. An inferential programming environment, similarly, should not force general retreat when small problems develop. Rather, it must allow an *incremental* approach to the manipulation of program derivations, which is again a change of attitude.

Very little success has been experienced in applying formal techniques from programming logics to reasoning about *very large programs*, and indeed a very natural worry about the inferential programming paradigm is its ability to scale up. Will inferential programming systems ever be sold other than in the toy shops? The answer lies partly in the development and use of powerful *modularization* techniques for both programs *and* program derivations. Modularization is an important concern of systems builders, and such a facility for inferential programming would allow individual parts of a large program to be derived independently and then combined together in such a way that all possible interactions can be anticipated. Here is a case where the right attitudes are already familiar, except they have not been applied sufficiently to deduction.

By anticipating interactions, *version control* becomes a much more precise activity — another very critical concern in large-scale programming. Current systems for version control, as seen in GANDALF and MasterScope, must apply a necessarily conservative strategy to the task. Because they are unable to make inferences about the *effects* of changes, *any* change must be treated as a *major* change, and analysis and perhaps recompilation is necessary for those modules that might *possibly* be affected. If a semantic component were added to these systems, then changes would need to be propagated only to those modules that were truly affected. We feel this would encourage more extensive experimentation.

Type checking, both at compile time (*static* checking) and at run time (*dynamic* checking), is among the essential mechanisms that programming languages have provided to help programmers protect lines of *abstraction* in their programs. The trend towards *self-documenting programs* has brought a variety of new kinds of abstraction facilities in programming language designs, along with correspondingly complex languages of types and type-inference algorithms. In program derivations, the lines of abstraction are drawn *between* programs rather than *within* them, so the need for complex typing mechanisms may diminish appreciably. But, we will still have to devote considerable effort to the design of the typing mechanisms used in the *formal* language in which the program derivations themselves are expressed. Like formal proofs — and indeed any formal objects (even programs) — program derivations have many, many internal consistency requirements, and a suitably rich typing mechanism can make the process of checking and maintaining these consistency requirements largely mechanical. Programmers who want to study and use strongly typed languages can still be accommodated, in any case, even if we feel we can shift part of the burden of type checking in program development to other phases of the process.

Developing Heuristics

An understanding of *meaning* does not necessarily imply a command of *technique*. Students can develop a reasonably deep understanding of the foundations of calculus without developing any skill at solving integrals. Similarly, many students are quite adept at integration, but have little understanding of the fundamentals, so we must conclude conversely that a command of technique does not imply a deep understanding of meaning. Though these remarks are truisms, they suggest that if we are to design a useful deductive facility, we must provide methods for introducing not only new knowledge to the database, but also information regarding how the facts are to be *used*.

Inferential programming tools will become more applicable as the heuristic knowledge they embody increases. Programmers will be tinkering constantly with their personal stores of heuristic knowledge in order to make them more powerful and flexible. It is necessary, however, to protect the database of *facts* from this constant tinkering, so our deductive mechanisms must be designed to keep correctness issues separate from the heuristic mechanism.

As we remarked in Section 2, the bulk of programming activity — whether in the modification of existing programs or in the creation of new programs — is carried out, often consciously, by analogy with past experience. Analogical inference is a fundamentally heuristic activity, involving search and pattern recognition. A system that supports it must store representations of past experience, aid programmers in finding useful analogies with derivation patterns in the store, help them select the most fruitful analogies, and finally allow them to adapt the store of knowledge as needs and understanding change [10]. As our grasp of this kind of heuristic reasoning improves, our tools will become better at helping programmers find not only new analogies but also new *kinds* of analogies. The heuristic mechanism of an inferential programming system must facilitate this kind of reasoning.

A simple example will illustrate the sort of reasoning that might go on. Consider the derivation of a program for numbering the nodes of a tree in preorder. A specification, say for generalized tree traversal, is committed to visit tree nodes in preorder. From this preorder enumeration algorithm, an algorithm for explicit preorder *numbering* is then derived. Now, the most natural way to derive the *postorder* numbering program is to follow this derivation, but with a slightly different commitment. These derivations are closely related because, although the commitments are different, the pattern of simplication steps is essentially the same. That is, at some level of abstraction, the same simplification activity is being performed. This could be the case even if the new program structure that results may not have any obvious resemblance with the original program.

Roughly, an analogy exists between two phenomena if there is a 'close' general phenomenon that captures essential qualities of both. If we are to reason effectively by analogy, then we will need to develop a language for program derivations that has an abstraction mechanism that is rich enough to *express* these generalizations. Thus, we must not only introduce conceptualizations concerning the fundamental program derivation steps, but about common *patterns* of their usage. Present programming languages do not, in general, have sufficiently rich abstraction mechanisms even to express directly the various kinds of analogies that can exist among *programs*. Were these analogies expressible, they still would not be nearly as useful (and would not reflect our intuitive thinking nearly as closely) as the sorts of analogies that exist among *derivations*. The issue boils down to this: Can we find program derivation *abstractions* that can capture the common patterns of programming activity?

5. PROGRAMS OF THE FUTURE

Just as twenty years ago we learned to move away from the details of object code by thinking about control and data structures more abstractly, we are learning now to move away from the details of algorithm, representation, and implementation by thinking instead about the qualities we desire of them and how they might be chosen. Thus, rather than leading to programs we can no longer understand, the use of inferential programming techniques will lead to a different view of how programs are to be presented.

Stripped down to essentials, our claim is that the 'programs' of the future will in fact be descriptions of program derivations. Documentation methods based on stepwise-refinement methodologies are already strong evidence that there is movement toward this approach. These documentation methods also provide support for the hypothesis that program derivations offer a more intuitive and revealing way of *explaining* programs than do conventional proofs of correctness. The conventional proofs may succeed in convincing the reader of the correctness of an algorithm without giving him any hint of *why* the algorithm works or how it came about. On the other hand, a derivation may be thought of as an especially well-structured 'constructive' proof of correctness of the algorithm, taking the reader step by step from an initial abstract algorithm he accepts as meeting the specifications of the problem to a highly connected and efficient implementation of it.

We shall not arrive at inferential programming overnight, however, because the very act of producing a complete derivation requires a programmer to express some of his previously unexpressed intuitions. Thus, it may often be harder to produce a complete program derivation than simply to write code for an implementation. The additional effort is justified by the fact that the explicit representation of the derivation sequence facilitates analysis, proof, and, most importantly, eventual modification of the programs derived. Many tools remain to be built to make this kind of programming possible. We believe, nevertheless, that the first comprehensive steps are becoming feasible, and we hope, further, that the arguments we have put forward in this paper will make the outcome seem worth the effort.

Department of Computer Science,
Carnegie-Mellon University, Pittsburgh, U.S.A.

NOTE

* This research was supported in part by the Defense Advanced Research Projects Agency (DOD), ARPA Order No. 3597, monitored by the Air Force Avionics Laboratory under Contract F33615-81-K-1539, and in part by the U.S. Army Communications R&D Command under Contract DAAK80-81-K-0074. The views and conclusions contained in this document are those of the authors and should not be interpreted as representing the official policies, either expressed or implied, of the Defense Advanced Research Projects Agency or the U.S. Government.

REFERENCES

1. Balzer, R.: 1981, 'Transformational Implementation: An Example', *IEEE Transactions on Software Engineering* SE-7(1), 3—14.
2. Barstow, D. R.: 1980, 'The Roles of Knowledge and Deduction in Algorithm Design', *Yale Research Report* 178, April.
3. Bates, J. L.: 1979, 'A Logic for Correct Program Development', Ph.D. Thesis, Cornell University.
4. Bates, J. L. and Constable, R. L.: 1982, 'Proofs as Programs', *Cornell University Technical Report*.
5. Bauer, F. L. *et al.*: 1981, 'Programming in a Wide Spectrum language: A Collection of Examples', *Science of Computer Programming* 1, 73—114.
6. Bauer, F. L.: 1982, 'From Specifications to Machine Code: Program Construction through Formal Reasoning', *Sixth International Conference on Software Engineering*.
7. Beckman, L., Haraldsson, A., Oskarsson, Ö, and Sandewall, E.: 1976, 'A Partial Evaluator and its Use as a Programming Tool', *Artificial Intelligence* 7, 319—357.
8. Broy, M. and Pepper, P.: 1981, 'Program Development as a Formal Activity', *IEEE Transactions on Software Engineering* SE-7(1), 14—22.
9. Burstall, R. M. and Darlington, J.: 1977, 'A Transformation System for Developing Recursive Programs', *Journal of the ACM* 24(1), 44—67.
10. Carbonell, J.: 1982, 'Learning by Analogy: Formulating and Generalizing Plans from Past Experience', *Carnegie-Mellon University Technical Report*.
11. Cheatham, T. E. and Wegbreit, B.: 1972, 'A Laboratory for the Study of Automatic Programmming', *AFIPS Spring Joint Computer Conference* 40.
12. Cheatham, T. E., Townley, J. A., and Holloway, G. H.: 1979, 'A System for Program Refinement', *Fourth International Conference on Software Engineering*, 53—63.
13. Clark, K. and Darlington, J.: 1980, 'Algorithm Classification through Synthesis', *Computer Journal* 23(1).
14. Clocksin, W. F. and Melish, C.S.: 1981, *Programming in PROLOG*, Springer-Verlag.
15. De Millo, R. A., Lipton, R. J., and Perlis, A. J.: 1979, 'Social Processes and Proofs of Theorems and Programs', *Communications of the ACM* 22(5), 271—280.

16. Detlefsen, M. and Luker, M.: 1980, 'The Four-Color Theorem and Mathematical Proof', *The Journal of Philosophy* **77**(12), 803—820.
17. Dijkstra, E. W.: 1971. 'Notes on Structured Programming', in: *Structured Programming*, O. J. Dahl, E. W. Dijkstra, C. A. R. Hoare (Eds.) Academic Press.
18. Dijkstra, E. W.: 1976, *A Discipline of Programming*, Prentice-Hall.
19. Ershov, A. P.: 1978, 'On the Essence of Compilation', in: *Formal Descriptions of Programming Concepts*, E. J. Neuhold (Ed.), North-Holland.
20. Feather, M. S.: 1982, 'A System for Assisting Program Transformation', *ACM Transactions on Programming Languages and Systems* **4**(1), 1—20.
21. Floyd, R. W.: 1979, 'The Paradigms of Programming', *Communications of the ACM* **22**(8), 455—460.
22. Goad, C.: 1982, 'Automatic Construction of Special-Purpose Programs', *6th Conference on Automated Deduction*.
23. Gordon, M. J., Milner, A. J., and Wadsworth, C. P.: 1979, *Edinburgh LCF*. Springer-Verlag Lecture Notes in Computer Science.
24. Green, C. C. and Barstow, D. R.: 1978, '*On Program Synthesis Knowledge*', *Artificial Intelligence* **10**, 241.
25. Green, C. *et al.*: 1981, 'Research on Knowledge-Based Programming and Algorithm Design', *Kestrel Institute Technical Report*.
26. Gries, D.: 1981, *The Science of Computer Programming*, Springer-Verlag.
27. Haken, W., Appel, K. and Koch, J.: 1977, 'Every Planar Map is Four-Colorable', *Illinois Journal of Mathematics* **21**(84), 429—567.
28. Knuth, D. E.: 1974, 'Structured Programming with Goto Statements', *Computing Surveys* **6**(4), 261—301.
29. Kriesel, G.: 1981, 'Neglected Possibilities of Processing Assertions and Proofs Mechanically: Choice of Problems and Data', in: *University-Level Computer-Assisted Instruction at Stanford: 1968—1980*. Stanford University.
30. Manna, Z. and Waldinger, R.: 1979, 'Synthesis: dreams ⇒ programs', *IEEE Transactions on Software Engineering* **SE-5**(4).
31. Martin-Löf, P.: 1979, 'Constructive Mathematics and Computer Programming', *6th International Congress for Logic, Methodology and Philosophy of Science*.
32. Paige, R. and Koenig, S.: 1982, 'Finite Differencing of Computable Expressions', *ACM Transactions on Programming Languages and Systems* **4**(3), 402—454.
33. Reif, J. and Scherlis, W. L.: 1982, 'Deriving Efficient Graph Algorithms', *Carnegie-Mellon University Technical Report*.
34. Rich, C. and Shrobe, H.: 1978, 'Initial Report on a Lisp Programmer's Apprentice', *IEEE Transactions on Software Engineering* **SE-4**(6), 456—467.
35. Scherlis, W. L.: 1981, 'Program Improvement by Internal Specialization', *8th Symposium on Principles of Programming Languages*, pp. 41—49.
36. Schwartz, J. T.: 1973, 'On Programming, an Interim Report on the SETL Project', Courant Institute of Mathematical Sciences, New York University.
37. Schwartz, J. T.: 1979, 'On Correct Program Technology', Courant Institute of Mathematical Sciences, New York University.
38. Sintzoff, M.: 1980, 'Suggestions for Composing and Specifying Program Design Decisions', *International Symposium on Programming*, Springer-Verlag Lecture Notes in Computer Science.

39. Swart, E. R.: 1980, 'The Philosophical Implications of the Four-Color Problem', *American Mathematical Monthly* **87**(9), 697—707.
40. Swartout, W. and Balzer, R.: 1982, 'On the Inevitable Intertwining of Specification and Implementation', *Communications of the ACM* **25**(7), 438—440.
41. Teller, P.: 1980, 'Computer Proof', *The Journal of Philosophy* **77**(12), 803—820.
42. Tymoczko, T.: 'The Four-Color Problem and its Philosophical Significance', *The Journal of Philosophy* **66**(2), 57—83.
43. Wand, M.: 1980, 'Continuation-Based Program Transformation Strategies', *Journal of the ACM* **27**(1), 164—180.
44. Waters, R. C.: 1981. 'A Knowledge Based Program Editor', *7th International Joint Conference on Artificial Intelligence*, Vancouver.
45. Wile, D. S.: 1981, 'Program Developments as Formal Objects', *USC/Information Sciences Institute Technical Report*.
46. Wirth, N.: 1971, 'Program Development by Stepwise Refinement', *Communications of the ACM* **14** (4), 221—227.

C. A. R. HOARE

MATHEMATICS OF PROGRAMMING

Mathematical Laws Help Programmers Control the Complexity of Tasks

Note: This article contains the text of the speech given by Tony Hoare at the Boston Computer Museum on the occasion of *BYTE*'s 10th Anniversary celebration. For practical reasons, some of the author's original notation has been modified slightly.

I hold the opinion that the construction of computer programs is a mathematical activity like the solution of differential equations, that programs can be derived from their specifications through mathematical insight, calculation, and proof, using algebraic laws as simple and elegant as those of elementary arithmetic. Such methods of program construction promise benefits in specifications, systems software, safety-critical programs, silicon design, and standards.

PRINCIPLES

To substantiate this opinion, I begin with four basic principles:
1. Computers are mathematical machines. Every aspect of their behavior can be defined with mathematical precision, and every detail can be deduced from this definition with mathematical certainty by the laws of pure logic.
2. Computer programs are mathematical expressions. They describe, with unprecedented precision and in the most minute detail, the behavior, intended or unintended, of the computer on which they are executed.
3. A programming language is a mathematical theory that includes concepts, notations, definitions, axioms, and theorems. These help a programmer develop a program that meets its specification and prove that it does.
4. Programming is a mathematical activity. Like other branches of applied mathematics and engineering, its successful practice requires the determined and meticulous application of traditional methods of mathematical understanding, calculation, and proof.

135

Timothy R. Colburn et al. (eds.), Program Verification, 135—154.
© 1993 *Kluwer Academic Publishers.*

HOWEVER . . .

These are general philosophical and moral principles, but all the actual evidence is against them. Nothing is as I have described it, neither computers nor programs nor programming languages nor even programmers.

I find digital computers of the present day to be very complicated and rather poorly defined. As a result, it is usually impractical to reason logically about their behavior. Sometimes, the only way of finding out what they will do is by experiment. Such experiments are certainly not mathematics. Unfortunately, they are not even science, because it is impossible to generalize from their results or to publish them for the benefit of other scientists.

Many computer programs of the present day are of inordinate size — many thousands of pages of closely printed text. Mathematics has no tradition of dealing with expressions on this scale. Normal methods of calculation and proof seem wholly impractical to conduct by hand, and 15 years of experience suggest that computer assistance can only make matters worse.

Programming languages of the present day are even more complicated than the programs you write with them and the computers on which they are intended to run. Valiant research has attempted to formulate mathematical definitions of these standard languages. But the size and complexity of those definitions make them impractical in deriving useful theorems or proving relevant properties of programs.

Finally, many programmers of the present day have been educated in ignorance and fear of mathematics. Of course, many programmers are mathematical graduates who have acquired a good grasp of topology, calculus, and group theory. But it never seems to occur to them to take advantage of their mathematical skills to define a programming problem and search for its solution.

Our present failure to recognize and use mathematics as the basis for a programming discipline has a number of notorious consequences. They are the same as you would get from a similar neglect of mathematics in drawing maps, marine navigation, bridge building, air-traffic control, and exploring space. In the older branches of science and engineering, the relevant physical and mathematical knowledge is embodied in a number of equations, formulas, and laws, many of which are simple enough to be taught to schoolchildren. The practicing scientist or engineer will be intimately familiar with these laws and will use them explicitly, or even instinctively, to find solutions to otherwise intractable problems.

What then are the laws of programming that would help programmers control the complexity of their tasks? Many programmers would be hard-pressed to name a single one.

ARITHMETIC

But the laws of programming are as simple, as obvious, and as useful as the laws in any other branch of mathematics, for example, elementary arithmetic. Consider the multiplication of numbers. Figure 1 shows some of the relevant algebraic laws: Multiplication is associative; its identity (or unit) is the number 1; it has the number 0 as its zero (or fixed point); and, finally, it distributes through addition. Figure 2 gives the defining properties of an ordering relation ($\leqslant$) like comparison of the magnitude of numbers. Such an order is reflexive, antisymmetric,

$$x \times (y \times z) = (x \times y) \times z$$
$$x \times 1 = 1 \times x = x$$
$$x \times 0 = 0 \times x = 0$$
$$(x + y) \times z = (x \times z) + (y \times z)$$

Fig. 1. Some algebraic laws relevant to the multiplication of numbers.

$$x \leqslant x$$
$$x \leqslant y \; \& \; y \leqslant x \to x = y$$
$$x \leqslant y \; \& \; y \leqslant z \to x \leqslant z$$

Note: $\leqslant$ can be interpreted as "is a subset of"; $\to$ denotes "implies"; & denotes "and."

Fig. 2. Defining properties of an ordering relation or a partial ordering.

and transitive. These laws hold also for a partial ordering like the inclusion relation between sets.

Figure 3 describes the properties of the least upper bound, or LUB, of an ordering, denoted by the cup notation ($\cup$). These laws are equally valid, whether the LUB is the union of two sets or the greater of two numbers. The first law states the fundamental property that the LUB is an upper bound on both its operands, and it is the least of all such bounds. The remaining laws are derived from the fundamental law by the properties of ordering. They state that the LUB operator is idempotent (i.e., $x \cup x = x$), symmetric, and associative. Finally, the partial ordering can itself be defined in terms of LUB.

Figure 4 shows some additional laws that hold for natural numbers or non-negative integers. Here, the LUB of two numbers is simply the greater of them. If you multiply the greater of x or y by z, you get the same result as multiplying both x and y by z and choosing the greater of the products. This fact is clearly and conveniently stated in the laws of distribution of multiplication through the LUB. An immediate con-

$$(x \cup y) \leqslant z \leftrightarrow x \leqslant z \,\&\, y \leqslant z$$
$$x \cup x = x$$
$$x \cup y = y \cup x$$
$$x \cup (y \cup z) = (x \cup y) \cup z$$
$$x \leqslant y \leftrightarrow x \cup y = y$$

Note: $\leftrightarrow$ denotes "is equivalent to."

Fig. 3. Properties of the least upper bound of an ordering.

$$x \cup y = \text{the greater of } x \text{ and } y$$
$$(x \cup y) \times z = (x \times z) \cup (y \times z)$$
$$z \times (x \cup y) = (z \times x) \cup (z \times y)$$
$$w \leqslant y \,\&\, x \leqslant z \rightarrow w \times x \leqslant y \times z$$

Fig. 4. Some additional laws for natural numbers or nonnegative integers.

sequence of these laws is that multiplication is a monotonic operator, in the sense that it preserves the ordering of its operands. If you decrease either factor, the product can only decrease, too, as stated in the last law of Figure 4.

In the arithmetic of natural numbers, multiplication does not in general have an exact inverse. Instead, we commonly use a quotient operator — which approximates the true result from below. It is obtained from normal integer division by just ignoring the remainder. Thus, the result of dividing y by nonzero z is the largest number such that when you multiply it back by z, the result still does not exceed y. This fact is clearly stated in the first law of Figure 5. The same fact is stated more simply in the second law, which I will call the fundamental law of division.

Other properties of division can be easily proved from the fundamental law. For example, the third law of Figure 5 is proved by substituting y divided by z for x in the first law. The last law states that division by a product is the same as successive division by its two factors. A proof is given in Figure 6. The proof shows that any w that is bounded by the left-hand side of the equation is bounded also by the right-hand side, and vice versa; it follows by the properties of partial ordering that the two sides are equal. The only laws used in the main part of the proof are the associativity of multiplication and the fundamental law of division, which is used three times to move a divisor from one side of the inequality to the other.

PROGRAMS

I have selected these laws to ensure that computer programs satisfy the same laws as elementary arithmetic. I will write programs in a mathe-

$$
\begin{array}{l}
if\ z \neq 0 \\
y \div z = \max\{x \mid x \times z \leqslant y\} \\
x \leqslant (y \div z) \leftrightarrow (x \times z) \leqslant y \\
(y \div z) \times z \leqslant y \\
x \div (y \times z) = (x \div z) \div y
\end{array}
$$

Fig. 5. Properties of division of natural numbers.

$$\text{Given } y \neq 0 \text{ and } z \neq 0$$
$$w \leqslant x \div (y \times z)$$
$$\leftrightarrow w \times (y \times z) \leqslant x$$
$$\leftrightarrow (w \times y) \times z \leqslant x$$
$$\leftrightarrow w \times y \leqslant x \div z$$
$$\leftrightarrow w \leqslant (x \div z) \div y$$

Fig. 6. A proof of the last law of division given in Figure 5.

matical notation first introduced by Edsger W. Dijkstra. Some of the commands are summarized as follows:

• The SKIP command terminates but does nothing else. In particular, it leaves the values of all variables unchanged.

• The ABORT command is at the other extreme. It may do anything whatsoever, or it may fail to do anything whatsoever. In particular, it may fail to terminate. A computer that has 'gone wrong' or a program that has run wild, perhaps by corrupting its own code, can behave this way. You would not want to write an ABORT command; in fact, you should take pains to prove that you have not created such a situation by accident. In such proofs, and in the general mathematics of programming, ABORT plays a valuable role. However much we dislike it, there is ample empirical evidence for its existence.

• The sequential composition of two commands x and y is written $(x \oplus y)$. This starts behaving like x. If and when x terminates, y starts in an initial state equal to the final state left by x. The composition $(x \oplus y)$ terminates when y terminates but fails to terminate if either x or y fails to do so.

The basic algebraic laws for sequential composition are given in Figure 7. The first law of associativity states that if three commands are combined sequentially, it does not matter in which way they are bracketed. The second law gives SKIP as the unit of composition. It states that a command x remains unchanged when it is either followed or preceded by SKIP. The third law gives ABORT as a zero for composition. You cannot recover from abortion by preceding it or following it by any other command. These three algebraic laws for composition are the same as those for multiplication of numbers.

$x \oplus (y, z) = (x \oplus y) \oplus z$
SKIP $\oplus x = x = x \oplus$ SKIP
ABORT $\oplus x =$ ABORT $= x \oplus$ ABORT

Note: $\oplus$ denotes "composition."

Fig. 7. The basic algebraic laws for sequential composition of commands.

The next important feature of programming is the conditional. Let b be a logical expression that in all circumstances evaluates to a logical value, true or false. If x and y are commands, the notation

$$x \lessdot b \gtrdot y \quad (x \text{ if } b \text{ else } y)$$

denotes the conditional command. If the logical expression b is true, the command x is obeyed and y is omitted. If the result is false, y is obeyed and x is omitted. This informal description you will find summarized in the first law of Figure 8.

Interpreting the if symbol $\lessdot$ and the else symbol $\gtrdot$ to be brackets surrounding the logical expression b, the notation $\lessdot b \gtrdot$ appears as an infix operator between two commands x and y. The reason for this novel notation is that it simplifies the expression and use of the relevant algebraic laws. For example, the conditional $\lessdot b \gtrdot$ is idempotent (i.e., $x \lessdot b \gtrdot x = x$) and associative, and it distributes through $\lessdot c \gtrdot$ for

$(x \lessdot \text{true} \gtrdot y) = x = (y \lessdot \text{false} \gtrdot x)$
$(x \lessdot b \gtrdot x) = x$
$x \lessdot b \gtrdot (y \lessdot b \gtrdot z) = (x \lessdot b \gtrdot y) \lessdot b \gtrdot z$
$\qquad = x \lessdot b \gtrdot z$
$x \lessdot b \gtrdot (y \lessdot c \gtrdot z) = (x \lessdot b \gtrdot y) \lessdot c \gtrdot (x \lessdot b \gtrdot z)$
$(x \lessdot b \gtrdot y) \oplus z = (x \oplus z) \lessdot b \gtrdot (y \oplus z)$

Note: $\lessdot$ denotes "if"; $\gtrdot$ denotes "else."

Fig. 8. The conditional operator.

any logical expression c. Finally, sequential composition distributes leftward (but not rightward) through a conditional.

Figure 9 shows a picture of the conditional as a structured flowchart. Such pictures are useful in first presenting a new idea and in committing it to memory. But pictures are quite unsuitable for expressing algebraic laws or for mathematical reasoning. Unfortunately, some problems are so widespread and so severe that flowcharts must be recommended and actually welcomed as their solutions.

Listing 1 shows the expression of the structure of a conditional in BASIC. Programming in BASIC is like doing arithmetic with roman numerals. For simple tasks like addition and subtraction, roman numerals are easier than arabic because you do not first have to learn 100 facts about the addition and subtraction of the 10 digits, and you avoid most of the complications of carry and borrow.

```
410 IF b THEN GO TO 554
411        .
           .
           .
550 GO TO 593
554        .
           .
           .
593 REM
```

Listing 1. The expression of the structure of a conditional in BASIC.

The disadvantages of roman numerals become apparent only in more complex tasks like multiplication, or worse, division for which the only known technique is trial and error. You have to guess the solution, test it by multiplying it by the divisor and comparing the dividend, and make a new guess if you are wrong. This is the way we teach beginners in BASIC, but division of roman numerals is much easier because the fundamental law of division tells you whether the new guess should be smaller or greater than the last.

Thankfully, arabic numerals have displaced roman ones in our schools, and the effective algorithm for long division has replaced the roman method of trial and error with an orderly process of calculation;

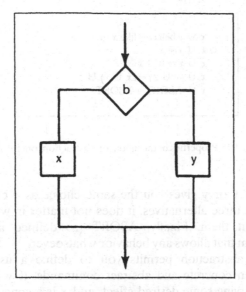

Fig. 9. A picture of the conditional as a structured flowchart.

when carefully executed, it leads invariably to the correct solution. In cases of doubt, the answer can still be checked by multiplication; but if this discovers an error, you do not try to debug the digits of your answer by trial and error. You go back over the steps of the calculation and correct them — or else you start again.

ABSTRACTION

An abstract command is one that specifies the general properties of a computer's desired behavior without prescribing exactly how that behavior is to be achieved. A simple example of an abstract command is the union, or LUB $(x \cup y)$, of two commands x and y, which may themselves be either abstract or concrete. The union command can be obeyed by obeying either of its operands. The choice between them is left open and can be made later by the programmer, the compiler, or even by some device in the computer during program execution. For this reason, abstract programs are sometimes called non-deterministic.

The properties of the union operator (Figure 10) are exactly what you would expect. A command to do x or x leaves you no choice but

$x \cup y$ behaves like x or y
$x \cup x = x$
$x \cup y = y \cup x$
$x \cup (y \cup z) = (x \cup y) \cup z$
$x \cup \text{ABORT} = \text{ABORT}$

Fig. 10. Properties of the union, or abstraction, operator.

to do x. To do x or y gives you the same choice as y or x. And in a choice between three alternatives, it does not matter in what order you choose between them. Finally, ABORT (as defined above) is the abstract program that allows any behavior whatsoever.

Introducing abstraction permits you to define a useful ordering relation between concrete and abstract commands. If y is an abstract command specifying some desired effect, and x is a concrete command that achieves this effect, you can say that x satisfies y and use the notation used here for a partial order: $x \leqslant y$.

The command x can also be abstract. If it is, the ordering relation means that x is the same as y or that x is more specific, concrete, or deterministic than y. In either case, x meets the specification y because every possible execution of x is described and therefore allowed by y. As stated in Figure 11, the satisfaction relation is a partial order, and the abstraction operator is its LUB.

$x \leqslant y \leftrightarrow x \cup y = y$
$\leftrightarrow x$ satisfies specification y
$x \leqslant x$
$x \leqslant y \,\&\, y \leqslant x \to x = y$
$x \leqslant y \,\&\, y \leqslant z \to x \leqslant z$
$(x \cup y) \leqslant z \to x \leqslant z \,\&\, y \leqslant z$

Fig. 11. The satisfaction relation ($\leqslant$), a partial order, with the abstraction operator ($\cup$) as its LUB.

Abstract commands can be combined by all the same operators as concrete commands. Figure 12 shows that sequential composition distributes through abstract choice in both directions. It follows that composition is monotonic in both its arguments. In fact, all the operators of a programming language are monotonic in this sense. There are good theoretical reasons for this, and there are also very beneficial consequences for the practical solution of programming problems.

$$(x \cup y) \oplus z = (x \oplus z) \cup (y \oplus z)$$
$$z \oplus (x \cup y) = (z \oplus x) \cup (z \oplus y)$$
$$w \leqslant y \ \& \ x \leqslant z \rightarrow w \oplus x \leqslant y \oplus z$$

Fig. 12. Sequential composition of commands distributes through abstract choice in both directions.

REFINEMENT

The task of a programmer can be described as a problem in mathematics. You start with an abstract description of what you want the computer to do, carefully checking that it is an accurate description of the right problem. This is often the most difficult part of the task and requires the most powerful tools. So in the specification y, you should take advantage of the full range of concepts and notations of mathematics, including concepts that cannot be represented on a computer and operations that could not be implemented in a programming language.

You must then find some program x that solves the inequality $x \leqslant y$, where y is the specification of the program. Mathematics provides many formulas and methods for solving equations (and inequalities), from linear and quadratic to differential and integral. In all cases, the derivation of a solution can use the full power of mathematics, but the solution itself must be expressed as a formula in some more restricted notation. The same is true in programming, where the eventual solution must be expressed in the restricted notations of an implemented concrete programming language.

The most powerful general method of solving a complicated problem is to split it into simpler subproblems, which can then be solved independently. The same method can be applied again to the subproblems until they are simple enough to solve directly by some other more direct method. In the case of computer programming, this is often called *top-down development* or *stepwise refinement* (see Figure 13). You start with the problem of finding some command x (expressed in a concrete programming language) that meets the specification y (expressed in the abstract language of mathematics). The first step requires the insight to split this task into two subproblems and the skill to specify them as abstract programs v and w. Before proceeding further, you prove the correctness of your design so far by showing that the sequential composition of v and w meets the original specification y, or more formally, $v \oplus w \leqslant y$.

Now these two subproblems v and w can be solved sequentially or concurrently by a single programmer or by two teams of programmers, according to the size of the task. When both subproblems are solved, you will have two commands, t and u, expressed in the restricted notations of your chosen programming language, each meeting their respective specifications: $t \leqslant v$ and $u \leqslant w$. All that remains is to deliver their sequential composition $(t \oplus u)$ as a solution to the original problem y. Correctness of the solution has been established not by the traditional laborious and ultimately unsound method of integration testing and debugging after the components have been constructed, but rather by a mathematical proof, which was completed on the very first step, even before the construction of components began.

The validity of the general method of top-down development depends on monotonicity of the composition operator and transitivity of the abstraction ordering. The method can therefore be applied to any other operator of a concrete programming language. It has been treated at length in many learned articles and books. A characteristic of the simplifying power of mathematics is that the whole method can be described, together with a proof of its validity, within the seven lines of Figure 13.

I have drawn an analogy between the multiplication of natural numbers and the sequential composition of commands in programming. This analogy extends even to division. As with division of natural numbers, the quotient of two commands is not an exact inverse. However, it is uniquely defined by the same fundamental law, as shown

Problem: find x such that $x \leqslant y$
Step 1: find v, w such that $v \oplus w \leqslant y$
Step 2a: find t such that $t \leqslant v$
Step 2b: find u such that $u \leqslant w$
Step 3: deliver $t \oplus u$
Proof: $t \oplus u \leqslant v \oplus w$ $\oplus$ is monotonic
    $\therefore t \oplus u \leqslant y$ $\leqslant$ is transitive

Fig. 13. Top-down development or stepwise refinement.

in Figure 14. The quotient of y divided by z is the most abstract specification of a program x, which, when followed by z, is sure to meet the specification y. As a consequence, the quotient itself, when followed by z, meets the original specification. And, finally, when the divisor is the composition of two commands, you can calculate the quotient by successively dividing by these two commands in the reverse order. Since the composition of commands is not symmetric, the reversal of order is important here.

In factoring large numbers, division saves a lot of effort: You only have to guess one of the factors and obtain the other one by calculation. The division of commands offers the same advantages in the factorization of programming problems. In refinement, it replaces the guesswork required in discovering two simpler subtasks by the discovery of only the second of them, as shown in Figure 15. Furthermore, the proof obligation in step 1 of Figure 15 has been eliminated. It is replaced by a formal calculation of the weakest specification that must be met by the first operand of the composition. Reducing guesswork and proof to

$$(x \oplus z) \leqslant y \leftrightarrow x \leqslant (y \div z)$$
$$(y \div z) \oplus z \leqslant y$$
$$x \div (y \oplus z) = (x \div z) \div y$$

Fig. 14. The fundamental law of the quotient of commands. Like the division of natural numbers, the quotient of two commands is not an exact inverse of their multiplication.

Problem: find x such that $x \leqslant y$
Step 1: choose suitable z
Step 2a: find t such that $t \leqslant y \div z$
Step 2b: find u such that $u \leqslant z$
Step 3: deliver $t \oplus u$
Proof: $t \oplus u \leqslant (y \div z) \oplus z$ $\oplus$ is monotonic
$\quad (y \div z) \oplus z \leqslant y$ property of $\div$
$\quad \therefore t \oplus u \leqslant y$ $\leqslant$ is transitive

Fig. 15. A simplification of Figure 13, using division of commands.

calculation is how mathematicians simplify their own tasks, as well as those of users of mathematics — the scientist, the engineer, and now the programmer.

The quotient operator for commands was discovered in a slightly restricted form by Dijkstra, who called it the weakest precondition (see [1]). It is one of the most effective known methods for the design and development of correct algorithms, as shown in numerous examples by David Gries in [2].

PROGRAM MAINTENANCE

In my description of the task of a programmer, I have concentrated on the more glamorous part of that task: specifying, designing, and writing new programs. But a significant portion of a programmer's professional life is spent making changes to old programs. Some of these changes are necessitated by the discovery of errors, some by changes in the specification of the program's desired behavior. The program and the specification may be so large that it is not practical to write a new program from scratch; when only a small part of the specification is changed, you hope that only a small part of the program will need changing to meet it.

Of course, such a hope is not always justified. Consider again the analogy of the division of numbers. A small change in the least significant digits of the dividend results in a small change in the least significant digits of the quotient and can be achieved by a small amount of recalculation. But a small change in the most significant digit of either operand requires that you start the calculation again, leading to a completely different result. In the case of programs, it is often difficult to know which small changes in a large specification will require major changes to the program.

It is, therefore, most important for the original programmer to decide which parts of a specification are most likely to be changed and to structure the program design so that a change to one part of the specification requires a change to only one part of the program. The programmer should then document the program with instructions on how to carry out the change. This too can be done in a rigorous mathematical fashion (see Figure 16). Let y be that part of a complete specification $g(y)$ that is liable to change. Let x be that command in a big program $f(x)$ that is designed to change when y changes. The problem now is to change x to x' so that $f(x')$ meets the changed specification $g(y')$.

The problem of program maintenance is most easily solved when the structure of the program f is the same as the structure of the specification g; in this case, it will be sufficient to ensure that the modified component meets the modified specification.

However, it is not always possible to preserve the structure of a specification in the design of a program. A specification is often most clearly structured with the aid of logical operators like negation and conjunction, which are not generally available in an implemented programming language. Nevertheless, mathematics can often help. Sometimes, the program f has an approximate inverse (f^{-1}), defined in the same way as for the quotient; it is then possible to calculate the proof obligation of the modified program as $x' \leqslant f^{-1}(g(y'))$.

Given: $f(x) \leqslant g(y)$
Problem: find x' such that $f(x') \leqslant g(y')$
Case 1: $f = g$
    solve $x' \leqslant y'$
Case 2: f has approximate inverse f^{-1}
    solve $x' \leqslant f^{-1}(g(y'))$

Fig. 16. Mathematical treatment of program maintenance.

IN REALITY

I must inject a note of realism into my mathematical speculations. Many operators of a programming language do not have suitable approximate inverses, and even if they do, their calculation is impossibly cumbersome. Would you like to calculate the inverse of f, when f is a million lines of FORTRAN code? The problem of the size of mathematical formulas is exactly the same problem that limits the use of mathematics in other branches of science and engineering. Many scientists believe as fervently as I do in the principle that the whole of nature is governed by mathematical laws of great smplicity and elegance, and brilliant scientists have discovered many laws that accurately predict the results of experiments conducted in a rigorously controlled laboratory environment. But when the engineer tries to apply the same laws in practice, the number of uncontrollable variables is so much greater than in the laboratory that a full calculation of the consequences of each design decision is hopelessly impractical. As a result, the engineer uses experience and understanding to formulate and apply various rules of thumb, and uses flair and judgment to supplement (though never to replace) the relevant mathematical calculations.

Experienced programmers have developed a similar intuitive understanding of the behavior of computer programs, many of which are now remarkably sophisticated and reliable. Nevertheless, I would suggest that the skills of our best programmers will be even more productive and effective when exercised within the framework of understanding and applying the relevant mathematics. The mathematics has been demonstrated on small examples, as it were, in the laboratory. The time has come to start the process of scaling up to an industrial environment. At Oxford University, we have a number of collaborative projects with industry to attempt this technology transfer. Preliminary indications are quite promising, both for advances in theory and for benefits in practice.

But it will be no easier to change the working habits of programmers than those of any other profession. I quote from Lord Hailsham, the British minister in charge of our judicial system (see [3]). He is conducting a review on the improvement of judicial practices in the legal profession. He first lists the faults of our current system, including delay, complexity, and obscurity. He continues: "A change in working methods is of course immensely difficult. Habit, interest, training all militate in favour of the status quo. People must be persuaded, taught, if necessary possibly even leaned on — or at least assisted."

Pilot schemes may be necessary to test new methods. They may be unpopular, "but if the alternative is to let all cases to continue subject to the anomalies of a malfunctioning system, we may have to contemplate unprecedented arrangements to allow new methods to be tested."

Changes in professional practice will be just as difficult for programmers as they are for lawyers and judges. They will occur first in the areas of greatest necessity, where the lack of mathematical precision leads to the heaviest costs. In particular, I suggest five such areas: specifications, systems software, standards, silicon structures, and safety.

SPECIFICATIONS

In the initial specification and design of a large-scale software product, the use of mathematics has been found to clarify design concepts and enable you to explore a wide variety of options at a time when less successful alternatives can be cheaply discarded. As a result, the final agreed-upon specification may enjoy that simplicity and conceptual integrity that characterizes the highest quality in design. Furthermore, user's manuals, tutorials, and guides that are based on mathematical specifications can be better structured, more complete, and more comprehensible, even to users who have no knowledge of the underlying mathematics. This promises to mitigate the greatest single cause of error, inconvenience, and customer frustration in the use of sophisticated software products, that is, failure to read and understand the user's manual.

SYSTEMS SOFTWARE

A computer's basic systems software includes items like an operating system, language compilers, utilities, transaction-processing packages, and database management systems. These programs are written by large teams of programmers and are delivered to thousands or millions of customers, who use them daily, hourly, or even continuously. In the years after delivery of such software, many thousands of errors are discovered by the customers themselves; each error must be laboriously analyzed, corrected, and retested; and the corrections must be delivered to and installed by the customer in the next release of the software. A reduction in the number of corrections needed would be cost-effective for the supplier and convenient for the customer. A reduction in the number of errors discovered before delivery of the software could also save a lot of money. No method by itself can guarantee absolute reliability. However, in combination with management control, a mathematical approach looks promising. Even when mistakes are made, they can be traced to their source, and steps can be taken to ensure they do not happen again.

STANDARDS

The standardization of languages and interfaces in hardware and software is a vital precondition for free competition and for the propagation of technical advances. The construction of a mathematical specification for these standards offers the same promise of improved quality in design that I have just described; it also offers a hope of reducing the ambiguities and misunderstandings which lead to errors and incompatibility in various implementations of the standard, and which have prevented us from obtaining full benefit from the existing standards.

SILICON STRUCTURES

The advance of technology in very large-scale integration now makes it possible to build hardware as complex as software. As a result, we are beginning to detect as many design errors during the manufacture and testing of complex devices as we have been finding with software. But now each error costs many thousands of dollars to remove; what is worse, by delaying the introduction of a new device onto the market, these errors can prevent an innovative company from assuming market leadership or even prejudice its survival. As a result, many products are delivered with known design errors, for which the customer will never obtain a correction or restitution. Fortunately, mathematical methods similar to those for sequential programs can be adapted to check logic designs. These methods are especially valuable when the design involves concurrency or parallelism.

SAFETY

Computer programs are increasingly used in systems that are critical to the safety of the general public — control of railway signaling, airplane engines, and chemical and nuclear processes. The engineering techniques used in these systems are subject to the most rigorous analysis and control, often enforced by law. Mathematical methods offer the best hope of extending such control to computer software, and when this happens, a program could gain a reputation as the most reliable component of any system in which it resides.

CONCLUSION

These are the five areas in which I believe the introduction of mathematical methods gives the greatest promise of rapid economic return. But their implementation will require a significant change in working habits and working attitudes, not only of individual programmers but of whole teams and departments. It requires the enthusiastic participation of coders, systems analysts, project leaders, and managers at all levels. It involves reforming the curricula in training departments, colleges, and universities. It even requires educating the general public to demand and expect as much — or more — quality and design integrity from computer programs as they do from other products critical to their economic and physical well-being.

That is why I am pleased and grateful for this opportunity to address you on the 10th anniversary of the foundation of BYTE magazine. This is a magazine that has a deservedly wide readership in our profession, ranging from theoretical computer scientists through lecturers, teachers, managers, engineers, and students to home computer enthusiasts. But it is also a magazine that takes seriously its responsibility to propagate improved methods and higher standards in the programming profession. I salute its ideals and achievement and wish it continued prosperity and progress.

Computing Laboratory,
Oxford University, U.K.

154 C. A. R. HOARE

REFERENCES

1. Dijkstra, Edsger W.: 1976, *A Discipline of Programming*. Englewood Cliffs, NJ: Prentice-Hall.
2. Gries, David: 1981, *The Science of Programming*. New York: Springer.
3. *The Guardian*, July 10, 1985, London, England.

ON FORMALISM IN SPECIFICATIONS

Specification is the software life-cycle phase concerned with precise definition of the tasks to be performed by the system. Although software engineering textbooks emphasize its necessity, the specification phase is often overlooked in practice. Or, more precisely, it is confused with either the preceding phase, definition of system objectives, or the following phase, design. In the first case, considered here in particular, a natural-language *requirements document* is deemed sufficient to proceed to system design — without further specification activity.

This article emphasizes the drawbacks of such an informal approach and shows the usefulness of formal specifications. To avoid possible misunderstanding, however, let's clarify one point at the outset: We in no way advocate formal specifications as a *replacement* for natural-language requirements; rather, we view them as a *complement* to natural-language descriptions and, as will be illustrated by an example, as an aid in *improving* the quality of natural-language specifications.

Readers already convinced of the benefits of formal specifications might find in this article some useful arguments to reinforce their viewpoint. Readers not sharing this view will, we hope, find some interesting ideas to ponder.

THE SEVEN SINS OF THE SPECIFIER

The study of requirements documents, as they are routinely produced in industry, yields recurring patterns of deficiencies. Table I lists seven classes of deficiencies that we have found to be both common and particularly damaging to the quality of requirements.

The classification is interesting for two reasons. First, by showing the pitfalls of natural-language requirements documents, it gives some weight to the thesis that formal specifications are needed as an intermediate step between requirements and design. Second, since natural-language requirements are necessary whether or not one accepts the thesis that they should be complemented with formal specifications, it provides writers of such requirements with a checklist of common

155

Timothy R. Colburn et al. (eds.), Program Verification, 155—189.
© 1993 *Kluwer Academic Publishers.*

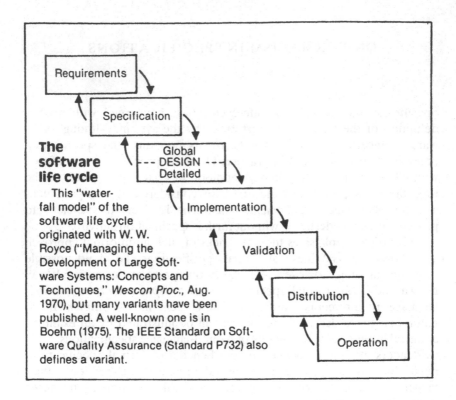

The software life cycle

This "waterfall model" of the software life cycle originated with W. W. Royce ("Managing the Development of Large Software Systems: Concepts and Techniques," *Wescon Proc.*, Aug. 1970), but many variants have been published. A well-known one is in Boehm (1975). The IEEE Standard on Software Quality Assurance (Standard P732) also defines a variant.

mistakes. Writers of most kinds of software documentation (user manuals, programming language manuals, etc.) should find this list useful; we'll demonstrate its use though an example that exhibits all the defects except the last one.

A REQUIREMENTS DOCUMENT

The reader is invited to study, in light of the previous list, some of the software documentation available to him. We could do the same here and discuss actual requirements documents, taken from industrial software projects, as we did in a previous version of this article [1]. But such a discussion is not entirely satisfactory; the reader may feel that the examples chosen are not representative. Also, one sometimes hears the remark that nothing is inherently wrong with natural-language

TABLE I
The seven sins of the specifier.

Noise:	The presence in the text of an element that does not carry information relevant to any feature of the problem. Variants: *redundancy: remorse.*
Silence:	The existence of a feature of the problem that is not covered by any element of the text.
Overspecification:	The presence in the text of an element that corresponds not to a feature of the problem but to features of a possible solution.
Contradiction:	The presence in the text of two or more elements that define a feature of the system in an incompatible way.
Ambiguity:	The presence in the text of an element that makes it possible to interpret a feature of the problem in at least two different ways.
Forward reference:	The presence in the text of an element that uses features of the problem not defined until later in the text.
Wishful thinking:	The presence in the text of an element that defines a feature of the problem in such a way that a candidate solution cannot realistically be validated with respect to this feature.

specifications. All one has to do, the argument continues, is to be careful when writing them or hire people with good writing skills. Although well-written requirements are obviously preferable to poorly written ones, we doubt that they solve the problem. In our view, natural-language descriptions of any significant system, even ones of good quality, exhibit deficiencies that make them unacceptable for rigorous software development.

To support this view, we have chosen a single example, which, although openly academic in nature, is especially suitable because it was explicitly and carefully designed to be a 'good' natural-language specification. This example is the specification of a well-known text-processing problem. The problem first appeared in a 1969 paper by Peter Naur where it was described as reproduced here in Figure 1.

Naur's paper was on a method for program construction and program proving; thus, the problem statement in Figure 1 was accompanied by a program and by a proof that the program indeed satisfied the requirements.

Given a text consisting of words separated by BLANKS or by NL (new line)
characters, convert it to a line-by line form in accordance with the following
rules:
   (1) line breaks must be made only where the given text has BLANK or
        NL;
   (2) each line is filled as far as possible, as long as
   (3) no line will contain more than MAXPOS characters.

Fig. 1. Naur's original statement of a well-known text-processing problem.

The problem appeared again in a paper by Goodenough and Gerhart,
which had two successive versions. Both versions included a criticism
of Naur's original specification.

Goodenough and Gerhart's work was on program testing. To explain
why a paper on program testing included a criticism of Naur's text, it is
necessary to review the methodological dispute surrounding the very
concept of testing. Some researchers dismiss testing as a method for
validating software because a test can cover only a fraction of significant cases. In the words of E. W. Dijkstra [2], "Testing can be a very
effective way to show the presence of bugs, but it is hopelessly inadequate for showing their absence." Thus, in the view of such critics,
testing is futile; the only acceptable way to validate a program is to
prove its correctness mathematically.

Since Goodenough and Gerhart were discussing test data selection
methods, they felt compelled to refute this *a priori* objection to any
research on testing. They dealt with it by showing significant errors in
programs whose "proofs" had been published. Among the examples
was Naur's program, in which they found seven errors — some minor,
some serious.

Our purpose here is not to enter the testing-versus-proving controversy. The Naur—Goodenough/Gerhart problem is interesting, however, because it exhibits in a particularly clear fashion some of the
difficulties associated with natural-language specifications. Goodenough
and Gerhart mention that the trouble with Naur's paper was partly due
to inadequate specification; since their paper proposed a replacement
for Naur's program, they gave a corrected specification. This specification was prepared with particular care and was changed as the paper

was rewritten. Apparently somebody criticized the initial version, since the last version contains the following footnote:

Making these specifications precise is difficult and is an excellent example of the specification task. The specifications here should be compared with those in our original paper.

Thus, when we examine the final specification, it is only fair to consider it not as an imperfect document written under the schedule constraints usually imposed on software projects in industry, but as the second version of a carefully thought-out text, describing what is really a toy problem, unplagued by any of the numerous special considerations that often obscure real-life problems. If a natural-language specification of a programming problem has ever been written with care, this is it. Yet, as we shall see, it is not without its own shadows.

Figure 2 (see p. 11) gives Goodenough and Gerhart's final specification, which should be read carefully at this point. For the remainder of this article, numbers in parentheses — for example, (21) — refer to lines of text as numbered in Figure 2.

ANALYSIS OF THE SPECIFICATION

The first thing one notices in looking at Goodenough and Gerhart's specification is its length: about four times that of Naur's original by a simple character count. Clearly, the authors went to great pains to leave nothing out and to eliminate all ambiguity. As we shall see, this over-zealous effort actually introduced problems. In any case, such length seems inappropriate for specifying a problem that, after all, looks fairly simple to the unprejudiced observer.

Before embarking on a more detailed analysis of this text, we should emphasize that the aim of the game is not to criticize this particular paper; the official subject matter of Goodenough and Gerhart's work was testing, not specification, and the prescription period has expired anyway. We take the paper as an example because it provides a particularly compact basis for the study of common mistakes.

Noise

'Noise' elements are identified by solid underlines in Figure 2. Noise is not necessarily a bad thing in itself; in fact, it can play the same role as

1	The program's input is a stream of characters whose end is
2	signaled with a special end-of-text character, ET. There is exactly
3	one ET character in each input stream. Characters are classified
4	as:
5	• Break characters – BL (blank) and NL (new line);
6	• Nonbreak characters – all others except ET;
7	• the end-of-text indicator – ET.
8	A *word* is a nonempty sequence of nonbreak characters. A
9	*break* is a sequence of one or more break characters. Thus, the
10	input can be viewed as a sequence of words separated by breaks,
11	with possibly leading and trailing breaks, and ending with ET.
12	The program's output should be the same sequence of words
13	as in the input, with the exception that an oversize word (i.e. a
14	word containing more than MAXPOS characters, where MAXPOS
15	is a positive integer) should cause an error exit from the program
16	(i.e. a variable, Alarm, should have the value TRUE). Up to the
17	point of an error, the program's output should have the following
18	properties:
19	1. A new line should start only between words and at the
20	beginning of the output text, if any.
21	2. A break in the input is reduced to a single break character in
22	the output.
23	3. As many words as possible should be placed on each line
24	(i.e., between successive NL characters).
25	4. No line may contain more than MAXPOS characters (words
26	and BLs).

Fig. 2. Goodenough and Gerhart's final specification of the original problem statement in Figure 1. Analysis of this text, overprinted, is according to the following key:

Noise	―――	Ambiguity	⬰
Remorse	= = = = =	Overspecification	· · · · ·
Contradiction	⟳	Forward reference	⟹

comments in programs. Often, however, noise elements actually obscure the text. When first encountering such an element, the reader thinks it brings new information, but upon closer examination, he realizes that the element only repeats known information in new terms. The reader must thus ask himself nonessential questions, which divert attention from the truly difficult aspects of the problem.

Here, a fraction of a second is needed to realize that a 'nonempty sequence' of characters (8) is the same thing as 'one or more' characters (9). These two expressions appear within a line of each other; the authors' aim was, presumably, to avoid a repetition. One is indeed taught in elementary writing courses that repetitions should be avoided, and no doubt this is a good rule as far as literary writing is concerned. In a technical document, however, the rule to observe is exactly the opposite — namely, the same concept should always be denoted by the same words, lest the reader be confused.

An interesting variant of noise is **remorse**, a restriction to the description of a certain specification element made not where the element is *defined* but where it is *used*, as if the specifier suddenly regretted his initial definition. An example here is 'the output text, if any' (20). Up to this point, the specification freely used the notion of output text (12, 17); nowhere was there any hint that such a text might not exist. If the reader wondered about this problem, the specification did not provide an answer. Now, suddenly, when the discussion is focusing, on something else, the reader is 'reminded' that there might be no such thing as an output text, but no precise criterion is given as to when there is and when there isn't.

Another instance of remorse is the late definition of the 'line' concept (24), to which we will return. We will meet again the tendency to say too much, which generates noise, as a source of contradiction and ambiguity.

Silence

In spite of all his efforts, the specifier often leaves, along with over-documented elements, undefined features. Commonly, these features are fairly obvious to a community of application specialists, who are close to the initial customers, but they will be more obscure to those outside this circle. An example is the concept of 'line', which is not really defined except in a parenthetical bit of remorse toward the end of the text (24), where it is described as a sequence of characters 'between successive NL characters'. (By the way, are those characters part of the line?)

An interesting point here is the cultural background necessary to understand this concept. In ASCII-oriented environments, 'New Line' is a character; thus, people working on ASCII environments (DEC machines, for example) will probably understand easily the specification's basic hypothesis — namely, that NL is treated as an ordinary character upon input but triggers a carriage return upon output. These concepts are foreign, however, to somebody working in an EBCDIC environment, especially on IBM OS systems, on which files are made up of a sequence of 'records' (corresponding, for example, to lines), each made up of a sequence of characters. A person coming from such an environment would not have written the above specification and will probably have trouble understanding it.

Besides, the late definition of line is plainly wrong. It applies only to lines that are neither at the very beginning nor at the very end of the text. In both these cases, a line is not 'between successive NL characters' but between the beginning of the file and an NL, or between an NL and the end of the file — that is, between an NL and an ET. If we accept the author's definition, the first and last lines of the output may be of arbitrary length; in fact, an output containing *no NL at all* is acceptable regardless of its length, since it does not have lines according to the definition given! This is obviously absurd and not what the authors had in mind, but the use of natural language leads naturally to such slips of the pen.

Another interesting silence concerns the variable Alarm. Line 16 specifies that this variable should be set to TRUE in case of an error, but nothing is said about what happens to it in other cases. The answer is obvious, of course; but the matter can only be brushed aside as minor by programmers who have never run into a bug due to an uninitialized variable. . . .

It must be pointed out that Goodenough and Gerhart corrected a notable silence in Naur's original description. Naur's text does not explain what should be done with consecutive groups of more than one break character; this is one of the seven errors analyzed in Goodenough and Gerhart's paper. Their specification corrects it by requiring that such groups be reduced to a single break character in the output. Although something had to be done about the problem, note that this solution is, to some extent, obtained at the expense of simplicity. Eliminating redundant break characters and dividing a text into lines are two unrelated problems; merging them into a single specification complicates the whole affair.

It is probably better to deal with these two requirements separately, and this is what we do in the formal specification given below. Some of the current trends in programming methodology emphasize this approach — most notably under the influence of the Unix programming environment, which, at least in principle, favors tools that are simple and composable rather than large and multipurpose.

Contradictions

There is another problem with the concept of line. Given a type t, one should distinguish between the types **seq**[t], whose elements are finite sequences of objects of type t, and **seq** [**seq**[t]], whose elements are sequences of sequences of objects of type t. Such a confusion can be found in Figure 2, where we are first told (1) that the input is a 'stream', or sequence, of characters and later (10) that it 'can be viewed' as a sequence of words and breaks. As any Lisp programmer knows, the sequences

$\langle a\,b\,a\,c\,c\,a \rangle$
[sequence of objects]

and

$\langle\langle a \rangle\,\langle b\,a \rangle\,\langle c\,c\,a \rangle\rangle$
[sequence of sequences of objects]

are not the same. Note that the same problem with respect to the output is redeemed only by ambiguity; the type of the output is not clear:

• Is it **seq**[$CHAR$] as (21—22) seems to imply?
• Is it **seq**[$WORD$] — that is, **seq**[**seq**[$CHAR$]] — as (12—13) indicates?
• Or is it even **seq**[LINE] — that is, **seq**[**seq**[**seq**[$CHAR$]]] — if we consider a line as a sequence of words and breaks?

Thus, a sentence that at first appears to be only noise (9—11) yields a contradiction within a few lines (13—14): 'The program's output should be the same sequence of words as in the input'. This last comment is remarkable since *neither the input nor the output* is a sequence

of words. Worse yet, even if we parse the input into a sequence of
words, this sequence is not sufficient to determine the output — one
also needs two binary informations: whether there is a leading and/or a
trailing break.

The same sentence (9—11), in its overzealous effort to leave no
stone unturned, ends up introducing another contradiction. An unbiased
reader would be puzzled. How can the input 'end with [the character]
ET' (11) and at the same time have a 'trailing break' (11)? 'Trailing',
precisely, means 'at the end'! What's the last character if there is a
'trailing' break: ET or a break character?

A more experienced reader, such as a programmer, will have no
difficulty resolving this contradiction; his experience will tell him that
'end' markers follow 'trailing' characters. But this reliance on intuition
and knowledge of the application domain can be particularly damaging
when transposed to large requirements documents, which will be handed
down to a group of system designers and implementors of diverse backgrounds and abilities.

Overspecification

Overspecification in requirements can be annoyingly close to silence.
The reader is told too much about the *solution* while he is desperately
trying to grasp the *problem* and figure out — by himself — features not
covered by the text. Overspecification is typically, although certainly
not exclusively, found in requirements documents written by programmers. Psychologically, this is understandable. An implementation-level
concept is good, concrete, technical stuff, whereas true requirements
deal with much less tangible material. To a computer specialist, a stack
is easier to visualize than, say, the flow of information in a company or
the needs of a radar operator. Thus, many specifiers have a natural
tendency to cling to programming concepts. There is a price to pay for
this: Implementation decisions taken too early may turn out to be
wrong, and important problem features can be overlooked.

The example text contains an overspecification right from the first
sentence: the notion of the end-of-text character ET. The only reason
for the presence of this notion is Goodenough and Gerhart's desire to
correct Naur's original program. Input—output facilities of the version
of Algol 60 used by Naur (and, for fairness, by Goodenough and
Gerhart) do not provide for end-of-file detection when reading, so one

must assume the presence of a special character at the end of the file to make up for this deficiency. But ET is an implementation detail and should not be included in an abstract specification. Conceptually, the input is a finite sequence of characters; it should be transformed into an output that is a sequence of lines or, depending on the interpretation chosen, a sequence of characters. It is a programmer's vice to insist that finite sequences be specially marked at the end.

Why does the ET character receive such emphasis in Goodenough and Gerhart's specification? The reason is one of the errors in Naur's original program, which would go into an infinite loop unless the input was incorrect (that is, contained an oversize word). Upon closer examination, however, a case can be made for Naur's solution (without the other errors, of course). It is not so unrealistic to consider the required program as a potentially infinite process, which takes characters as input and produces lines as output, working somewhat like a device handler (for instance one that drives a printer) in an operating system. Such an interpretation should, of course, be clearly described in the specification, which was not the case with Naur's text. That decision would be less arbitrary than the one taken by Goodenough and Gerhart: their inclusion of ET changes the data structure at the specification level to accommodate the programming language used at the implementation stage.

The unacceptability of the change is further evidenced by the fact that the output does not satisfy the requirement on the input. Is it realistic to expect an existing file to be terminated by an explicit marker? If it is, the output produced by the program should satisfy that condition; however, examination of the specification, which is not completely clear on this matter, and, as a final criterion, of the proposed program, shows that ET will *not* be passed on to the output file. Assume that we want to write another program, for, say, right-justifying the text, that will take Goodenough and Gerhart's output (in 'pipe' mode *à la* Unix). In designing that program, we will not be able to make the same assumption on its input. Thus, the overspecification has opened the way to serious inconsistencies.

Another overspecification in the text is the concept of 'error exit' (16), which causes a 'variable', Alarm, to have the value TRUE. Clearly, the notion of a variable belongs to the world of programs, not specifications. This piece of overspecification would have been less shocking if the problem had been defined as the task of writing a *procedure*, with

	1	2	3	4	5	6	7	8	9	10		
1	U	N	I	X		I	S		A			
2	T	R	A	D	E	M	A	R	K			
3	O	F		B	E	L	L					
4	L	A	B	O	R	A	T	O	R	I	E	S
	1	2	3	4	5	6	7	8	9	10		

Fig. 3. Output requirement (MAXPOS = 10).

Alarm as one of its parameters, or as one of the 'exceptions' (in the sense of Clu or Ada) it might raise. A variable is internal to the program unit to which it belongs, whereas the specification of a parameter or an exception can be given relative to the environment of that unit.

The problem of the Alarm variable is less innocuous than it seems. One reason for shock at meeting the reference to this variable in a sequential reading of the text is that the definition of the error case (the one in which there is an oversize word) looks like over-specification until one sees the *last* sentence (25—26), 10 lines down, which gives the basic line-size constraint, MAXPOS. The world is really standing upside down here. Clearly, the constraint on word size is a consequence of the constraint on line size, and the definition of the error case cannot be understood until the latter constraint has been introduced.

We see here one of the major deficiencies plaguing requirements documents of more significant size: early inclusion of detailed descriptions of error handling, interwoven with descriptions of normal cases, which are usually much simpler. Here the matter is even worse; error processing is described before the reader has had a chance to recognize the problem — that is, before gaining an understanding of normal processing. Failure to clearly separate normal cases from erroneous ones makes the document much harder to understand.

Mathematically, a program that performs an input-to-output transformation often corresponds to the implementation of a partial function, which is not defined for some arguments of the input domain. Error processing then consists in 'completing' the function with alternate results, such as error messages, for those arguments. This completion should not be confused with the definition of the function in its normal cases. Here, as we'll see later in a formal specification, failure to accommodate words larger than MAXPOS is a consequence of the requirements for normal processing, which can be *proved*, as a theorem, from the definition of the function.

Ambiguities

Error processing raises an ambiguity in the example text (Figure 3). The requirement that the output text satisfy properties 1 to 4 'up to the point of an error' is susceptible to at least two interpretations.

The text says that up to (and presumably including) the point of the error, the program's output should correspond to the input. But where is the 'point of the error' in Figure 3? Is it [line 4, column 10], last acceptable letter, or [3, 7], end of the last acceptable word? Nothing in the text allows the reader to decide between these two interpretations.

Another interesting ambiguity is connected with the basic constraint on acceptable solutions (23): 'As many words as possible should be placed on each line.' If we have, say, MAXPOS = 10 and the input text

WHO WHAT WHEN

there are two equally correct two-line solutions (WHAT may be on either the first or second line). This ambiguity may be acceptable since neither solution appears superior to the other; the specification as such is nondeterministic. We suspect (perhaps wrongly) that this nondeterminism was not intentional and that there was an implicit overspecification in the authors' minds: they considered it obvious that the input would be processed sequentially, so any ambiguity, as in the example above, would be solved by placing as many words as possible on the earlier line (giving line WHO WHAT followed by line WHEN). In this interpretation, property 3 (23—24) actually means, 'As many words as possible should be placed on each line as *it is encountered in the sequential construction of the output.*' If this is the case, the specification should state it precisely.

Another potential source of ambiguity is the use of imprecise or poorly defined terms — for example, the use of 'stream' (1) rather than the more standard 'sequence'. The expression 'error exit' (15), stemming from the overspecification seen above, is ambiguous, and the reader is not comforted by the explanation that follows it ('i.e., a variable, Alarm, should have the value TRUE'); the notion of assigning a value to a variable does not by itself imply the idea of an 'exit', which also means that the program stops in some fashion. We have seen that the concept of 'line' is not well defined (24). Also note that the expression 'new line' is to be parsed as a single entity (the *new line* character) in its first appearance (5) and as separate words ('a new *line* should start . . .') in its second (19).

Forward References

In a requirements document, not all forward references are bad. Some, corresponding to a top-down presentation of the concepts ('the notion of . . . will be studied in detail in section . . .'), might even be considered good practice, provided there are not too many. But *implicit* forward references (that is, uses of a concept that come before definition of the concept, without particular warning to the reader) can present much more of a problem. They make a document extremely hard to read, especially in the absence of the technical apparatus (index, glossary, etc.) that should be a part of all requirements specifications and other software documents.

Here, of course, the text is very short, so the annoyance caused by forward references is nowhere near what it can be with full-size documents. Note, however, that ET is used three times (2, 3, 6) before it is defined (7), that the notion of line, defined not quite satisfactorily (24), has been used earlier (19—20), and that MAXPOS is used just before its definition (14).

So What?

In dissecting Goodenough and Gerhart's specification, we identified a significant number of problems in a text that may seem innocuous to a superficial observer. Not all the problems were equally serious, and the reader may have felt that we were a bit pedantic at times. We submit, however, that one must be pedantic in dealing with such matters. Inconsistencies, ambiguities, and the like may not warrant the gallows when the problem is to split up a sequence of characters into lines. But keep in mind how the above defects transpose to more serious matters — a nuclear reactor control system, a missile guidance system, or even just a payroll program. The computer that executes the code resulting from a faulty specification is more pedantic than any human referee could ever be.

Thus, we should consider Goodenough and Gerhart's specification not only as an object of study in itself but also, and more importantly, as a microcosm for conveniently observing deficiencies typical of more meaningful requirements documents. Although the text was written with great care, we have witnessed how the authors, who started out to improve upon Naur's terse but simple text, sentence after sentence became a little more entangled in their own rosary of caveats. This says a lot about why interminable manuals occupy so much shelf space in programmers' offices and computer rooms.

In our opinion, the situation can be significantly improved by a reasoned use of more formal specifications. But again, let's emphasize that such specifications are a complement to natural language documents, not a replacement. In fact, we'll show how a detour through formal specification may eventually lead to a better English description. This and other benefits of formal approaches more than compensate for the effort needed to write and understand mathematical notations.

We will now introduce such notations, which will allow us to give a formal specification of the Naur—Goodenough/Gerhart problem.

ELEMENTS FOR A FORMAL SPECIFICATION

Many formal specification languages have been designed in recent years (see Bibliography). Choosing one of these languages would force the reader to learn its particular notation and would obscure the essential fact — namely, that their underlying concepts are, for the most part, well-known mathematical notions like sets, functions, relations, and sequences. We thus prefer to use a more-or-less standard mathematical notation. The style of exposition will be similar to that found in mathematical texts; translation to a specific formal specification language should not be hard, provided the language supports the relevant concepts.

Overview

Perhaps the only difficult part of the Naur—Goodenough/Gerhart problem is that the processing to be performed on the text involves three aspects: reducing breaks to a single break character, making sure no line has more than MAXPOS characters, and filling lines as much as possible. If these three requirements are separated, things become much simpler. Consequently, we will define the problem formally by considering two simple binary relations, called *short_breaks* and *limited_length*, and a function called *FEWEST_LINES*. (Throughout the discussion of the formal specification, the reader may wish to refer to Figure 4 for a picture of the overall structure of the relations and functions involved.)

Relation *short_breaks* holds between two sequences of characters *a* and *b* if and only if *b* is identical to *a*, except that breaks in *a* (i.e., successive break characters) have been reduced to single break characters in *b*.

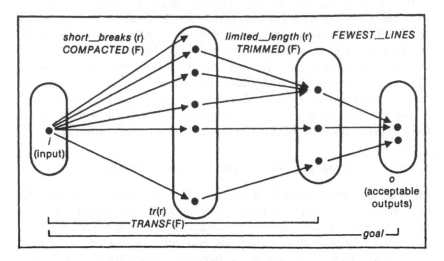

Fig. 4. Overall structure of the specification: (r) indicates a relation, (F) a function.

Relation *limited_length* holds between two sequences of characters *b* and *c* if and only if *c* is a 'limited length version' of *b*: that is, no line in *c* has length greater than MAXPOS, and *c* is identical to *b* except that some blanks may have been replaced with new lines and/or some new lines with blanks.

By applying these two relations successively, we associate with any sequence of characters *a* all sequences of characters that are 'made of the same words', separated only by single breaks, and fit on lines no longer than MAXPOS. Given such a set of sequences, say, *SSC*, then

A reminder on functions and relations
Consider two sets — for example, *INPUT* and OUTPUT. A
binary relation between these two sets is a set of pairs

$$\{\langle i_1, o_1\rangle, \langle i_2, o_2\rangle, \ldots\}$$

A relation.

where each i_i belong to set *INPUT* and each o_k belongs to
set *OUTPUT*. Such a relation is represented pictorially at
right. If *goal* is a relation, then we write *goal* (i, o) to express that the pair $\langle i, o\rangle$ belong to the relation.

The **domain** of such a relation, written **dom** (*goal*), is the subset of *INPUT*
containing only those elements i such that *goal*(i, o) holds for **at least** one element
o in *OUTPUT*. Thus, in the example pictured, i_1, i_2, and i_4, but not i_3, belong to the
domain of the relation.

A **function** is a relation f such that for any i there is **at most** one o for which $f(i,$
$o)$ holds; if o exists, then one may write $o = f(i)$. The relation pictured above is not
a function, since i_1, for instance, has two buddies o_1 and o_2. Note that the domain of
a function is made of those elements of *INPUT* for which there is **exactly** one
corresponding element in *OUTPUT*.

FEWEST_LINES (*SSC*) is the subset of *SSC* containing those sequences that consist of a minimum number of lines and thus are acceptable outputs for the program.

We'll now define these notions formally, but a few simple conventions are needed first.

Basic Form of the Specification

As a general convention, we use uppercase for sets and for functions
whose results are sets and lowercase for other functions, elements of
sets (except for MAXPOS, which we write in upper-case as in the
original specification), sequences, and relations.

The program to be written is the implementation of a function

$$sol: INPUT \rightarrow OUTPUT$$

where *INPUT* and *OUTPUT* are the sets of possible inputs and
outputs, which we will describe below as sets of sequences. Function
sol must satisfy certain constraints, which it is the role of the specification to express.

As noted above, there may be more than one correct output for a given input; in other words, a truly general specification of the problem should be nondeterministic. We will represent this fact by defining a binary relation between sets *INPUT* and *OUTPUT*. We call this binary relation *goal*; then a function *sol* will be a correct solution if and only if the following two conditions are satisfied (readers who are not so sure about functions and relations are referred to the refresher in the box):

- function *sol* is defined wherever relation *goal* is defined — that is, $sol(i)$ exists for any i in the domain of *goal*;
- for any i for which *goal* is defined, then $sol(i)$ yields a 'solution' to *goal* — that is, $goal(i, sol(i))$ holds.

This definition is expressed in mathematical notation by writing that *sol* is an acceptable function if and only if

$$\forall i \in \mathbf{dom}(goal),$$
$$i \in \mathbf{dom}(sol) \text{ and } goal(i, sol(i))$$

where $\mathbf{dom}(sol)$ is the domain of function *sol*. Note that there may be some inputs for which there is no acceptable solution (those not in the domain of *goal*), so *sol* may be a partial function. Also, in more concise notation, the above property can simply be expressed by writing that the domain of *sol* is at least as large as the domain of *goal*, and that *sol* is **included** in *goal* (both being defined as sets of pairs):

$$\mathbf{dom}(goal) \subset \mathbf{dom}(sol) \text{ and } sol \subset goal$$

This way of presenting a specification is of very general applicability for programs performing input-to-output transformations. Such a program may be viewed as the implementation of a certain function (*sol*) which must ensure that a certain relation (*goal*) is satisfied between its argument and its result; in mathematical terms, the function is included in (is a subset of) the relation. To specify the problem is to define the relation; to construct the program is to find an implementable function *sol* satisfying the above conditions. [3]

Characters and Sequences

The principal set of interest in our problem is the set of characters, which we denote by *CHAR*. The only property of *CHAR* that matters here is that *CHAR* contains two elements of particular interest, *blank*

Basic set and logic notations

The definitions marked (*) introduce predicates, that is, expressions which may have value 'true' or 'false'.

$\{a, b, c, \ldots\}$: the set made up of elements $a, b, c, \ldots$

$x \in A$: x is an element of A (*).

$x \notin A$: x is not an element of A (*).

$A \subset B$: A is a subset of B (all elements of A are elements of B) (*).

$\{x \in A \mid P(x)\}$: The (possibly empty) subset of A made up of those elements x which satisfy property P.

$\forall x \in A, P(x)$: All elements x of A, if any, satisfy property P (or: no element of A violates P); holds in particular whenever A is empty (*).

$\exists x \in A, P(x)$: There is at least one element x in A which satisfies property P; may only hold if A is nonempty (*).

$a \Rightarrow b$: a implies b.

$a \ldots b$: the integer interval containing all the integers i such that $a \leqslant i \leqslant b$; empty if $a > b$. This notation is borrowed from Pascal.

The symbol $\equiv$ means 'is defined as'.

---

and *new_line*. We call *BREAK_CHAR* the subset of *CHAR* consisting of these two elements:

$$BREAK\_CHAR \equiv \{blank, new\_line\}$$

The basic concept in this problem is that of sequence. If X is a set, we denote by **seq**[X] the set whose elements are finite sequences of elements of X. Such a sequence is written, for example, as

$$\langle a, b, a, c, c, d \rangle$$

and has a length that is a nonnegative integer; thus, *length* is a function from **seq**[X] to the set of natural numbers. Elements are numbered starting at 1; the ith element of a sequence s (for $1 \leqslant i \leqslant length(s)$) is written $s(i)$. A **subsequence** of s is a sequence made of zero or more of the elements of s, in the same order as in s; for example, if s is the above sequence, then some of its subsequences are

$$\langle a, b, c, d \rangle$$
$$\langle b, c, c \rangle$$

On the other hand, $\langle b, d, c \rangle$ is not a subsequence of s because the original order of its elements in s is not preserved.

The set of subsequences of s will be written $SUBSEQUENCES$ (s).

The concept of sequences is well known, and we rely on the reader's understanding here. A formal definition of sequences and of the above notions is given in the box on page 177.

Minima and Maxima

If X is a set, and f is a function from X to the set of natural numbers,

$$MIN\_SET(X, f)$$

denotes the subset of X consisting of the elements for which the value of f is minimum. For example, if X is the following set, containing four sequences

$$X \equiv \{\langle a, c, b, a \rangle, \langle a, b \rangle, \langle b, a, b \rangle, \langle c, c \rangle\}$$

and f is the *length* function on sequences, then $MIN\_SET(X, f)$ will be the set consisting of the shortest of these sequences, namely, the second and last.

In the same fashion, we denote by

$$MAX\_SET(X, f)$$

the subset of X consisting of the elements for which the value of f is maximum; thus, in the above case, $MAX\_SET(X, f)$ is the set $\{\langle a, c, b, a \rangle\}$, containing just one sequence.

MAX_SET, however, is not always defined; we have to be careful to apply it only to sets X which are finite; otherwise, there might be no maximum value for f. Note that the results of MIN_SET and MAX_SET are a subset of X rather than a single element, since there may be more than one element with minimum or maximum f value. These subsets are non-empty if and only if X is nonempty.

We will also need a way to denote the minimum and maximum elements of a set of natural numbers SN. They will be written, in the usual fashion, $min(SN)$ and $max(SN)$. Thus, if SN is the set

$$SN \equiv \{341, 7, 3, 654\}$$

then $min(SN)$ is 3 and $max(SN)$ is 654. Note that min and max, contrary to MIN_SET and MAX_SET, yield a natural number, not a set. Also in contrast to MIN_SET and MAX_SET, which are defined for empty sets (they yield an empty result), both min and max are defined only if the set *SN* is not empty; *max* further requires that *SN* be finite. It is essential to check for these conditions whenever using these functions.

Input and Output Sets

In the problem at hand, the input is a sequence of characters; we choose to describe the output as a sequence of characters as well. Thus, we define the two sets:

$$INPUT \equiv \mathbf{seq}[CHAR]$$
$$OUTPUT \equiv \mathbf{seq}[CHAR]$$

Note that, as mentioned above, another interpretation could have defined the set of possible outputs as **seq**[*LINE*], with *LINE* itself being defined as **seq**[*CHAR*] (or possibly **seq**[*WORD*] with *WORD* ≡ **seq**[*CHAR*], plus information on leading and trailing breaks).

We will now define the relations *short_breaks* and *limited_length* and the function *FEWEST_LINES*.

THE FORMAL SPECIFICATION

Short Breaks

Let a be a sequence of characters. We define *SINGLE_BREAKS* (a) as the set of subsequences of a such that no two consecutive characters are break characters:

$$SINGLE\_BREAKS\,(a) \equiv \{s \in SUBSEQUENCE\,(a)\,|$$
$$\forall i \in 2 \ldots length\,(s),\, s(i-1) \in BREAK\_CHAR$$
$$\Rightarrow s(i) \notin BREAK\_CHAR\}$$

Note that we use the Pascal notation, $a \ldots b$, to denote the (possibly empty) set of integers i such that $a \leqslant i \leqslant b$.

Next, we define *COMPACTED* (a) as the subset of *SINGLE_BREAKS* (a) containing those sequences of maximum length:

$$COMPACTED\,(a) \equiv MAX\_SET\,(SINGLE\_BREAKS\,(a),$$
$$length)$$

As stated above, *MAX_SET*(*X*, *f*) may be be undefined if *X* is an infinite set. This cannot occur here, however, since *SINGLE_BREAKS* (*a*) is a subset of *SUBSEQUENCES* (*a*) which, for any sequence of characters *a*, is finite.

Note that any sequence *b* in *COMPACTED* (*a*) must have retained from *a* all nonbreak characters (if such a character had been omitted, it could be inserted into *b* and yield a longer element of *SINGLE_BREAKS* (*a*)), and has a single break character where *a* had one or more consecutive break characters.

Thus, the relation *short_breaks* (*a*, *b*), which holds between *a* and *b* if and only if *a* and *b* are made of the same sequences of words and breaks but the breaks in *b* consist of a single break character, can be expressed simply by

$$short\_breaks\,(a, b) \equiv b \in COMPACTED\,(a)$$

Limited Length

The relation *limited_length* (*b*, *c*) holds between sequences *b* and *c* if and only if
- *c* is the same sequence as *b*, except that it may have a *new_line* wherever *b* has a *blank*, or conversely; and
- the maximum line length of *c*, defined as the maximum number of consecutive characters none of which is a *new_line*, is less than or equal to MAXPOS.

This is expressed more precisely as follows:

$$limited\_length\,(b, c) \equiv c \in TRIMMED\,(b)$$

where

$$TRIMMED\,(b) \equiv \{s \in EQUIVALENT\,(b)\mid$$
$$max\_line\_length\,(s) \leqslant MAXPOS\}$$

$$EQUIVALENT\,(b) \equiv \{s \in \mathbf{seq}[CHAR]\mid$$
$$length(s) = length\,(b) \text{ and } (\forall i \in 1 \ldots length\,(b),$$
$$s(i) \neq b(i) \Rightarrow$$
$$s(i) \in BREAK\_CHAR \text{ and } b(i) \in BREAK\_CHAR)\}$$

$$max\_line\_length\,(s) \equiv max(\{j - i \mid$$
$$0 \leqslant i \leqslant j \leqslant length\,(s) \text{ and } (\forall k \in i + 1 \ldots j,$$
$$s(k) \neq new\_line)\})$$

A few explanations may help in understanding these definitions. If *s*

A definition of sequences

The following presentation is based on the formal specification of sequences given in the Z reference manual [11].

N will denote the set of natural numbers.

DEFINITION: **seq[X]**, the set of finite sequences of elements of X, is defined as the set of partial functions from N to X whose domains are intervals of the form $1 \ldots n$ for some natural number n.

So a sequence is defined as a partial function; for example, the sequence $s = \langle a, b, a, c \rangle$ is the function defined for arguments 1, 2, 3, and 4 only, and whose value is a for 1 and 3, b for 2, and c for 4. The following is a pictorial representation of s:

$$
\begin{array}{ccccccccc}
 & 1 & 2 & 3 & 4 & 5 & 6 & 7 & \ldots & N \\
s & \downarrow & \downarrow & \downarrow & \downarrow & & & & & \\
 & a & b & a & c & & & & & X
\end{array}
$$

Note that the above definition allows $n = 0$ (empty interval, thus empty function — that is, empty sequence) and that it justifies the notations $s(i)$ for the ith element of sequence s (which is the result of applying function s to element i).

The **length** of a sequence is defined as the largest integer for which the associated partial function is defined (i.e. n in the above definition).

Now let s be a sequence of elements of X and g be a (total) function from X to some set Y. The composition

$$g \cdot s$$

is a partial function from the set of natural numbers to Y, which has the same domain as s; thus, it is a sequence of elements of Y, with the same length as s. This sequence is obtained from s by applying g to all the elements of s. Again, a picture may help (we set $g(a) = a'$, etc.):

$$
\begin{array}{ccccccccc}
 & 1 & 2 & 3 & 4 & 5 & 6 & 7 & \ldots & N \\
s & \downarrow & \downarrow & \downarrow & \downarrow & & & & & \\
 & a & b & a & c & & & & & X \\
g & \downarrow & \downarrow & \downarrow & \downarrow & & & & & \\
 & a' & b' & a' & c' & & & & & Y
\end{array}
$$

Now take for X the set N of natural numbers. A **sorted sequence** of natural numbers is an element s of **seq[N]** such that

$$\forall i \in 2 \ldots length\ (s), s(i-1) \leqslant s(i)$$

With this definition, it becomes easy to formally define the notion of **subsequence** used in the text.

DEFINITION: Let s be an element of **seq[X]** for some set X. A subsequence of s is a sequence of the form $s \cdot u$ where u is a sorted sequence of natural numbers.

The following picture shows how $\langle a\ a\ b\ c \rangle$ is obtained as a subsequence of $\langle a\ b\ a\ a\ b\ d\ c\ d \rangle$ using the above definition. The sorted sequence u of natural numbers used here is $\langle 3\ 4\ 5\ 7 \rangle$; $\langle 1\ 3\ 5\ 7 \rangle$ or $\langle 1\ 4\ 5\ 7 \rangle$ would also work.

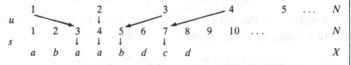

is a sequence of characters, max_line_length (s) is the maximum length of a line in s, expressed as the maximum number of consecutive characters, none of which is a new line. In other words, it is the maximum value of $j - i$ such that $s(k)$ is not a new line for any k in the interval $i + 1 \ldots j$. (We will have more to say about this definition below.) $EQUIVALENT$ (b) is the set of sequences that are 'equivalent' to sequence b in the sense of being identical to b, except that new_line characters may be substituted for $blank$ characters or $vice\ versa$. Finally, $TRIMMED$ (b) is the set of sequences which are 'equivalent' to b and have a maximum line length less than or equal to MAXPOS.

Fewest Lines

Let SSC be a set of sequences of characters. These sequences can be interpreted as consisting of lines separated by new_line characters. We define the set $FEWEST\_LINES$ (SSC) as the subset of SSC consisting of those sequences that have as few lines as possible:

$$FEWEST\_LINES\ (SSC) \equiv$$
$$MIN\_SET\ (SSC, number\_of\_new\_lines)$$

where the function $number\_of\_new\_lines$ is defined by:

$$number\_of\_new\_lines\ (s) \equiv$$
$$\textbf{card}\ (\{i \in 1 \ldots length\ (s)\ |\ s(i) = new\_line\})$$

and $\textbf{card}$ (X), defined for any finite set X, is the number of elements (cardinal) of X.

The Basic Relation

The above definitions allow us to define the basic relation of the problem, relation $goal$, precisely. Relation $goal$ $(i,\ o)$ holds between input i and output o, both of which are sequences of characters, if and only if

$$o \in FEWEST\_LINES\ (TRANSF\ (i))$$

$TRANSF$ (i) is the set of sequences related to i by the composition of the two relations $short\_breaks$ and $limited\_length$:

$$TRANSF\ (i) \equiv \{s \in \textbf{seq}[CHAR]\ |\ tr(i,\ s)\}$$

with

$$tr \equiv limited\_length \cdot short\_breaks$$

The dot operator denotes the composition of relations (see box). A look at Figure 4 may help explain the role of the various functions and relations in the above specification.

Existence of Solutions

Once we have a formal specification, what can we do with it? Relying on the specification as a basis for the next stages of the software life cycle — program design and implementation (e.g., translating ∀s into loops) is the most obvious use. However, we'd like to emphasize two others. One use, studied in the next section, is as a starting point for better natural-language requirements. The other, to which we now turn, is querying the specification to learn as much as possible about properties of the problem and valid solutions.

What can the given specification teach us about the Naur—Goodenough/Gerhart problem and its solution? First, let's determine when solutions do exist. It is trivial to prove that, given a sequence of characters a, there is always at least one sequence b such that relation *short_breaks* (a, b) holds. Given b, however, the necessary and sufficient condition for the existence of at least one sequence c such that *limited_length* (b, c) holds is that b contains no word (i.e., contiguous subsequence of non-break characters) of length greater than MAXPOS. This follows from the definitions of *TRIMMED* and *max_line_length* used in the definition of *limited_length*. Thus, the domain of definition of the relation *tr*, which is also the domain of the function *TRANSF* and thus of the relation *goal*, is the set of input texts containing no word longer than MAXPOS. This can be formulated as a theorem:

$$\mathbf{dom}\,(goal) = \{s \in \mathbf{seq}[CHAR] \mid$$
$$\forall i \in 1 \ldots length\,(s) - \text{MAXPOS},$$
$$\exists j \in i \ldots i + \text{MAXPOS},$$
$$s(j) \in BREAK\_CHAR\}$$

The property expressed by this theorem is that the domain of relation *goal* consists of sequences such that, if a character c is followed by MAXPOS other characters, at least one character among c and the other characters must be a break.

An important problem, not addressed here, is how the specification deals with erroneous cases — that is, with inputs not in the domain of the *goal* relation — like sequences with oversize words. Clearly, a robust and complete specification should include (along with *goal*) another relation, say, *exceptional_goal*, whose domain is *INPUT*—**dom** (*goal*) (set difference); this relation would complement *goal* by defining alternative results (usually some kind of error message) for erroneous inputs. Formal specification of erroneous cases falls beyond the scope of this article, but a discussion of the problem and precise definitions of terms such as 'error', 'failure', and 'exception' can be found in a paper by Cristian [4].

Discussion

What we have obtained is an abstract specification — this is, a mathematical description of the problem. It would be difficult to criticize this specification as being oriented toward a particular implementation: if followed to the letter, the specification would lead to a program that (as illustrated in Figure 4) would first generate all possible distributions of the input over lines of length less than or equal to MAXPOS and then search the resulting list for solutions with minimum number of *new_line* characters — not a very efficient implementation!

An element that does seem to point toward a particular implementation technique is the composition of relations *short_breaks* and *limited_length*, which seems to imply a two-step process (first remove break characters, then cut into lines). A first design could indeed use a two-step solution. The steps could then be merged using coroutine-like

---

Composition of relations

Let r and t be two relations; r is from X to Y and t is from Y to Z (see figure).

The composition of these two relations, written $t \cdot r$ (note the order), is the relation w between sets X and Z such that $w(x, z)$ holds if and only if there is (at least) one element y in Y such that both $r(x, y)$ and $t(x, y)$ hold.

Thus, in the example illustrated, w holds for the pairs $\langle x_1, z_1 \rangle$, $\langle x_1, z_2 \rangle$, and $\langle x_5, z_3 \rangle$ (and for these pairs only).

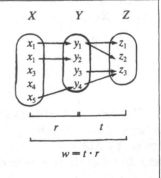

concepts, such as the Unix notion of pipe or the 'program inversion' idea of Jackson's program design method [5].

We chose to model the problem's object and operations with very simple mathematical notions (sets, relations, functions, sequences). Because of the specific nature of this problem, another approach would have been to rely on a more advanced theory, such as the theory of regular languages. As emphasized below, a realistic specification system should permit reuse of existing theories [6].

Starting from the above definition, the specification should of course be refined, taking into account the physical form of the data structure (including, for example, the end-of-file marker) and the particular response that should be given by the program in case of erroneous input.

CONCLUSION

Although natural language is the ideal notation for most aspects of human communication, from love letters to introductory programming language manuals, there are cases [7] where it is not appropriate. Software specifications, for example, require more rigorous formalism.

The use of formal notation does not, however, preclude that of natural language. In fact, mathematical specification of a problem usually leads to a better natural-language description. This is because formal notations naturally lead the specifier to raise some questions that might have remained unasked, and thus unanswered, in an informal approach.

Mathematical Definition

Formal specifications help expose ambiguities and contradictions because they force the specifier to describe features of the problem precisely and rigorously. The problem studied in this article contains many examples of this. For example, let us try to redefine the function *max_line_length* using the definition of 'line' taken from Goodenough and Gerhart's specification (line 24: 'between successive NL characters'). Writing this definition mathematically, we obtain something like

$$max\_line\_length\ (s) \equiv$$
$$max(\{line\_length\ (s, i) \mid 1 \leqslant i \leqslant length\ (s)\ \textbf{and}$$
$$s(i) = new\_line\})$$

where *line_length* (s, i), the length of the line beginning after the *new_line* at position i in sequence s, may be defined as a minimum:

$$line\_length~(s, i) \equiv$$
$$min(\{k \mid 0 \leqslant k < length~(s-i)~\text{and}~s(i + k + 1) = new\_line\})$$

However, as mentioned above, the maximum or minimum of a set of natural numbers is defined if and only if this set is nonempty and, in the maximum case, finite; so using mathematical notation prompts us to check for these conditions. Finiteness presents no problem, but we see immediately that the set whose maximum is sought in the definition of *max_line_length* will be empty if the sequence s does not contain any *new_line* character. Even if it contains one, *line_length* (s, i), itself a minimum, will not be defined if there is no other *new_line* further in the sequence. This prompts us to look for a better definition.

A fairly natural reaction at this point is to see that we really don't need to define the concept of 'line', only that of **maximum line length**. Once we have noticed this, it's easy to come up with a correct definition: *the maximum number of consecutive characters, none of which is a new line*. This is the definition that was given above:

$$max\_line\_length~(s) \equiv$$
$$max(\{j - i \mid 0 \leqslant i \leqslant j \leqslant length~(s)~\text{and}~(\forall k \in i + 1 \ldots j, s(k) \neq new\_line)\})$$

Note that we have been careful to apply *max* to a set that always contains at least one value (zero, obtained for $i = j = 0$), even if s is an empty sequence (see box).

Natural Language Definition

Once such a mathematical definition has been produced, it may in return influence the natural language definition. In this example, the formal definition suggests that we should refrain from trying to define the concept of 'a line in the text' which, although intuitively clear, is slightly tricky when one attempts to specify it precisely, as Goodenough and Gerhart's text shows. Instead, we should focus on the notion of 'maximum line length', which is always defined, even for a text consisting of *new_line* characters only. Once we have obtained the specifi-

The reasoning behind formal specifications:
the example of *max_line_length*

How does one obtain a formal expression such as the one defining *max_line_length*? Let's analyze the different steps involved.

We want to express the fact that *max_line_length* (*s*) is the maximum length of a line in *s*. A definition that avoids the pitfalls mentioned in the analysis of Goodenough and Gerhart's text is, informally, 'the maximum number of consecutive characters, none of which is a new line'.

To translate this definition into a formal description, we have to express the notion of a contiguous subsequence of *s* that does not contain a *new_line*. A contiguous subsequence can be given by its end indices, say, *i* and *j*. The sequence comprising the elements between indices *i* and *j* will have length $j - i + 1$; if it is to yield a line length, then $s(k)$ should be a character other than *new_line* for any *k* between *i* and *j*, inclusive. Thus, a first try might yield

$$max\_line\_length\ (s) \equiv max(LINE\_LENGTHS)$$

where the set *LINE_LENGTHS* is defined as

$$LINE\_LENGTHS \equiv \{j - i + 1 \mid 1 \leqslant i \leqslant j \leqslant length\ (s)\ \textbf{and}$$
$$(\forall k \in i \ldots j, s(k) \neq new\_line)\}$$

But beware! One should only apply *max* to nonempty sets. With the above convention, we can end up with *LINE_LENGTHS* being empty if *s* is an empty sequence or all its characters are *new_line*; in either case, no *i*, *j* pair satisfies the condition. Now, if we write a program for the Naur–Goodenough/Gerhart problem and put it into a library, sooner or later someone will apply it to a sequence that is empty or entirely made of *new_line* characters, so we had better deal with these cases in a clean fashion.

The culprit is the condition $i \leqslant j$, which prevents us from finding a satisfactory *i* and *j* in the borderline cases mentioned. The problem disappears, however, if we replace this condition by $i - 1 \leqslant j$. Then, for a sequence having only *new_line* characters or no character at all, the set *LINE_LENGTHS* will contain one element, 0, obtained for $i = 1$ and $j = 0$. For these values, the interval $i \ldots j$ is empty; thus, the $\forall \ldots$ clause is true. (Remember that a property of the form $\forall x \in E, P(x)$ is always true when the set *E* is empty, regardless of what property *P* is.) Thus, we obtain the following replacement:

$$LINE\_LENGTHS \equiv \{j - i + 1 \mid 0 \leqslant i - 1 \leqslant j \leqslant length\ (s)\ \textbf{and}$$
$$(\forall k \in i \ldots j, s(k) \neq new\_line)\}$$

(The first condition has been written $0 \leqslant i - 1$ instead of $1 \leqslant i$.)

We have chosen to simplify slightly the writing of this condition by a change of variable (use *i* for $i - 1$, thus eliminating $+ 1$ and $- 1$ terms):

$$LINE\_LENGTHS \equiv \{j - i \mid 0 \leqslant i \leqslant j \leqslant length\ (s)\ \textbf{and}$$
$$(\forall k \in i + 1 \ldots j, s(k) \neq new\_line)\}$$

This new version is defined in all cases.

It should be noted that this kind of analysis, which at first sight might seem quite remote from programmers' concerns, is in fact closely connected to typical patterns of reasoning about programs. Anyone who has tried to debug a loop that sometimes goes one iteration too few or too many, or works improperly for empty inputs or other borderline cases, will recognize the line followed in the above discussion. It is our contention, however, that such analysis is better performed at the specification level, dealing with simple and well-defined mathematical concepts, than at program debugging time, when the issues are obscured by many irrelevant details, implementation-dependent features, and idiosyncrasies of programming languages.

cation of *max_line_length*, we can build on it and include it in the English problem definition a sentence such as

The maximum number of consecutive characters, none of which is a *new_line*, should not exceed MAXPOS.

This sentence, a direct translation from the formal definition, is not, admittedly, of the most gracious style; but it is easy to remove the double negation, yielding

Any consecutive MAXPOS + 1 characters should include a *new_line*.

The main advantage of natural language texts is their understandability. One should concentrate on this asset rather than trying to use natural language for precision and rigor, qualities for which it is hopelessly inadequate. Understandability is seriously hindered when natural language requirements become ridiculously long in a vain attempt to chase away silence, ambiguity, contradiction, etc. Such attempts, as shown by the text studied here, only make matters worse. The length of many requirements documents found in actual industrial practice, often extending over hundreds or even thousands of pages, is due to such misuse of natural language. Natural language descriptions should remain reasonably short; the exact description of fine points, special cases, precise details, etc., should be left to a formal specification.

The advantages of brevity cannot be overemphasized. It could even be argued that Naur's specification, once the problems of termination and consecutive break characters are tackled properly, is preferable to Goodenough and Gerhart's because it is shorter and doesn't fuss unnecessarily.

New Specification

It would be fair game for the reader at this point to ask what natural-language specification we have to offer in lieu of both Naur's and Goodenough and Gerhart's texts. To answer such a request, we'd try to capitalize on the lessons gained from writing the mathematical definition. We'd propose something like the text in Figure 5, which is directly deduced from that definition (see in particular its relation to Figure 4).

No doubt this text deserves some criticism of its own. In particular, it still needs to be refined. For example, the implementor must know how to 'report the error' before embarking upon detailed design and coding; he must know what the allowable characters are apart from *blank* and *new_line*, etc. Also note that this text avoids defining specific concepts (e.g., line length, word) explicitly; rather, it substitutes the definition for the concept when needed. Although this device can lead to interesting literary experiments [8], it is certainly not recommended for large requirements documents where one must repeatedly refer to the same basic concepts.

It seems to us, however, that the above statement of the requirements embodies the essential elements of the problem and achieves a reasonable tradeoff between the imprecision of Naur's and the verbosity of Goodenough and Gerhart's specifications. (Its length is in fact slightly more than double the former's and half the latter's.) Its most important feature is that it draws heavily from the lessons gained in writing the formal specification, while retaining (we hope) clarity and simplicity.

End-Users

An objection that is often voiced against formal specifications relates to the needs of end-users, who request easily understandable documents. Such an objection, we think, is based on an incorrect assessment of what specification is about. There is a need for requirements documents that must be read, checked, and discussed by noncomputer scientists, but there is also a need for technical documents used by computer professionals. The difference is the same as that between user requirements and engineering specifications in other engineering disciplines. Of course, there must be a way to communicate back the contents of technical specifications (for example, in the case of changes). As we have seen, the existence of a good mathematical specification is a great asset for improving a natural-language description.

Other ways can be found for translating formal elements into forms that are more easily understood. Many people like graphical descriptions, which play a basic role in such (non-formal) specification methods as SADT [9] or SREM [10]. A picture may be worth a thousand words at times, but it can also be dangerously misleading. On the other

hand, a pictorial explanation of a well-defined concept certainly does
no harm. If the picture

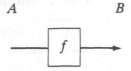

is considered more understandable than the function definition

$$f : A \rightarrow B$$

they why not have graphics tools generate the picture from the formula
for the benefit of those who want it? There is certainly a great need for
software tools of this kind in specification systems.

Techniques

The last point we want to emphasize is that formal specification *is
not necessarily difficult*. The reader who is familiar with specification
techniques will have noted that the example did not rely (at least
explicitly) on such notions as abstract data types, finite-state machines,
and attribute grammars. In fact, it used only very simple notions from
elementary set theory and logic. These notions are no more difficult
than the basic core of college calculus, even if most of today's university
students are regrettably less at ease dealing with such concepts as sets,
relations, partial functions, composition, and predicate calculus than
with other mathematical objects and operations that are better established in the traditional curriculum.

Of course, the example studied here is a small problem. Experience
with the Z language [11, 12] and subsequent work prompted by this experience [13—15] shows, however, that the same basic concepts can be
carried through to the description of much more complex systems. The
main limitation of the problem studied here is that it is defined by a
simple input-to-output relation, whereas most significant programs can
be characterized, in our view, as *systems* that offer various services in
response to possible user requests. We are currently working on methods, notations, and tools for the modular specification of such systems.
[16]

Given are a nonnegative integer MAXPOS and a character set including two 'break characters' *blank* and *new_line*.

The program shall accept as input a finite sequence of characters and produce as output a sequence of characters satisfying the following conditions:

- it only differs from the input by having a single break character wherever the input has one or more break characters;
- any MAXPOS + 1 consecutive characters include a *new line*;
- the number of *new_line* characters is minimal.

If (and only if) an input sequence contains a group of MAXPOS + 1 consecutive nonbreak characters, there exists no such output. In this case, the program shall produce the output associated with the initial part of the sequence, up to and including the MAXPOS-th character of the first such group, and report the error.

Fig. 5. Yet another statement of the requirements.

Reuse

An essential requirement of a good specification formalism is that it should favor reuse of previously written elements of specifications. For example, the notion of sequence and the associated operations should be available as predefined specification elements. Languages Z and Affirm, among others, provide for such libraries of basic specifications. More work is needed to share and reuse the work of formal specifiers. Along with the availability of simple and efficient software tools, this is one of the conditions that must be met before formal specifications become for software engineers what, say, differential equations are for engineers in other fields.

ACKNOWLEDGMENTS

I learned most of what I know about specification from Jean-Raymond Abrial. Much of the material was contained in an earlier article, written in French and published in 1979 in a newsletter [1]. I am grateful to Axel van Lamsweerde for reminding me of the existence of that article and suggesting that it might be of interest to a wider audience (and to him and Jean-Pierre Finance for some heated discussions on specification). I also thank Flaviu Cristian for important comments on a previous version. The referees' comments were also useful.

This article also benefited from the involuntary contributions made by the authors of all the system requirements and other software documentation I have had to struggle with over a number of years.

Department of Computer Science,
University of California, Santa Barbara, U.S.A.

REFERENCES

1. Meyer, Bertrand: 1979, 'Sur le Formalisme dans les Spécifications', *Globule, Newsletter of the AFCET* (French Computer Society) *Working Group on Software Engineering*, No. 1, 81—122.
2. Dijkstra, Edsger W.: 1972, 'The Humble Programmer', *Comm. ACM* **15**(10), 859—866.
3. Meyer, Bertrand: 1980, 'A Basis for the Constructive Approach to Programming', in: *Information Processing 80 (Proc. IFIP World Computer Congress*, Tokyo, Japan, Oct. 6—9, 1980), S. H. Lavington (Ed.), North-Holland, Amsterdam, pp. 293—298.
4. Cristian, Flaviu: 1985, 'On Exceptions, Failures and Errors', *Technology and Science of Informatics* **4**(1), Jan.
5. Jackson, Michael A.: 1975, *Principles of Program Design*, Academic Press, London.
6. Burstall, Rod M. and Goguen, Joe A.: 1977, 'Putting Theories Together to Make Specification', *Proc. Fifth Int'l Joint Conf. Artificial Intelligence*, Cambridge, Mass., pp. 1045—1058.
7. Hill, I. D.: 1972, 'Wouldn't It Be Nice If We Could Write Computer Programs In Ordinary English — Or Would It?' *BCS Computer Bulletin* **16**(6), 306—312.
8. Oulipo: 1967, *Ouvroir de Littérature Potentielle*, Gallimard, Paris.
9. Ross, Douglas T. and Schoman, Kenneth E. Jr.: 1977, 'Structured Analysis for Requirements Definitions', *IEEE Trans. Software Engineering* **SE-3**(1), 6—15.
10. Alford, Mack W.: 1977, 'A Requirements Engineering Methodology for Real-Time Processing Requirements', *IEEE Trans. Software Engineering* **SE-3**(1), 60—69.
11. Abrial, Jean-Raymond, Schuman, Stephen A. and Meyer, Bertrand: 1980, 'A Specification Language', in: *On the Construction of Programs*, R. McNaughten and R. C. McKeag, (Eds.), Cambridge University Press.
12. Abrial, Jean-Raymond: 1980, 'The Specification Language Z: Syntax and Semantics', Oxford University Computing Laboratory, Programming Research Group, Oxford.
13. Abrial, Jean-Raymond and Schuman, Stephen A.: 1979, 'Specification of Parallel Processes', in: *Semantics of Concurrent Computation* (Proc. Int'l Symp., Evian, France, July 2—4, 1979), Gilles Kahn (Ed.), Springer-Verlag, Berlin-New York.
14. Morgan, Carroll and Sufrin, Bernard: 1984, 'Specification of the Unix File System', *IEEE Trans. Software Engineering* **SE-10**(2), 128—142.

15. Sufrin, Bernard: 1982, 'Formal Specification of a Display-Oriented Text Editor', *Science of Computer Programming* 1(2).

16. Meyer, Bertrand: 1984, 'A System Description Method', in: *Int'l Workshop on Models and Languages for Software Specification and Design*, Robert G. Babb II and Ali Mili (Eds.), Orlando, Fla., pp. 42—46.

References on the Naur-Goodenough/Gerhart Problem

Original reference, Naur:
  Naur, Peter: 1969, 'Programming by Action Cluster', *BIT* 9(3), 250—258.
First version, Goodenough and Gerhart:
  Goodenough, John B. and Gerhart, Susan: 1975, 'Towards a Theory of Test Data Selection', *Proc. Third Int'l Conf. Reliable Software*, Los Angeles, 1975, pp. 493—510. Also published in *IEEE Trans. Software Engineering* **SE-1**(2), 1975, 156—173.
Revised version, Goodenough and Gerhart:
  Goodenough, John B. and Gerhart, Susan: 1977, 'Towards a Theory of Test: Data Selection Criteria', in: *Current Trends in Programming Methodology, Vol. 2*, Raymond T. Yeh (Ed.), Prentice-Hall, Englewood Cliffs, N.J. pp. 44—79.
Another paper that uses the same problem as an example:
  Myers, Glenford J.: 1978, 'A Controlled Experiment in Program Testing and Code Walkthroughs/Inspections', *Comm. ACM* **21**(9), 760—768.

References on Formal Specification

Many formal specification languages have been designed in recent years. A few are listed here, without any claim to exhaustivity.

Abrial, Jean-Raymond, Schuman, Stephen A. and Meyer, Bertrand: 1980, 'A Specification Language', in: *On the Construction of Programs*, R. McNaughten and R. C. McKeag (Eds.), Cambridge University Press.

Burstall, Rod M. and Goguen, Joe A.: '1977, 'Putting Theories Together to Make Specifications', *Proc. Fifth Int'l Joint Conf. Artificial Intelligence*, Cambridge, Mass, pp. 1045—1058.

Jones, Cliff B.: 1980, *Software Development: A Rigorous Approach*, Prentice-Hall, Englewood Cliffs, N.J.

Locasso, R., Scheid, John, Schorre, Val and Eggert, Paul R.: 1980, 'The Ina Jo Specification Language Reference Manual', Technical Report TM-(L)-/6021/001/00, System Development Corporation, Santa Monica, Calif.

Musser, David R.: 1980, 'Abstract Data Type Specification in the AFFIRM System', *IEEE Trans. Software Engineering* **SE-6**(1), 24—32.

Robinson, L. and Roubine, Olivier: 1980, *Special Reference Manual*, Stanford Research Institute.

PETER NAUR

FORMALIZATION IN PROGRAM DEVELOPMENT

1. INTRODUCTION

In recent years several authors concerned with the methodology of program development have claimed that the use of formalized modes of expression in certain particular manners offers considerable benefits to the program development activity. The claims are related to a few key notions, most conspicuously formal specifications of programs and abstract data types. The present study is an analysis of some of the arguments given in support of these claims. It will be found that several of these arguments are invalid, and it will be maintained that the emphasis on formalization in program development may in fact have harmful effects on the results of the activity.

The present analysis is not an attempt at a treatment of the complete literature related to it. It will deal with only a few selected contributions, chosen so as to cover both the discussion of general principles and their illustrative application to specific instances of program development. General principles will be discussed primarily as they are expounded by Liskov and Zilles [11] and by Jones [9]. Applications will be taken from Henhapl and Jones [7] and Liskov [10].

2. SPECIFICATIONS, FORMALIZATION, AND MATHEMATICS

In this essay specification will mean a description of a problem to be solved with some aid from a program running on a computer, the specification to be produced by a programmer in addition to the program. This definition of specification is explicitly designed to exclude descriptions of a form that is used to control the computer directly, by means of automatic translation or interpretation, the reason being that in the terminology employed here such a description is a program. The present definition of specification is an attempt to capture the notion given to this word by Jones [9] and by Liskov and Zilles [11], but identity of meaning cannot be assured owing to their

Timothy R. Colburn et al. (eds.), Program Verification, 191—210.

addition of constraints to the notion of the word, such as the requirement that the description be formal, in some sense.

The arguments for the program development techniques to be
considered here are based to a large extent on claims about the
techniques employed in mathematics and about techniques characterized as 'formal'. Thus Liskov and Zilles [11] on page 6 say:

Formality. A specification method should be formal, that is, specifications should be
written in a notation which is mathematically sound . . . formal specification techniques
can be studied mathematically. . . .

Jones [9] on page 8 writes:

This book shows how precise specifications for programs can be written for data-
processing problems. To achieve this, formal definition methods are explained and then
shown to provide a basis for precise, though not fully formal, specifications. A similar
approach is taken with correctness arguments. For these to be sound (or rigorous) the
writer needs to understand what would comprise a formal proof.

Goguen [6] expresses himself similarly.

In spite of this stress on formalization, none of the authors quoted
makes any attempt to clarify what it is to be formal. As an approach to
clarifying this concept, here are the relevant explanations given in the
Oxford Concise Dictionary:

Formal: concerned with the form, not the matter, of reasoning; valid
in virtue of its form, explicit and definite, not merely tacit.

Formalize: give definite shape to; make precise, or rigid; imbue with
formalism.

These definitions are too vague to be used as basis of a discussion of
the formalization in program development referred to in the quotations
above. In these contexts the concept appears to be understood in a
more specific sense, essentially this:

Formal: expressed purely by means of symbols given a specialized
meaning.

Here the restriction to specialized meaning is intended to exclude the
symbolism used in natural language writing. Even with this clarification,
however, the formality or otherwise of many common forms of expression, for example numerals expressed in decimal notation, is unclear
according to the definition. This vagueness of the concept seems to be
unavoidable, and accords with the main point of the following discus-

sion, which asserts that the formal mode is merely an extension of the informal one, not a replacement of it.

On the other hand, the special notations used for example in arithmetic and in so-called formal systems of mathematical logic are clearly formal according to the definition.

It will here be argued that the uses of the word formal in the contexts formal specification and formal proof of the correctness of a program rest on a misunderstanding of the nature of formalization and therefore are misleading. In essence what Liskov and Zilles [11] and Jones [9] suggest is that the full meaning of an expression given in an informal mode can be conveyed by a formal expression. What will be argued here is that in reality the meaning of any expression in formal mode depends entirely on a context which can only be described informally, the meaning of the formal mode having been introduced by means of informal statements.

The dependence of formal expressions upon informal ones holds in mathematics as much as anywhere else. As illustration, consider the following passage from Gauss [5], appearing as part of a proof to which he himself attached extraordinary importance, his third proof of the law of quadratic reciprocity:

We can shorten the following discussion considerably by introducing certain convenient notations. Let the symbol (k, p) represent the number of products among

$$k, 2k, 3k, \ldots, \frac{p-1}{2} k$$

whose smallest positive residues modulo p exceed $p/2$. Further if x is a non-integral quantity we will express by the symbol $[x]$ the greatest integer less than x so that $x - [x]$ is always a positive quantity between 0 and 1. We can readily establish the following relations:

I. $[x] + [-x] = -1.$

This passage says clearly that the reason for introducing the formalizations is brevity and convenience. Here and elsewhere in the proof the formalizations are used in the first instance as convenient ways of stating certain facts. In some cases the derivation of a new fact from given ones can be accomplished by purely formal manipulation, but for the most part the derivations depend directly on the reader's intuitive understanding of the facts, whether these are expressed formally or informally.

The use of the formal mode, on the other hand, clearly has nothing to do with the validity of the various steps in the argument. Throughout the proof Gauss passes back and forth between the two modes with perfect ease. However, this ease cannot conceal the pre-eminence of the informal mode. As a verification of this point one might make the experiment of rewriting Gauss's proof using exclusively one or the other of the modes. This would show immediately that a purely informal formulation would be long and inconvenient, but perfectly possible, while a purely formal one could not conceivably make sense.

The pre-eminence of the informal mode of expression, with particular regard to program development, has already been asserted very strongly by Zemanek [16] on page 14:

No formalism makes any sense in itself; no formal structure has a meaning unless it is related to an informal environment [. . .] the beginning and the end of every task in the real world is informal.

This holds as much in mathematics and automatic theorem proving as anywhere else. Even in the most formalized mathematical argument the justification of each argument step in terms of a rule of inference must finally rest with the author's intuitive, informal acceptance that the rule applies and has been used correctly. In an automatic theorem prover the validity finally rests with the program author's intuitive, informal acceptance that the actions of his theorem prover match the adopted rules of inference properly, or with the user's equally intuitive and informal trust that the author has realized the rules that the user wants to employ.

Instead of regarding the formal mode of expression as an alternative to the informal mode we must view it as a freely introduced part of the basic informal mode, having sometimes great advantages, mostly for the expression of highly specialized assertions, but having also great disadvantages, first and foremost in being limited to stating facts while being inapplicable to the many other shades of expression that can be reached in the informal mode. With this view of the formal mode of expression many of the turns of phrase employed by Liskov and Zilles [11] turn out to be logically improper and therefore misleading. Thus 'formal specification' and 'formal proof' suggest a contrast to 'specification' and 'proof' which does not exist. At most one might distinguish between specifications or proofs that make more or less use of formal modes of expressions. It follows that the explanation of the first criterion for evaluating specification methods in Liskov and Zilles's section 1.2, which says that "A specification method should be formal ... This criterion is mandatory if the specifications are to be used in conjunction with proofs of program correctness", is meaningless. It also contradicts the explanation on the previous page which suggests that "code reading becomes an informal proof technique." This whole contradistinction between formal and informal techniques is a discussion of a pseudo-problem.

For similar reasons the whole argument in [11] that purports to justify the need for a separate specification that somehow makes the connection between what is called a concept and a corresponding program is void. Assertions may be proved true whether they are expressed informally or formally, and it is perfectly possible to prove that the result of a program satisfies a requirement stated informally. In fact, many of the proofs of the properties of programs given in such a treatise as Aho, Hopcroft, and Ullman [1], conform to this pattern, which again is quite similar to the usual mathematical style, as illustrated in the quotation from Gauss given above. Thus the claim of Liskov and Zilles that formal specifications, whatever they are, are necessary as basis for proofs of programs cannot be substantiated.

3. CORRECTNESS AND ABSTRACTION IN PROGRAMMING

The dominating reason given for insisting on formalization in program development is that supposedly formalization is necessary for proofs of correctness, and correctness supposedly must have absolute priority over any other concerns. Having already dismissed the first part of this argument it might seem less important to consider the second part of it. However, as an illustration of the argumentation used in support of formalization the question of program correctness and its importance is in need of examination.

Liskov and Zilles [11] say on page 2:

Although we are coming to realize that correctness is not the only desirable property of reliable software, it is surely the most fundamental: If a program is not correct, then its other properties (e.g. efficiency, fault tolerance) have no meaning since we cannot depend on them.

This is a remarkable statement since it is obviously proved false every day, in large scale, in probably most of the activities in which computers are used. Surely many programs that are used have errors in them, and are thus not correct, but even so provide useful service, and their efficiency and fault tolerance are issues of meaningful concern.

The claim that correctness of programs is an issue of absolute priority seems to be connected with the adoption of a scale of values that also attaches high importance to what is called abstraction, while it tends to dismiss as unimportant and trivial such issues as the format of data used in interfaces to users. In order to see some of the manifestations and consequences of such a scale of values, consider the example of a specification of a program, with particular illustration of the manner in which input and output will be described, as given by Jones [9], Chapter 19, 'Input/Output Statements'. Jones in his Figure 79 gives a specification of a program for producing as an ordered output list the first n primes, where n must be given in the input to the program. The figure gives the specification first at the highest level, consisting of eight lines of formulae, and then decomposed into more detailed descriptions, consisting of 18 lines. Most of the substance of the description is concentrated in two lines of the description at the highest level:

> **let** $n =$ **hd** il
> **elems** $ol = \{i \mid 1 \leq i \leq n \land is\text{-}prime(i)\} \land is\text{-}ordered(ol)$

These lines refer to an input list, il, and an output list, ol, and express that when the program has finished execution, if n denotes the head element of the input list then the output list is such that the set of its elements is the same as the set of the primes between 1 and n, and in addition the output list is ordered.

This specification reflects very clearly a scale of values that gives high priority to such abstract characteristics of the output as its inclusion of all primes, its exclusion of non-primes, and its being ordered, while such issues as the output medium, the line and page format, and the number representation, are ignored completely. Such a scale of values can only be maintained consistently if the responsibility of the program designer is defined so as to embrace only the strictly computer controlled part of the activity, while excluding the part of the data processing that includes interaction with humans. If a wider responsibility is assigned to the program designer this scale becomes unacceptable. As illustration, consider a solution of the production of primes that produces output in a smudged, printed form that makes it difficult to distinguish the digits 3, 5, 6, 8, and 9, or one that uses octal or hexadecimal number representation. Let us be concerned with the perception of this output by a human, and in particular the error rate of that perception. In this situation the correctness of the internal production in the computer might be less important than the characteristics of the physical form of the output.

This reflection points to a deep inconsistency in the arguments of those who in the interest of reliability advocate the use of so-called formal specifications and at the same time insist on concentrating on what is called the abstract aspects of the problems. These arguments must assume that reliability is highly dependent on the forms of expressions used in *interfacing* with those people who do the program development, while it is independent of the forms used in interfacing with the eventual users of the programs. Clearly such a perspective on program development leaves important reliability issues uncovered.

4. NOTES ON A DESCRIPTION OF ALGOL 60

As one example of the results of following the methods based on so-called formal specifications, the description of Algol 60 given by Henhapl and Jones [7] (referred to as the HJ-report in what follows) will in this section be subject for some critical analysis. The HJ-report will be compared with the official description of the same language given in the Modified report on the algorithmic language ALGOL 60 [4], referred to as the Mod-report in what follows.

As the first point in the comparison of the two reports, the HJ-report presents itself not as an independent description of a programming language but as a demonstration of a description technique. Thus the introduction of the Mod-report, which places the language in the context of history and related techniques and gives an overview of its characteristics, is replaced in the HJ-report by one that accounts for special technical points in which the description deviates from earlier related descriptions. Because of this difference in aim of the two descriptions the following comparisons will be confined to those parts of the two reports that describe the same matters.

As the second point of the comparison, the Mod-report includes examples of uses of the language that are not found correspondingly in the HJ-report. A fair comparison of the sizes of the descriptions will therefore have to omit the examples from the Mod-report.

Third, the formal notation used for the description of the language is defined in the Mod-report as Section 1.1, having 32 lines, while the corresponding basis of the HJ-report must be sought elsewhere, for example in Jones [8], a description of 60 pages. While the enormous difference in size of this background material must be kept in mind, both descriptions will be omitted from the size comparison below.

Fourth, with the omissions stated in the previous points, the sizes of the two descriptions of the language, counting lines of formal and informal mode statements equally, are as follows:

Mod-report   1310 lines
HJ-report   1110 lines.

Fifth, as the first substantial item in the comparison of the overall characteristics of the two language descriptions, the HJ-report does not define the language Algol 60, but a language, let us call it VDMgol, defined by a so-called abstract syntax. The relation between the language Algol 60 and VDMgol is not specified, indeed is only hinted at in Section 1.2 of the HJ-report. This means that the number of lines of the HJ-report given above is entirely misleading, since they correspond to only a fragment of the description of Algol 60. There can be no doubt that a complete description of the language obtained by completing the HJ-report would be considerably longer than the Mod-report.

Sixth, the formulations of Section 1.2 of the HJ-report, which are informal notes concerning the relation between Algol 60 and VDMgol, are remarkable for their unclarity and ambiguity. They rely heavily on the reader's thorough background knowledge of Algol 60 and on his willingness to guess an intended meaning behind obscure phrases. Thus they mostly leave it unclear whether a word such as expression, block, or body, refers to one or the other of the two languages, and they refer to the body of a **for** statement, a concept defined in neither of the languages.

The relation between the descriptions of language details in the HJ-report and the Mod-report will be illustrated here only by some examples taken from the part of the HJ-report reproduced in Figure 1. This part of the HJ-report is the first one to present actual language rules. It describes in lines 1 to 9 the restrictions that must be satisfied by programs, and gives in lines 10 to 23 part of the logic that goes into the check of blocks. The following notes are brief indications of details of Figure 1 that either in themselves or in a comparison with the Mod-

```
Line
1       is-wf-program(mk-program(b)) =
2          * for all type-decl, array-decl's within b. their s-oid is unique * &
3.         (let oads = {d | within (d. b) & is-array-decl(d) & s-oid(d) ≠ NIL}
4          * all expressions is s-bdl of elements of oads are integer constants */) &
5          (let env = [n → mk-type-proc(INT) | n ∈ Int-funct-names] ∪
6                     [n → mk-type-proc(REAL) | n ∈ Real-funct-names] ∪
7                     [n → PROC | n ∈ Proc-names]
8          is-wf-block(b. env))
9       type: Program → Bool
10      is-wf-block(mk-block(dcls. stl). env) =
11          let labl = /* list of all labels contained in stl without an intervening block */
12          is-uniquel(labl) &
13          is-disjoint(⟨elems labl. {s-id(d) | d ∈ dcls}⟩) &
14          (let renv = env\{s-id(d) | d ∈ dcls}
15          let lenv = [s-id(d)                           → (cases d:
16                      mk-type-decl(,,tp)       → tp
17                      mk-array-decl(,,tp,)     → mk-type-array(tp)
18                      mk-switch-decl(,)        → SWITCH
19                      mk-proc-decl(,PROC,,,,)  → PROC
20                      mkproc-decl(,tp,,,,)     → mk-type-proc(tp))
21                         | d ∈ dcls] ∪
22                      [lab → LABEL | lab ∈ elems labl]
23          let nenv = renv ∪ lanv
```

Fig. 1. Excerpt from Henhapl and Jones [7], Section 2.1.

report are incorrect or problematic. In this comparison many issues are uncertain, owing to the lack of a description of the relation between Algol 60 and VDMgol.

(1) Line 2 is incompatible with the Mod-report or is misplaced, depending on what is held in the *s-oid* fields of type and array declarations, a matter left to be guessed by the reader. If the *s-oid* fields hold the original identifiers in the case of own declarations then the uniqueness is inconsistent with the Mod-report. If, on the other hand, the *s-oid* fields hold internal identifiers generated in the translation from Algol 60 to VDMgol then the rule has nothing to do with the language Algol 60, being merely a requirement on the translator.

(2) Lines 3 to 4, requiring the bound expressions of own array declarations to be integer constants, should be compared with the following sentence from the Mod-report, Section 5.2.4.2: "The bounds

of an array declared as *own* may only be of the syntactic form integer
(see Section 2.5.1)." An examination of what is meant by integer
constant in the HJ-report and integer in Section 2.5.1 in the Mod-
report shows that they cannot be identical, as demonstrated by some
examples:

Bound expression	Accepted by Mod-report	Accepted by VDMgol
(34)	No	No
−45	Yes	Yes

(3) Lines 11—12 check that no identifier is used more than once as a
label in one block, and line 13 checks that no label identifier is otherwise declared in the block. However, there is no check that an identifier
is not declared explicitly more than once. For this reason lines 15 to 21
may form an inconsistent mapping. What lines 11 to 13 of Figure 1
thus fail to convey properly is expressed in the Mod-report, Section 5,
as follows: "No identifier may be declared either explicitly or implicitly
(see Section 4.1.3) more than once in any block head."

(4) In line 14 the mapping *renv* is formed from the global environment by removing the items corresponding to identifiers that are
redeclared in the block. However, items corresponding to identifiers
used as local labels are not thus removed. For this reason the operation
used in line 23 is incompatible with the specification language.

Other remarks on the formulations used in Figure 1 are made in
Section 5 below.

The remarks above are concerned only with a small fraction of the
1110 lines of the HJ-report. The remaining part of the report has not
been similarly analyzed. However, even the present limited enquiry
provides examples that indicate the problematic nature of several of the
claims made in support of the use of formalization. First, items (3) and
(4) above both indicate flaws in the purely formal part of the HJ-report
that, in addition to failing to describe Algol 60 properly, entail formal
inconsistencies related solely to the rules of the description language
employed. Thus, in spite of the insistency that formalization should be
employed in order to make proofs possible, the consistency of the
formulations of the HJ-report cannot have been proved by its authors.
As to the arguments employed in the present discussion, they have
been based on the intuitive understanding obtained by means of careful
reading of the formulae and thus are quite independent of the fact that
some of the matters involved have been expressed formally.

In connection with the statement (1) above it is relevant to compare directly with the corresponding formulation of the Mod-report, Section 5.1.3:

A variable declared **own** behaves as if it had been declared [. . .] in the environmental block, except that it is accessible only within its own scope. Possible conflicts between identifiers, resulting from this process, are resolved by suitable systematic changes of the identifiers involved.

In discussing this passage, which incidentally was added to the Revised Report on the Algorithmic Language ALGOL 60 [3] when forming the Mod-report, it must be noted that strictly speaking the last sentence is unclear since "this process" is undefined. The sentence seems to have been formed by thoughtless paraphrasing of Section 4.7.3.3 of the same report. Presumably the intended meaning is that which is expressed if the first part of the passage is rewritten so as to define the meaning in terms of a process of program transformation. With this understanding of the passage, the meaning of **own** is explained by means of a virtual program transformation. Line 2 of Figure 1 appears to be an attempt to express this notion formally, and the flaw noted in statement (1) indicates the failure of this attempt.

In summary, the description of Algol 60 given in Henhapl and Jones [7] fails to confirm the claims made for the advantages of using a formal specification language on all counts. Compared with the standard description of the language the more formal description is quite incomplete with roughly the same size; an examination of a small part of it has revealed numerous errors and inconsistencies, and the proof of its consistency appears to be impossible, or at least so impractical that it has not been done by its authors.

5. NEGLECT OF INFORMAL PRECISION AND DISDAIN OF INTUITION

While the development and use of formal modes of description in itself are at worst harmless, some of the current discussion and argumentation that tries to promote the use of formal modes most likely is harmful to effective program development. As the most prominent harmful effect, the claim that formal modes can be regarded as a superior alternative to informal ones inevitably will influence the authors arguing for formal modes into suggesting or implying that there are no such things as principles and practice of good informal expression, or, at least, that such matters are not worthy of attention.

Examples of unclear informal formulations can be found abundantly in the writings of the authors who make a special point of the alleged superiority of the formal mode. Thus much of the criticism of the argumentation in Liskov and Zilles [11] presented above can be viewed as directed against inconsistent use of words such as 'formal', 'proof', and 'specification'. Examples of unclear informal formulations that are directly involved in specifications of programs can be found in Liskov [10], Section 3.1 Problem Specification, an excerpt of which is shown in Figure 2. The section gives in one page an informal, high-level specification of a text formatter, supposedly as it might result from an analysis of the initial problem description. Quite apart from the fact that, as may be seen from sentences S6 and S12 of Figure 2, this specification is heavily oriented towards sequential processing and thereby in direct contradiction to the idea of a specification of 'what' and not 'how' it is supposed to illustrate, it gives rise to many questions of detail. Thus 'space' is used ambiguously, in S4, S6, and elsewhere, to denote a character, in S16 to denote whatever is between words, and in the obscure sentence S5 to denote, one may guess, position. The order of items explained is confusing; the paragraph from S6 gives certain rules about treatment of tab and space characters in the input, while the following paragraph, from S9, tells that the formatter has two modes of operation, and that in one of these the previous paragraph does not apply (presumably, the matter is not clear since the meaning of 'modification' in S10 is not given). The attentive reader will become frustrated from lack of clear answers to simple questions; for example, are the 10 spaces of indentation mentioned in S1 considered to be a modification as in S10, and thus not produced in 'nofill' mode? Are these spaces included in the 60 characters produced as an output line in the 'fill' mode according to S11? Are the 5 lines of the header mentioned in S2 included in the 50 text lines that form a page according to S1?

As other examples of negligent informal formulations made by authors who make a point of advocating the use of formal modes, consider the excerpt from Henhapl and Jones [7] given as Figure 1. Line 11 is defined in terms of "all labels contained in stl without an intervening block". Here the word 'label' is used inconsistently; the abstract syntax definition of the structure of 'stl' contains an item called label only as the operand of go-to-statements, while what must be meant to make sense at all in the context are labels appearing before colons before statements, identified in the abstract syntax, one may venture to guess, by the selector s-lp. Again, the phrase 'contained in stl [a statement list] without an intervening block' is likely to make sense only to readers with a good background knowledge of Algol 60.

Further examples of unclear informal formulations in Henhapl and Jones [7] are found abundantly in their Section 1.2, as already pointed out above.

The weakness of informal formulations shown as illustration above are caused predominantly by violations of a very elementary principle of expression. In a nut's shell the principle is that one should choose one's designations carefully and stick to them. Spelled out in more detail, the principle says

(1) that one should choose each specialized word and symbol carefully so as to identify unmistakably a concept that is useful and relevant in the context,

(2) that one should introduce each concept and designating word or symbol in the documentation where they are used, and

(3) that throughout the documentation each concept introduced must be referred to invariably by the same designation.

Obviously this principle is the basis of the use of symbols in mathematics. However, it is clearly just as important to clarity in informal expression. It is neither profound nor difficult to apply. When one finds that it nevertheless is grossly neglected by authors who wish to propagate formal modes, then it is hard to escape the conclusion that this neglect is influenced by the desire to demonstrate the alleged superiority of the formal mode.

Hand in hand with the claim that formal modes of expression are superior to informal ones goes an insistence that informal modes depend on unreliable intuition, without any mention that the same holds for formal modes. Liskov and Zilles [11] even talk of a superior "formal understanding", whatever that may be. In the view presented here, neglect of the quality of informal expression and disdain of intuition amount to intellectual suicide. As already stated above, any description of real matters of this world, and not just such partial aspects of them that have already been isolated so as to allow formal description, must fundamentally be based on an informal mode. If the informal basis is not clear and consistent then there is no hope for the relevance or usefulness of the description, irrespective of whether it makes use exclusively of an informal mode or also makes use of formalizations. And any argument, no matter how formally expressed, depends in the final analysis on intuitive insight. Intuitive insight comes in many degrees of certainty and doubt, and may actually be mistaken, but this does not imply that it has to be vague or ambiguous, nor that it is beyond discussion and refinement. In fact, an important part of any scientific activity is to express, discuss, criticize, and refine, intuitive notions. In this activity formalization may be very helpful, but it is, at best, just an aid to intuition, not a replacement of it.

6. NEGLECT OF SIMPLE FORMALIZATIONS

The misguided notion that a formal mode of expression can replace the informal mode entirely seems to be also the background of the special stipulation put on the use of formal modes by Liskov and Zilles [11] when they say that "specifications should be written in a notation which is mathematically sound" and that "the syntax and semantics of the language in which the specifications are written must be fully defined." With these demands formalization is made an awe inspiring, heavy tool that can only be employed by those who have been initiated into the appropriate mathematical methods.

Faced with these high demands one may first wonder whether the actual uses of formalization that one finds in the literature satisfy them. Does Gauss's introduction of the notation $[x]$ quoted above qualify? Does the introduction of the formalism for syntactic description used by Backus [2] and subsequently taken over in the Algol 60 report qualify? It has 16 lines of informal prose, one example of a meta-syntactic formula, and four examples of symbol sequences generated by it. Does the introduction of the Pidgin ALGOL employed by Aho, Hopcroft, and Ullman [1] throughout their book satisfy the demands? On page 34 they say explicitly that "No attempt is made to give a precise definition, as that would be far beyond the scope of the book."

In the view taken here Liskov and Zille's demand for full definition of the notation used in specifications is an unjustified takeover from the area of programming languages, where the demand is justified because such languages are designed for general use. The demand not only goes far beyond much established practice, but may have a harmful effect by discouraging useful formalization of limited scope, for special purposes. If we regard formalization, not as an overall replacement of the informal mode, but as a result of a freely made extension of the informal mode, then it will be obvious that one and the same description may employ any number of different formal notations side by side without contradiction, and there will be no objection to introducing a formal notation for a special purpose and defined only so as to make the meaning of that particular use of it clear. This is the attitude adopted by Aho, Hopcroft, and Ullman, who, as the continuation of the quotation above, say: "It should be recognized that one can easily write programs whose meaning depends on details not covered here, but one should refrain from doing so."

The unfortunate consequences of insisting on using only formalizations that are supported by heavy mathematical machinery and that supposedly replace the informal mode entirely are clearly visible in the illustration give by Liskov [10]. In the concluding discussion of that report it is said that "we have experimented with writing formal specifications and have found them very error prone and difficult." The reason for this experience can only be that the kinds of formalizations that have been tried have been inadequate. In fact, as already pointed out above the informal descriptions given in the report are quite unclear and incomplete. They could undoubtedly be improved considerably by the addition of a few simple formalized descriptions, thus in Section 3.1, 'Problem Specification', quoted in part in Figure 2, the handling of the various character classes in the input might be stated precisely by means of a table corresponding to viewing the input analysis as a finite-state machine. Similarly in Figure 3, 'Partial Specification of Streams', the description might be clarified by the addition of a table showing the reaction of a stream in dependence on the possible states of a stream along one dimension and the relevant operations along the other. Such tabular descriptions are the obvious means for helping to assure that in a certain situation all cases, or all combinations of cases, are considered and treated properly. Often they can be used directly as the basis for efficient and clear programming (see, e.g. Naur [12], pp. 88—95, and Naur [13]). That all of these advantages are disregarded, with no compensating gain whatsoever, is a telling comment on some of the argumentation and thinking going on around the use of formalization.

7. PSYCHOLOGY OF FORMALIZATION AND SPECIFICATION

With the view on formalization expressed above, the choice between informal and formal modes becomes, not a question of what can be expressed or the validity of what is expressed, but a matter of ease and effectiveness for author and reader, and thus to a large extent a psychological question. On the ease of the use of formalization no generally valid claim seems possible. Bertrand Russell, well known for his highly formalized studies of the foundations of mathematics, says in [15]: ". . . a good notation has a subtlety and suggestiveness which at times make it seem almost like a live teacher." On the other hand in [14], page 76, he says: "It is not easy for the lay mind to realize the importance of symbolism in discussing the foundations of mathematics, and the explanation may perhaps seem strangely paradoxical. The fact is that symbolism is useful because it makes things difficult." On this background there is reason for scepticism when Liskov and Zilles [11] claim that

because it is difficult to construct specifications using informal techniques, such as English, specifications are often omitted, or are given in a sketchy and incomplete manner. Formal specification techniques [. . .] provide a concise and well-understood specification or design language, which should reduce the difficulty of constructing specifications.

As an approach of clarifying the ease or difficulty of using formal modes, let us consider some concrete examples. If we have to multiply two hundred and seventy four by three hundred and forty one, most of us would immediately rewrite the problem formally as 274×341 and then proceed from there using some kind of formal manipulation. The obvious ease of the use of the formal mode in this case should not make us overlook, however, that this ease depends partly on the problem being of a highly specialized kind for which very effective formal tools have been developed, and partly on our having been trained in the use of these tools.

As a concrete example of formalization directly relevant to programming, consider the writing of programs in a programming language such as Fortran or Pascal. Probably most users would agree that for some purposes, such as describing certain arithmetic calculations, these languages make the required formalization easy, while for other purposes, such as describing the generation of elaborate reports as output, they impose difficulties. Thus it is not the particular form of formal description employed in programming languages that gives difficulties, but the requirement to express particular kinds of actions using particular given forms of expression. Clearly this consideration must have been a driving force in the development of the programming language Cobol during the years following its first introduction, characterized by the addition of built-in formalizations of actions such as sorting and report generation.

On the background of these examples one may suggest that for a particular kind of formalization to be convenient and helpful it must first have been designed specifically so as to cover a particular class of problems in a manner that is convenient to the people who have to use it, and, second, these people must have been trained in its use. It must further be inferred that in so far as established programming languages, such as Fortran and Pascal, are felt to be inconvenient to users who are well familiar with them, they must fail to cover certain classes of problems adequately.

This kind of failure of established programming languages seems to have arisen in at least two different manners: (1) the wealth and variety of the problem areas that are of significant practical interest in present-day programming are so large and change so rapidly that adequate coverage by a single programming language is incompatible with keeping the language small and economical; (2) the form of expression predominantly employed in the established programming languages, essentially strings generated by productions, excludes several useful forms of expression, such as tables laid out in two dimensions.

Faced with these difficulties in expressing required solutions using established programming languages, the advocates of so-called formal specifications offer what is claimed to be a help to the programmers. According to these ideas the programmer must first express the solution in a so-called formal specification language, second express the same solution in a programming language, and third prove that the two solutions are equivalent, in some sense. From the discussion above it should be clear that in the view presented here this approach offers no help to the programmer, but only adds to his or her burdens. Indeed, the two manners in which programming languages pose difficulties in covering problems may be expected to work in largely the same way with any given, pre-defined specification language. Thus with this approach the programmer has to produce not one but two formal descriptions of the solution, and in addition has to prove that they are equivalent.

From this discussion it follows that while formalization under certain circumstances is very helpful to the people who employ it, the formalization required when using any one given language for formal specifications cannot be expected to yield any particular advantage, but rather will be a burden on programmers.

8. SPECIFICATIONS FOR HUMAN UNDERSTANDING

While the discussion above must conclude in an invalidation of many of the arguments given for using formalization in particular manners in developing programs, it is not an attempt to argue for abolishing specifications or for avoiding formal modes. What will be argued is that for effective program development the criteria for selecting the form of specifications and for the use of formalization should be quite different from those advocated by the authors discussed above, and that consequently the kinds of specifications of programs that should be advocated are also quite different. According to the views adopted here specifications are sometimes a necessary evil, to be used for documentation of such aspects of programs that are not satisfactorily documented by the programs themselves. When specifications cannot be avoided they should be designed to bridge the gap between the user's intuitive understanding of each aspect of his or her problem and its solution on the one hand, and the programmed solution on the other. In general a program will have to be supported by part-specifications of several different forms, corresponding to its various aspects. The ideal specification of an aspect of a program is a description of what that aspect does such that on the one hand it is intuitively obvious to the user that it corresponds to his or her requirements, and on the other hand it is equally obvious to the programmer that it is realized in the program itself. In either case the intuitive understanding may need to be supported by an argument having several steps, a proof, but clearly it is preferable if such is unnecessary.

From this view of specifications it follows that the criteria for choosing the style and mode of each part-specification must combine ready understandability with the power to express whatever problem aspects are relevant. In addition the criteria must include clauses that ensure the ready, effective realization of the solutions described in terms of the programming language to be used. It follows that the most suitable form of a part-specification must depend on a variety of factors, including personal and environmental ones, to such an extent that the forms must be considered an open range of possibilities, unrestricted by any closed system of notions and techniques. Each part-specification must be connected to the environment by informal explanations, but may employ formalizations of any suitable form, such as tables for enumerating cases, decision tables, tables or graphs corresponding to finite-state algorithms, program skeletons, formulae of any kind, etc. A part-specification may employ a well-established description technique, or it may employ description forms that have been introduced specifically for a particular purpose.

When this is said it should be clear that these criteria correspond closely to the style and technique used in practice for traditional technical documentation. This, then, is the conclusion of the present discussion, that programs should be supported and specified by documentation of any kind, the overriding concern in producing this documentation being clarity to the people who have to deal with it. For achieving clarity any formal mode of expression should be used, not as a goal in itself, but wherever it appears to be helpful to authors and readers alike.

Datalogisk Institut,
Copenhagen, Denmark.

REFERENCES

1. Aho, A. V., Hopcroft, J. E., and Ullman, J. D.: 1974, *The Design and Analysis of Computer Algorithms*, Addison-Wesley, Reading, Mass.
2. Backus, J. W.: 1959, 'The Syntax and Semantics of the Proposed International Algebraic Language of the Zürich ACM-GAMM Conference, Proc. International Conf. on Information Processing', UNESCO, pp. 125—132.
3. Backus, J. W., Bauer, F. L., Green, J., Katz, C., McCarthy, J., Naur, P. (Ed.), Perlis, A. J., Rutishauser, H., Samelson, K., Vauquois, B., Wegstein, J. H., van Wijngaarden, A., and Woodger, M.: 1963, 'Revised Report on the Algorithm Language ALGOL 60', *Comm. ACM* **6**(1), 1—17; *Computer Journal* **5**, 349—367; *Num. Math.* **2**, 106—136.
4. de Morgan, R. M., Hill, I. D., and Wichman, B. A.: 1976, 'Modified Report on the Algorithmic Language ALGOL 60', *Computer Journal* **19**, 364—379.
5. Gauss, C. F.: 1876, *Theoremais arithmetici — demonstratio nova, Commentationes Societatis Regiae Scientiarum Gottingensis*, Vol. **16**, Göttingen, 1808; *Werke,*

1876, Bd. 2, pp. 1—8; *English translation in D. E. Smith, A Source Book in Mathematics*, Vol. 1, Dover, New York, 1959, pp. 112—118.

6. Goguen, J.: 1980, 'Thoughts on Specification, Design and Verification', *ACM SIGSOFT Software Engineering Notes* 5(3), 29—33.

7. Henhapl, W., and Jones, C. B.: 1978, 'A Formal Definition of ALGOL 60 as Described in the 1975 Modified Report', in D. Bjørner and C. B. Jones (Eds.), *The Vienna Development Method: The Meta-Language*, Springer, Lecture Notes in Computer Science 61. Berlin, Heidelberg, New York, pp. 305—336.

8. Jones, C. B.: 1978, 'The Meta-Language: A Reference Manual', in D. Bjørner and C. B. Jones (Eds.), *The Vienna Development Method: The Meta-Language*, Springer, Lecture Notes in Computer Science 61, Berlin, Heidelberg, New York, pp. 218—277.

9. Jones, C. B.: 1980, *Software Development: A Rigorous Approach*, Prentice-Hall, Englewood Cliffs, New Jersey.

10. Liskov, B.: 1980, 'Modular Program Construction Using Abstractions', in D. Bjørner (Ed.), *Abstract Software Specifications*, Springer, Lecture Notes in Computer Science 86, Berlin, Heidelberg, New York, pp. 354—389.

11. Liskov, B., and Zilles, S.: 1977, 'An Introduction to Formal Specifications of Data Abstractions', in R. T. Yeh (Ed.), *Current Trends in Programming Methodology*, Vol. 1, Prentice-Hall, Englewood Cliffs, New Jersey, pp. 1—32.

12. Naur, P.: 1974, *Concise Survey of Computer Methods*, Studentlitteratuur, Lund, Sweden.

13. Naur, P.: 1977, 'Control-Record-Driven Processing', in R. T. Yeh (Ed.), *Current Trends in Programming Methodology*, Vol. 1, Prentice-Hall, Englewood Cliffs, New Jersey, pp. 220—232.

14. Russell, B.: 1953, *Mysticism and Logic*, Penguin, Harmondsworth, England.

15. Russell, B.: 1922, 'Introduction', *in Wittgenstein, L.: Tractatus Logico-Philosophicus*, Routledge and Kegan Paul, London.

16. Zemanek, H.: 1980, 'Abstract Architecture', in D. Bjørner (Ed.), *Abstract Software Specifications*, Springer, Lecture Notes in Computer Science 86, Berlin, Heidelberg, New York, pp. 1—42.

PART III
CHALLENGES, LIMITS, AND ALTERNATIVES

BRUCE I. BLUM

FORMALISM AND PROTOTYPING IN THE SOFTWARE PROCESS*

1. INTRODUCTION

The September 1988 issue of the *Communications of the ACM* contained a paper by James Fetzer entitled 'Program Verification: The Very Idea' [1]. The title was adapted from a book by John Haugland on artificial intelligence [2]. One is accustomed to controversy in AI. One might not have expected an essay by a professor of philosophy to have aroused passions, but it did. The March 1989 issue of the *Communications* printed eight pages of technical correspondence on the paper plus three pages of discussion in the ACM Forum. A letter in the Forum, signed by ten researchers in the field, stated that Fetzer's paper was 'not a serious scientific analysis of the nature of verification' [3]. Fetzer replied by referring to the authors of the letter as 'The Gang of Ten' and said their response was an example of 'the ancient practice of killing the messenger when you do not like the message' [4]. Regrettably, the continuing debate has served to obscure rather than clarify the goals and benefits of formal methods.

The Hubble Space Telescope was to have been launched in October 1986. The Challenger accident delayed the launch date until December 1989 at the earliest. This postponement avoided the embarrassment of having to operate the satellite manually without the support of its automated Science Operations Ground System (SOGS). It is not that the SOGS software was mismanaged. Work on the requirements began in 1980, and a two-inch thick document was prepared. The software contract was let in 1981, and the tested components began to arrive in 1983. Unfortunately, as the *Science* reporter noted, "it was also becoming all too clear to the astronomers that 'meeting the requirements' was not the same thing as 'working as desired'" [5]. James Weiss, data systems manager for Space Telescope at NASA headquarters, remarked about the process used to develop SOGS, "It's the methodology that got us to Apollo and Skylab, but it's not getting us to the 1990s. The needs are more complex and the problems are more complex" [6]. Rapid prototyping is Weiss's solution.

213

Timothy R. Colburn et al. (eds.), Program Verification, 213—238.
© 1993 *Kluwer Academic Publishers.*

Two stories; two seemingly different solutions for resolving the 'software crisis'. The goal of program verification is to prove correctness. It treats programming as a form of mathematics and emphasizes the role of finding the correct formalisms for the problem domain. Prototyping, on the other hand is a type of experimentation with software — an unstructured approach to discovering 'what is desired'. On the surface, formalism and prototyping may seem to contradict each other. Yet, as I shall show, each is an integral part of the software process.

2. THE SOFTWARE PROCESS

'Software process' is a relatively new term. It includes all aspects of computer software creation and maintenance from problem identification to that of product retirement. Study of the software process is an active area of research in software engineering; it is believed that we can design better tools and environments only after we have understood the process they are to support. In any event, the software process provides a unifying framework for software engineering, which is the collection of disciplines dealing with the development and maintenance of software systems. In one sense, software engineering is a specialization of systems engineering; in another view it is a subset of computer science. These are murky distinctions, and the definitions of the discipline boundaries are undergoing change. Perhaps it will be better to start at the beginning.

Software engineering emerged as a distinct field as the result of a NATO Science Committee conference held in Garmisch, Germany, in 1968 [6]. The third generation computers were in operation, time sharing was available, high order languages such as FORTRAN were in widespread use, and structured programming had just been introduced. The challenge now was to make the building of large systems more manageable, predictable and reliable. The hope was that engineering techniques could be applied to software. Hence the conference title.

The first NATO conference generated great enthusiasm. There was agreement about the problem and hope for a solution. A second conference was held near Rome, Italy, the following year. The following is extracted from the introduction to that conference report [7].

The Garmisch conference was notable for the range of interests and experience represented among its participants . . . [They] found commonality in a widespread belief as to the extent and seriousness of the problems facing . . . 'software engineering'. This enabled a very productive series of discussions . . . [whose] goal was to identify, classify and discuss the problems, both technical and managerial, up to and including the type of projects that can only be measured in man-millennia. . . .

. . . [The Rome] conference bore little resemblance to its predecessor. The sense of urgency in the face of common problems was not so apparent as at Garmisch. Instead, a lack of communication between different sections of the participants became, in the editors' opinions at least, a dominant feature. Eventually the seriousness of this communication gap, and the realization that it was but a reflection of the situation in the real world, caused the gap itself to become a major topic of discussion.

One part of the problem was that software engineering, unlike the other engineering disciplines, was not derived from real world models; it was based on models of the artificial. Although many of the participants at the NATO conference were mathematicians who sought mathematical or logical solutions to the problem (in the sense that Algol offered a formal approach to the management of computer programming), there was no reality against which to calibrate their engineering practices. Consequently, software engineering was soon reduced to the study of those issues that affect the management of large software development projects. A waterfall life cycle model was proposed [8], and later refinements produced a software process model that was isomorphic to the hardware development model [9]. One had only to substitute 'code and debug' for 'fabricate'. Where structured programming enforced discipline in program construction, software engineering was intended to establish discipline in project management. One should not be able to write code before the specifications were accepted.

The software engineering of the early 1970s was adequate for the problems being addressed. Equipment was still relatively expensive, support for the programmers and analysts was still limited, and most of the projects had well defined goals (many of which could be stated with engineering precision). A decade later the equipment had become very inexpensive, the personal computers and work stations changed our perceptions of what computers could do, and new classes of very large, vaguely-defined projects were being undertaken. Of course, this is a byproduct of success. Like the Rome attendees, today's software engineers offer a diversity of solutions and the communication gaps persist.

If the software process is the collection of all software development and maintenance activities, then it can be characterized as a simple transformation from a need in the application domain to the realization of a software product that meets that need. As Lehman has shown, once the product is embedded in the application domain, the needs are modified, and changes to the product are required [10]. Thus, the transformation is but one step in an iterative process. In fact, it is widely recognized that some two-thirds of a software product's cost comes after it has been installed and that over half this cost involves product enhancement [11]. This leads some to refer to software development as a special case of software maintenance and others to describe software maintenance as software development with severe constraints. In what follows I will consider only software development. I choose this course for simplicity's sake; nevertheless, it is important to recognize that what is being addressed is only one iteration of the process.

The software life cycle that emerged in the 1970s formalized the classical problem-solving paradigm: decide what to do, decide how to do it, do it, and evaluate the result. This is the way that we build bridges and buildings, conduct wet-bench research and clinical trials, and carry out goal-directed learning. For the software process model, several phases were identified: system-level requirements analysis, software-level requirements analysis, allocation of functions, detailed design, code and debug, testing, and operations and maintenance. Although the phase titles were subject to variation, they were always organized as a sequence in which no phase could begin until the previous phase was complete in the sense that its output satisfied some verification or validation criteria. The flow was displayed diagrammatically as a series of cascading steps that suggested a waterfall. Backward-pointing dotted lines were used to show the need for backtracking. Obviously, conformance to this waterfall flow model would eliminate the tendency to program solutions before the problem was understood.

Figure 1 reduces the essence of the waterfall model reduced to three basic transformations:

— From the need in the real world to a problem statement that identifies a software solution for that need. This is the definition of what is to be done.
— From the problem statement to a detailed implementation statement that can be transformed into an operational system. This transformation includes the bulk of the waterfall activity. It encompasses the definition of how to build the software, its building, and its testing.

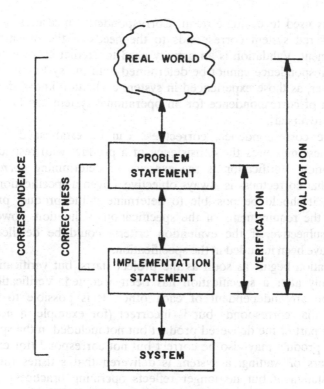

Fig. 1. The essence of the waterfall model.

The implementation statement includes the design descriptions, the source code, and the testing materials.
— From the implementation statement to a system that will satisfy the real-world need. This requires equipment, procedures, people, etc. It represents the embedding of the software product within the system.

The composite transformation is from a need to a software product that satisfies that need, i.e., it is the essential software process. Naturally, after the system is installed in the real world, the environment is modified thereby generating new software requirements. Thus, the figure represents only a temporal trace of one path of a continuing process.

The figure also displays two quality measures for the system with the processes used to evaluate them. Correspondence measures how well the delivered system corresponds to the needs of the organizational environment. Validation is an activity used to predict correspondence; true correspondence cannot be determined until the system is in place. (Moreover, as those experienced in system evaluation know, the determination of correspondence for an operational system can be difficult and controversial).

Unlike correspondence, correctness can be established formally. Correctness measures the consistency of a product with respect to its specification. Verification is the exercise of determining correctness. Notice that correctness is always objective. Given a specification and a product, it should be possible to determine if the product precisely satisfies the requirements of the specification. Validation, however, is always subjective; if the evaluation criteria could be detailed, they would have been included in the specification.

Validation begins as soon as the project starts, but verification can begin only after a specification has been accepted. Verification and validation are independent of each other. It is possible to have a product that corresponds but is incorrect (for example, a necessary report is part of the delivered product but not included in the specification). A product may also be correct but not correspond (for example, after years of waiting, a system is delivered that satisfies the initial design statement but no longer reflects operating practices). Finally, observe that when the specification is informal, it is difficult to separate verification from validation.

The distinction between the objectivity of verification and the subjectivity of validation suggests a boundary between the use of formalisms and prototyping. It is the difference between 'meeting the requirements' and 'working as desired'. Stating this more formally, the real world need of Figure 1 can be expressed as the application (A), the problem statement as the specification (S), and the implementation statement as the product (P). The essential software process can then be defined as the transformation $A \Rightarrow P$. Of course, this transformation is difficult to work with because it combines both subjective and objective evaluations. Consequently, it is more convenient to decompose the process into two transformations: $A \Rightarrow S$ and $S \Rightarrow P$.

Lehman, Stenning, and Turski hae described a software process model (LST) in these terms [12]. They call the transformation $A \Rightarrow S$ *abstraction* and the transformation $S \Rightarrow P$ *reification*. In a later work,

Turski opts for the terms *abstraction* and *satisfaction* [14]. Others use the terms *analysis* and *detailing* for these activities. In almost all cases, however, the researchers offer little insight regarding what happens before S exists. The emphasis is on the construction of a product P that satisfies its specification S.

The LST model provides one of the clearest statements of the reification process. It is represented as follows:

$$S \Rightarrow S_1 \Rightarrow S_2 \Rightarrow \ldots \Rightarrow S_i \Rightarrow S_{i+1} \Rightarrow \ldots \Rightarrow S_n \Rightarrow P.$$

Reification begins with some specification S. For a given problem, there are many potentially valid specifications that could be selected. Reification is not concerned with why S was selected. In the LST model the primary concern is that P be correct with respect to S. Presumably, this correctness can be demonstrated logically. Although there is an explicit validation obligation, it is recognized as an extralogical activity; validity cannot be proven. The central issue is how to preserve correctness.

The specification S is considered a theory in the sense that a set of axioms establishes a mathematical theory. Given a theory, the designers derive a model in a linguistic form that is closer to that of the desired product P. The result is a model of the theory. Many correct models for a theory exist, and the designers must select the one they consider to be the best. If the model is not derived by use of behavior-preserving transformations, then it must be verified. A validation obligation follows to ensure that no inappropriate behaviors have been introduced. Finally, the accepted model becomes the theory for another iteration of this canonical step.

Expressed in this fashion, it is clear that reification is a sequential process. The model S_{i+1} cannot be built until the model S_i is accepted as a theory. Once S_j has been accepted as a theory, any changes to the theories S_k, $k < j$, will invalidate the chain of logic from S_k forward. The goal is to maintain a logically correct trail until some model (S_n) exists that is isomorphic to the desired product P. The S_n represents a program that will be correctly transformed into the product.

Because it is unlikely that a single linguistic representation is appropriate for each level of modeling, the canonical steps are presented in terms of different linguistic levels. One may think of going from a top-level design to a detailed design to a program design language (PDL) to code. Of course, these are all informal models, and it is difficult to prove that one (model) is correct with respect to its parent (theory).

To guarantee rigor, one must have formalisms to express the descriptive theories of the application domain in ways that can be transformed correctly. For example, to the extent that FORTRAN expresses the scientist's intent, the FORTRAN program can be considered a descriptive theory. The compiled code will be a correct model of that theory. (Different compilers might produce alternative, but equally correct, models.)

Even when an appropriate formalism for the specification exists, the specification will not normally be complete. Maibaum and Turski observe that there are two reasons for this [14]. First, the sponsors seldom know enough about the application to define it fully at the start. Second, and from the LST model perspective even more importantly, there are many behaviors in the product P that are not important to the sponsor. If the specification S is complete, then S will be isomorphic to P. In short, the problem will be overspecified. What is necessary is that S contain only the essential behaviors. Any behaviors added to some S_i are permissive so long as they do not violate the behavior of S.

What started out as a clean description of a logical process turns out not to be so neat. The LST model assumed that S contains all the essential behaviors of the desired P. But we have just observed that the sponsor may not know what those behaviors are. Therefore, during reification it may be necessary to augment S as the problem becomes better understood. Changes to S, naturally, may invalidate the models that were derived from the original S. Consequently, the software process is not simply one of the logical derivation of a product from a fixed specification. Rather, it is best understood as one of problem solving in two domains using both subjective and objective criteria. Of course, that is what makes it so difficult.

This is summarized in Figure 2, which decomposes the software process activities by phase (abstraction and reification) and evaluation criteria (objective and subjective). The figure suggests the following interpretation of the LST model. Abstraction consists of a judgmentally-based requirements analysis activity that relies on formalisms and heuristics derived from domain experience. Where domain-formalisms exist, they describe properties and behaviors in the application domain, but they offer little insight into what the software should do. The outcome of the abstraction step is the specification S. The reification that follows is best described as an iteration of canonical steps, $S_i \Rightarrow S_{i+1}$, each of which culminates with (objective) verification followed by

	ABSTRACTION	REIFICATION
OBJECTIVE	DOMAIN EXPERIENCE	VERIFICATION
SUBJECTIVE	REQUIREMENTS ANALYSIS	VALIDATION

Fig. 2. Objectivity and subjectivity in the software process.

a (subjective) validation obligation. Much of the objective domain knowledge will be represented in S; it is the goal of the validation obligation to ensure that none of the introduced, permissive behaviors is inconsistent with the domain experience. Naturally, the boundary between the objective and subjective criteria is fuzzy. Nevertheless, the figure clearly shows the important role of the subjective and domain-specific (i.e., extralogical) decision making in the process.

3. THE ESSENTIAL SOFTWARE PROCESS

Earlier I made reference to an essential software process. The term is taken from the article 'No Silver Bullet' by Frederick Brooks, which is subtitled 'Essence and Accidents of Software Engineering'. Following Aristotle, Brooks examines the difficulties of software technology and divides them "into *essence*, the difficulties inherent in the nature of software, and *accidents*, those difficulties that today attend its production but are not inherent" [15]. He identifies the essential difficulties of software as its complexity, conformity (i.e., the software is constrained by a physical reality), changeability, and invisibility (i.e., it cannot be visualized). Some past breakthroughs that he cites are high-level languages, time-sharing and unified environments. His conclusion: accidental improvements cannot mitigate the essential difficulties of software. In the sense of the hygiene factors defined by Herzberg in his work on motivation [16], there are encumbrances that inhibit productivity but whose removal will have only a limited positive effect.

Using italics for emphasis, Brooks makes the following statement
[15]:

I believe the hard part of building software to be the specification, design, and testing of
this conceptual construct, not the labor of representing it and testing the fidelity of the
representation. We still make syntax errors, to be sure; but they are fuzz compared with
the conceptual errors in most systems.

This comment seems to place the emphasis on the subjective rather
than the objective part of the software process. Indeed, the recom-
mendations that he makes emphasize the need to control the scope of
the subjective problem solving. He suggests that we buy versus build,
that we use requirements refinement and rapid prototyping to enhance
our understanding, that we use incremental development to grow
(rather than build) systems, and that we foster the maturation of great
designers. He concludes that software construction is a *creative* process
and that the central question in the improvement of the software arts
centers on people.

If there can be an argument with his paper it is that it focuses on
software and not the software process, which I defined as being the
transformation from an application domain problem to a software
product that solves that problem. That is, it concentrates on the software engineer and not the software process. Of course, this is a reflection of the historical necessity of computer science. It grew out of a
need to create instructions for a new class of machine; the means for
specifying and transforming those instructions has been and remains
computer science's central concern. But if software engineering is to be
seen as a specialization of systems engineering, then it may be more
instructive to view the process from the perspective of the problem to
be solved rather than the product that represents the solution.

Figure 3 displays a model of the essential software process that
transforms the identification of a problem to be solved into a software
solution. I use the term essential because the constraints dictated by
accidental difficulties have been removed. I am concerned only with the
process of creating and validating the solution to the problem. I exclude
from this model any external considerations such as project management and sponsor involvement. It may seem foolish to ignore these
factors, but recall that the evolution of software engineering has been
guided by the need to manage large projects that integrate both hardware and software. Thus, the waterfall model, which was designed to

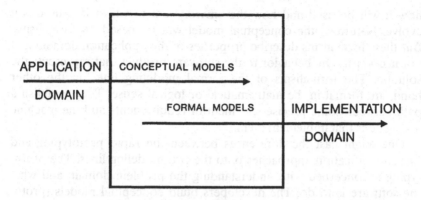

Fig. 3. The essential software process model.

control the accidental difficulties of managing 'man—millennia' projects, has come to define the software process that we use. I suggest that we must ignore these extraneous issues to understand what the essential software process is. Once this is done, we can address the accidental difficulties in a more rational way.

The figure shows this essential process as the transformation from the application domain to the implementation domain; from a need to a software solution. What was previously described as subjective and objective decision making is shown here as conceptual and formal modeling. The conceptual models are concerned with the application need and the specification of a product that will meet that need. Conceptual modeling begins in the application domain. Because of its orientation, conceptual models cannot prescribe the implementation. Formal models, on the other hand, are concerned with the reification process. This modeling line cannot extend out to the application domain, and every product must result from a formal modeling activity. (The computer program that is transformed automatically into the implementation represents the lowest level of the formal model.)

The names of the model types are somewhat misleading. In other papers I have described them as subjective and objective models: perhaps descriptive and prescriptive models would be a better choice. The term conceptual model is meant to imply the formulation of a problem statement that can be satisfied by a software solution. The model is a description of what the software is to do in the context of how it will be used and how the environment in which it is used will evolve. Naturally, the conceptual model will be based on formalisms. But these formalisms describe properties of the application domain and not necessarily the behavior to the software product that represents the solution. The formalisms of the formal modeling line, on the other hand, are formal in the mathematical or logical sense. They establish a path between some precise statement of requirements and the machine instructions that implement them.

One might cast the differences between the rapid prototyping and formal verification approaches onto the two modeling lines. The prototyping is concerned with understanding the problem domain and what the software is to do. The developers build conceptual models (prototypes) and experiment with them. Their goal is to derive some formal model that can be implemented. Formal methods, on the other hand, are concerned with the domain-independent formal modeling activity. Researchers in this area desire to extend the scope of the modeling line to the left and thereby provide assurance that the product preserves the properties stated in the specification. This is a neat mapping, but it obscures as much as it explains.

Software construction, as Brooks observes, is a creative process. The essential software process is a problem-solving activity based on defining the role of a software product for some application domain need and transforming that product concept into an implementation. There are two levels of creativity; two categories of problem solving. First, there is problem solving in the application domain. Obviously, this will be domain dependent; it will depend on the state of knowledge and experience, the application formalisms available, the constraints of the operational environment, etc. From the perspective of the implementation domain, however, it is a cognitive activity that leads to a specification, which prescribes the behavior of the implementation. The creation of the implementation begins with a precisely stated goal; it must proceed with rigor if the product objectives are to be achieved.

There is a fundamental tension between the conceptual and formal modeling activities as I have described them. The former involves problem solving in an application domain. Like most such processes, it entails learning and is managed in parallel incremental steps with backtracking. Except for domains in which there is considerable experience, there are few formal models to guide the determination of what the software should do. (This, of course, is quite different from the formalisms used to evaluate application-domain truths that are independent of any computer application.) The modeling activity can best be thought of as one of cognition. There is no single, unified model; its components are formulated from chunks that reflect reality, and experience with a potential solution refines the perceptions of its utility. Formal modeling, as expressed by the LST reification step, however, must be top down and sequential. It cannot begin until the initial specification exists. Changes to the specification tend to be discouraged because they require a reverification of the details linked to the modification. Both the conceptual and formal modeling activities are creative problem-solving tasks. The difference is that there are always objective criteria for evaluating the formal solutions.

There are several ways of managing this tension between the two modeling activities. One may accept the specification statement as valid and freeze it until an implementation exists to be tested. This is the method implicit in the water-fall model. It satisfies the need to be explicit about what is to be procured, i.e., the specification serves as a contractual device. Alternatively, one may allow the changing understanding of needs to drive the process. With the proper tools, it is now possible to have conceptual changes correctly integrated into the implementation during the development. In most cases, unfortunately, attempts to do this reduce confidence in the quality of the delivered product. Consequently, there is a growing move to recognize the dynamics of the development cycle by organizing work as small increments, with the lessons learned from one step providing guidance for the next step [17]. This is what Brooks implies in his advice to grow, not build, software. In what follows I examine some instances of the essential software process from the perspective of the paradigms in the title: formalism and rapid prototyping.

4. FORMALISM

It should be clear that success in computer science rests on formal methods. Instruction sequences in a computer can only be understood in terms of formal abstractions. Programming languages and their compilers are consistent and reliable to the extent we have good formal models for them. The issue is not one of having formalisms. It is one of identifying when and how they contribute to the software process and of understanding their limitations.

In a short note, Turski separates the conceptual and formal concerns that I have described. He puts it as follows [18]:

Logic — a calculus of formal systems — plays an important role in software development from specification to implementation. It can also play a minor role in validation of specifications, in so far as the validation is seen to include derivation of consequences. That logic plays no discernible role in descriptive theory formation — an act always serendipitous and contingent on invention — should not worry us too much, as long as we do not wish to claim the possession of the philosopher's stone.

Implicit in Turski's note is that the computer scientist is expected to develop linguistic levels for program (and system) specification so that domain specialists can express descriptive theories in a form that may be mapped easily into prescriptive theories (specifications) for computer software. That is, the software engineer should not be a domain specialist, but he should provide the high level languages for the specialist to define his needs. Here is the division of responsibility between the conceptual and formal models: to expect more implies mythological solutions.

This can be restated in terms of the essential software model shown in Figure 3. Notice that the two modeling lines can be expressed in terms of three transformations:
— From the application need to a (possible collection of) conceptual model(s).
— From the conceptual model(s) to a formal model.
— From the formal model to the implementation.
What Turski is suggesting is that computer scientists must establish languages such that the conceptual and formal models are conceptually close to each other. The domain specialist can then deal with the problem solving in his sphere of interest while the software engineer uses the formal methods of his profession to reify the specification and thereby produce the formal implementation. (Note that in this description of his job, the software engineer indeed becomes an engineer.)

When is it possible to restrict the software process to the reification activity? It seems to me that there are two categories of application. In one, the high level language (or environment) available to the domain specialist is powerful enough for him to express his conceptual model (i.e., domain theory). Examples are:
— The use of FORTRAN for the engineer to define a computation.
— The use of a fourth generation language (4GL) for an analyst to define a report.

— The use of spreadsheet program for an accountant to model his organization's finances.
— The use of a shell for a knowledge engineer to build the knowledge base of an expert system.

The second application category is that in which the formal model is the most appropriate representation for defining the desired behavior. Here the domain of concern can only be modeled formally using linguistic levels that also serve as specifications. Examples include:

— Implementation of algorithms as computer programs.
— Definition of programming languages, and by extension, the implementation of their compilers.
— Analysis of the abstract properties of automated systems such as software safety and computer security.
— Manipulation of well-defined interface conventions such as communication protocols.
— Experimentation with new computational environments, such as parallel computing, in which our intuitive foundations are weak and the rigor of a formalism is a prerequisite for a disciplined understanding.

The use of formalisms is not a new concept. Texts that explain methods for enforcing a formal approach have been available throughout this decade [19, 20]. It has been demonstrated that a reliance on formal methods can be used in a commercial setting and that the use of such methods improve both reliability and productivity. Indeed, it also has been shown that one can gain a sound understanding in certain of these complex domains only by recourse to formalisms.

The list of domains that can be served by formally-based tools is open ended. Considerable reseach is underway in the development of environments that will allow designers to specify a system and experiment with the behavior of the specification. This use of executable specifications, sometimes called the operational approach [21], is offered as an alternative model for system development. Here one defines the behavior of an application in a specification and then uses the specification as a simulation of the final system. Once the specification performs as desired, the specification is frozen and implementation can begin. Notice that the specification, even though it may be executable, is not the same as a program. The specification defines the behaviors of all possible implementations of the system; the implementation consists of programs optimized for a specific environment exhibiting the specified behavior. One can specify a system that one does not know how to implement. For example, the specification of a chess-playing program defines the moves and termination conditions for all implementations. There is an infinite number of possible implementations for this specification, and none has been found that will never lose.

The executable specification allows one to exercise the simulated behavior of the target system. In this sense, it allows the designers to validate that the system will 'work as desired'. Because the specification is defined formally, the process of definition can identify inconsistencies in the specification. That is, some categories of error can be eliminated at an early stage. It is common to think of testing as a process that increases our confidence in the correctness of a software system. But testing does not establish correctness; its focus in the identification of errors. A good test case is one that discovers a previously undetected error. In this context, a good test program is one in which the product is systematically examined to identify potential defects caused by oversight or carelessness. Those who favor formal methods assert that the use of a formalism reduces the number of failures resulting from oversight and provides a better mechanism for avoiding potential errors.

In a nondeterministic environment or when dealing with an infinite solution space, one cannot prove correctness experimentally. It can only be proven with logic. (In some cases, of course, it may only be possible to prove that an implementation is not correct.) And this is the center of the controversy about the very idea of program verification exposed by Fetzer's paper. The software is part of some larger system. One may prove the correctness of that software with respect to its specification, but what can be said about that specification with respect to the known (and unknown) system demands? Clearly, a formally verified program will be superior to one that is not. But at some level we must recognize that we are operating with some degree of uncertainty; we can never be too confident of our specifications, even when we can implement them. This is the substance of Parnas' cautions regarding the Strategic Defense Initiative [22].

5. RAPID PROTOTYPING

One of the limits of the formal methods is that their linguistic constructs may offer little insight into the application domain problem to be solved. In the context of the essential software process model, it may be difficult to map from the conceptual model into the formal model. That is, the formalism masks our understanding of the application intent. There are two ways to correct this deficiency. First, one may construct new languages (or representation schemes) that narrow this gap. Second, one may train analysts and software engineers to use the existing formalisms. (After all, familiarization with mathematical equations has led us to accept multiplication tables, algebra, and statistics as elementary tools. This was not true several hundred years ago.) However, until better formalisms and/or education are available, rapid prototyping offers an alternative.

Like the term 'formalism', rapid prototyping has many interpretations. Earlier, I placed rapid prototyping in opposition with formalism by suggesting that the former was concerned with conceptual modeling and the latter with formal modeling. Like all truths, this was only a partial statement of reality. As I shall soon demonstrate, some prototypes are formal models, and some have shown how prototyping and formal methods can fruitfully coexist in the same paradigm [23]. Nevertheless, the view of rapid prototyping as a method to establish valid requirements for a software application is a reasonable place to begin.

In the section on the software process I went to great lengths to separate the subjective validation activity (getting the right system) from the objective verification activity (getting the system right). I also pointed out that verification could not begin until there was a specification against which to verify a product. In the present context, therefore, I examine rapid prototyping as a technique to define (or refine) the initial specification. It is always based on subjective judgments rooted in the application domain; in the traditional waterfall flow, it is a requirements analysis subtask. Many variations of rapid prototyping have been identified, and in what follows I discuss some representative examples.

Although rapid prototyping was a topic of discussion in the 1970s, it received renewed attention in 1981 when Gomma and Scott presented their experience at the International Conference on Software Engineering [24]. Their goal was to specify the user requirements for a computer system to manage a semiconductor processing facility. The prototype method was selected because of shortcomings in the traditional requirements analysis methods. An interpretative tool with a report generator was used to build an incomplete demonstration prototype (SHARP APL). The prototype was shown to the potential users, and feedback from the exercise resulted in a refinement of both the developers' and users' understanding of the system needs. After the prototype was experimented with and the report formats were established, a requirements document was drawn up and a waterfall flow followed. The prototype was discarded. The total cost of the prototype exercise was 10% of the total project cost.

This view of the rapid prototype as a throwaway is a widely accepted notion. The prototyping activity is seen as a means for learning about the problem to be solved. In his *Mythical Man Month* Brooks has a chapter called 'Plan to Throw One Away'. Because the first system reflects an imperfect understanding of the problem, one ought to start with a pilot system. '*Plan to throw one away: you will anyhow*', he cautions the reader. The rapid prototype, however, typically involves considerably less than a pilot system. It is limited to those parts of the application that can best be understood through experimentation with a software product. The user interface is an excellent example of a requirement that can be effectively defined in this manner. The prototype need not address the details of input validation or database integrity. Thus, it may be constructed rapidly and — because of its incompleteness — ought to be destroyed after use. Naturally, not all the properties of an application can be prototyped (or stimulated, for that matter). Moreover, there is no guarantee that the processes accepted because of a prototype demonstration will be valid in an operational environment.

(It also is important to recognize that this type of prototyping activity can extend well beyond the initial requirements analysis task. As the software process progresses, the critical problems tend to migrate from the application to the implementation domain. In writing an Ada compiler, for example, even after the end product is clearly defined there may be uncertainty regarding how to construct certain features. Viewing the prototype as an experiment using software, the same techniques can provide a richer understanding of the implementation issues. In this sense, the prototype is used to test fesibility rather than establish initial requirements. Both modes of use are essentially the same. The point is that the software process is iterative, and the key problem-solving tasks shift as the project evolves. For large projects with very long development schedules, it is difficult to sustain user interest over long periods of apparently little progress. The Hubble Space Telescope SOGS requirements document reflected the users' very early understanding of their needs. Perhaps a prototype would have been a more effective way to elicit an understanding of their needs. As it turns out, however, it seems that the SOGS scheduling algorithm was its key deficiency; here a prototype in the sense of a feasibility-testing tool might have been more productive.)

Gomma and Scott identified four reasons for the success of their prototype. First, the prototype was limited to the user interface. Second the prototype was developed by 3 people whereas the full system required 12 people to be implemented. (Project studies have consistently shown the individual productivity is inversely proportional to project size; that is, smaller is better.) The final two reasons were testimonials to the language that they used for the prototype. In fact, it can be asserted that if SHARP APL were more efficient, then the prototype could be transformed into the operational implementation. This is the view of the operational prototype previously introduced in the discussion of formalism.

With the operational approach the goal is to define a valid specification. Balzer observes that the software process involves two activities: establishing what the product is to do, and optimizing its performance [26]. What makes software development so difficult is that the two processes are entwined. The solution, he and his colleagues suggest, is a modified development paradigm in which the specification and optimization activities are separated. One builds and evaluates a prototype specification. Once it has been accepted, behavior-preserving transformations are applied to improve the performance until an acceptable implementation emerges. Unlike the throw-away prototype, the specification is preserved, and maintenance is performed by testing changes to the prototype. Indeed, this is one of the major weaknesses of the throw-away technique. Because most of the system life-cycle cost is for product evolution, if the links cannot be preserved between the prototype and the product, then the utility of a prototype is severely circumscribed. For example, who would consider discarding a six-degree-of-freedom simulation just because the missile that it modeled was operational?

Although the operational approach has considerable appeal, most implementations are still in a pre-production state. Moreover, in many cases it is not possible to transform the specification into an implementation automatically. In these environments, the lessons learned from the prototype are incorporated into a textual requirements document, and the implementation proceeds with the standard waterfall flow. Nevertheless, it must be observed how operational prototypes are an expression of the essential software process model. The prototype is intended to act as a formal model of the desired application. The weakness of the approach is in our inability to manage the transformation from that initial problem specification to its implementation. The problems are too complex or too poorly understood to automate the process. Research in program transformations aims at reducing this barrier.

For some well understood application domains that rely on relatively mature software technology, it is possible to go directly from a prototype to an implementation. For example, in the previous section the spreadsheet program was used as an illustration of a high level language that permits the domain specialist to develop his conceptual models without regard for implementation details. In the context of the essential software process model, the computer environment is used to develop the conceptual model that can be transformed automatically into an implementation. That is, the software process is reduced to one of conceptual modeling, and no implementation domain problem solving is required. By way of contrast, in the formal approach, the goal is to provide a representation for building domain theories (conceptual models) thereby restricting the problem solving to implementation concerns.

At present, there are few domains in which the conceptual models can be transformed into implementations automatically. In very specialized disciplines, artificial intelligence can be used by domain specialists to solve problems that cannot be stated explicitly. For example, Barstow uses automatic programming techniques to support engineers in the analysis of oil tool data [27]. But this is an atypical application. A more common application is the use of 4GL tools to refine the understanding of the requirements. Although the process is sometimes referred to as prototyping, the 'prototype' products are retained as part of the delivered system. It is as if the SHARP APL programs of Gomma and Scott's prototype were of production quality and need not be thrown away. Stated another way, the linguistic level of the 4GL is high enough to satisfy descriptive theory building, and the transformations required to produce an implementation are understood well enough to automate its generation.

Blum has extended this form of prototyping in relatively large and complex clinical information systems [28]. He calls the technique systems sculpture (and not prototyping). The method is based on a sculpture metaphor, in which one starts with a concept, models iteratively, and — once the product is aesthetically pleasing — has a finished product. In this system the conceptual model is built as specifications that are retained in an integrated database along with descriptive text. A program generator maintains a system style that defines the default actions, user interface standards and rules for interpreting the specifications. Each generated program is complete in that it includes all functions stated explicitly in the specifications or implicitly in the system style. The system sculpture process proceeds by adding or revising specifications in the database, generating programs, and experimenting with the products. When the programs work as desired, they are ready for operational use. Because the specification database, maintenance is performed at the specification level. All problem solving is limited to the conceptual model; solutions to implementation domain problems are formalized in the system style.

The system sculpture method can also be described as the capture of knowledge about what the application is to do, and this is how prototypes generally are described in AI applications. Here the goal is to build a tool that applies knowledge in the solution of a narrow range of domain problems. The process of developing an expert system using a shell has been described as one of modeling in the application domain without concern for the implementation activity [29]. That is, the expert system shell provides the inference mechanism, and the knowledge engineer focuses on the representation of domain knowledge. This knowledge is formalized, modified and tested in the shell environment. During the early stages of this process, the system with an incomplete knowledge base is considered a prototype. As in the sculpture paradigm, once the knowledge base is rich enough to support an operational function, the development shell is replaced by a delivery shell, and the prototype label is removed. (Actually, the case study described in Ref. [29] is called a prototype because the end product is being used as a throw-away prototype in support of a waterfall flow development process.)

There are also modifications to the waterfall flow that use prototypes to reduce the risk of failure. Reference has already been made to the concept of iterative development in which a large project is broken down into many small increments, each with limited risk [17]. This approach requires an open architecture and the ability to be flexible in defining the tasks for the next iteration or build. Boehm has defined a spiral software process model in which there are many iterations (called prototypes) in the requirement definition phase [30]. Each cycle is designed to resolve some issues that have a high risk; once the requirements have been defined, implementation follows the traditional waterfall model flow.

Finally, it should be observed that there is one form of prototyping that cannot be recommended. Here the developers implement a partial system that exhibits some of the desired functionality. This is their prototype, which they then repair for delivery as the operational product. Notice how this practice perverts the process. System development before the Garmisch conference was plagued by the practice of code first and then define the problem. A rigid waterfall flow was established to eliminate this cause of defects, but subsequent experience has shown that the waterfall remedy is not without fault. Prototyping has been offered as a correction. This discussion has shown that there are many valid versions of the prototyping solution. However, a return to the hacking of the pre-Garmisch days is not one of them.

6. CONCLUSION

This essay has discussed formalism and experimentation (i.e., rapid prototyping) in the software process. Like the blind men describing the elephant, there are many partially-correct descriptions and an equal number of contradictions. It would be nice if there were a canonical formalism or a unified definition for prototyping. Alas, this is not the case. The software process is messy; formalisms and prototypes are but two projections from a holistic entity. What makes the software process so difficult is that it involves problem solving in two (or more) overlapping domains using two (or more) categories of modeling construct to solve problems that may have few (if any) physical constraints. Indeed, some of our problems may defy all software solutions.

In addition to the essential difficulties of the software process, there are those of carrying out the process in a real setting. These involve both systems engineering and management considerations. One of the major benefits of the waterfall flow is that it identifies a particular starting point (the requirements analysis that leads to the specification) and establishes the acceptance criteria for the finished product. Many who advocate the use of formal methods also favor a contract model for the software process. Here each phase begins with some specification that serves as a contract and ends when that contract has been satisfied. The outcome of a contract step may be the definition of lower-level contracts (for subsequent reification steps) or a correct implementation. Naturally, this contract approach provides for the negotiation of change, corrections after the discovery of errors, etc. The assumption is, however, that the initial specification is a good model that will be subject to only minor refinements. The advantage of the contract approach is that it provides traceability and well-defined development units.

Yet, as the discussion of prototyping has shown, for some projects we may not know what that initial specification ought to state. Here, a top-down, specify-and-decompose approach may be of too high a risk. In such cases it may be better to model the objects that are best understood and then compose a design from these fragments. But if the product is constructed in this manner, where then is the specification? How does the sponsor know what the product is to do or what has been purchased? It may be that the contract is simply a description of what has been delivered; ex post facto, the product is correct and valid. Of course, this is primarily a management concern. Both formalists and prototypers are concerned with the intellectual task of building a shared understanding between the client and software engineering. Different kinds of systems will require different techniques to arrive at that understanding.

As in the case of the conceptual and formal models, there is a tension between contracting for what is known to be needed and contracting for what is perceived to be desired. This is the management's version of the difference between the cognitive and mathematical domains. We are on the horns of a dilemma. We seek objective controls in a subjective universe. If it were not for the fact that we have made so many unprecedented advances, one might despair and consider the situation hopeless.

I began this paper by contrasting two views of a unified process; I shall close on a similar note. The following is a quotation from a widely-cited paper. It is out of context here; nevertheless, it provides a fitting conclusion [31].

Traditional formal logic is a technical tool for discussing either *everything that can be deduced from some data or whether a certain consequence can be so deduced*; it cannot discuss at all what ought to be deduced under ordinary circumstances. Like the abstract theory of Syntax, formal Logic without a powerful procedural semantics cannot deal with meaningful situations.

I cannot state strongly enough my conviction that the preoccupation with Consistency, so valuable for Mathematical Logic, has been incredibly destructive to those working on models of mind. At the popular level it has produced a weird conception of the potential capabilities of machines in general. At the 'logical' level it has blocked efforts to represent ordinary knowledge, by presenting an unreachable image of a corpus of context-free 'truths' that can stand almost by themselves. And at the intellect-modelling level it has blocked the fundamental realization that *thinking begins first with suggestive but defective plans and images that are slowly (if ever) refined and replaced by better ones.*

ACKNOWLEDGMENT

I would like to thank the following people for their helpful reviews of earlier versions of this paper: M. A. Ardis, F. P. Brooks, J. H. Fetzer, S. Gerhart, R. L. Glass, D. A. Gustafson, P. H. Loy, M. M. Tanik, R. L. Waddell, and the four *Information and Decision Technologies* referees. This is a much better paper because of their efforts.

Applied Physics Laboratory,
Laurel, MD, U.S.A.

NOTE

* This work was supported in part by the US Navy, Space and Naval Warfare Systems Command (SPAWAR) under contract N00039-89-C-3501, task VMAR7 with the Office of Naval Research (ONR), and by the Air Force Office of Scientific Research (AFOSR) under grant AFOSR-89-0080.

REFERENCES

1. Fetzer, J. H.: 1988, 'Program Verification: The Very Idea', *Comm. ACM* **31**, 1048–1863.
2. Haugland, J.: 1985, *Artificial Intelligence, The Very Idea*, Bradford Books, Boston, MA.
3. Ardis, M. A.: 1989, 'Editorial Process Verification', *ACM Forum. Comm. ACM* **32**, 287–288.

4. Fetzer, J. H.: 1989, 'Response from the Author', *ACM Forum. Comm. ACM* **32**, 288—289.
5. Waldrop, M. M.: 1989, 'Will the Hubble Telescope Compute?' *Science* **243**, 1437—1439.
6. Naur, P. and Randell, B. (Eds.): 1969, *Software Engineering*. Report on a conference sponsored by the NATO Science Committee. Science Affairs Division, NATO, Brussels.
7. Buxton, J. N. and Randell, B. (Eds.): 1970, *Software Engineering Techniques*. Report on a conference sponsored by the NATO Science Committee, Scientific Affairs Division, NATO, Brussels, p. 5.
8. Royce, W. W.: 1970, 'Managing the Development of Large Software Systems', *IEEE WESCON*, 1—9.
9. Boehm, B. W.: 1976, 'Software Engineering', *IEEE Trans. Comp.* **C-25**, 1226—1241.
10. Lehman, M. M.: 1980, 'Life Cycles and Laws of Program Evolution', *Proc. IEEE* **68**, 1060—1076.
11. Lientz, B. P. and Swanson, E. G.: 1980, *Software Maintenance Management*, Addison Wesley, Reading, MA.
12. Lehman, M. M., Stenning V. and Turski, W. M.: 1984, 'Another Look at Software Design Methodology', *ACM SIGSOFT SEN* **9**, p. 3.
13. Turski, W. M. and Maibaum, T. S. E.: 1987, *The Specification of Computer Programs*, Addison-Wesley, Workingham, England.
14. Maibaum, T. S. E. and Turski, W. M.: 1984, 'On What Exactly is Going On when Software is Developed Step-By-Step', *Seventh Int. Conf. S.E.*, pp. 528—533.
15. Brooks, F. P. Jr.: 1987, 'No Silver Bullet', *Computer* **20**(4), 10—19.
16. Herzberg, F., Mausner, B. and Sayderman, B. B.: 1969, *The Motivation to Work*, 2nd ed., Wiley, London.
17. Gilb, T.: 1988, *Principles of Software Engineering Management*, Addison Wesley, Reading, MA.
18. Turski, W. M.: 1985, 'The Role of Logic in Software Enterprise', *Proc. Int. Conf. S.E.*, p. 400.
19. Jones, C. B.: 1980, *Software Development, A Rigorous Approach*, Prentice Hall, Englewood Cliffs, NJ.
20. Gries, D.: 1981, *The Science of Programming*, Springer-Verlag, New York, NY.
21. Zave, P.: 1984, 'The Operational versus the Conventional Approach to Software Development', *Comm. ACM* **27**, 104—118.
22. Parnas, D. L.: 1985, 'Software Aspects of Strategic Defense Systems', *Comm. ACM* **28**, 1326—1335.
23. Hekmatpour, S. and Ince, D.: 1988, *Software Prototyping, Formal Methods and VDM*, Addison-Wesley, Reading, MA.
24. Gomma, H. and Scott, D. B. H.: 1981, 'Prototyping as a Tool in the Specification of User Requirements', *Proc. Intl. Conf. S.E.*, pp. 333—342.
25. Brooks, F. P. Jr.: 1975, *The Mythical Man-Month*, Addison-Wesley, Reading, MA.
26. Balzer, R.: 1985, 'A 15 Year Perspective on Automatic Programming', *IEEE Trans. S.E.*, **SE-11**, 1257—1268.
27. Barstow, D. R.: 1985, 'Domain-Specific Automatic Programming', *IEEE Trans. S.E.*, **SE-11**, 1321—1336.

28. Blum, B. I.: 1989, *TEDIUM and the Software Process*, MIT Press, Cambridge, MA.
29. Weitzel, J. R. and Kerschberg, L.: 1989, 'Developing Knowledge-Based Systems: Reorganizing the System Development Life Cycle', *Comm. ACM* **32**, 428—488.
30. Boehm, B. W. : 1988, 'A Spiral Model of Software Development and Enhancement', *Computer* **21**(5), 61—72.
31. Minsky, M.: 1981, 'A Framework for Representing Knowledge', in: J. Haugeland, (Ed.) *Mind Design*, MIT Press, Boston, MA, pp. 95—128.

CHRISTIANE FLOYD

OUTLINE OF A PARADIGM CHANGE IN SOFTWARE ENGINEERING

INTRODUCTION

This paper is a contribution of a special kind: rather than describing one specific result, I will attempt to give **a synopsis of an ongoing controversy** between rivalling ideas and attitudes underlying our scientific and technical work in software engineering. In doing so, I will argue for a shift of emphasis which I term a paradigm change. This change seems urgent in a situation where software as part of computer-based systems affects even more areas of the living human world, while the existing discipline of software engineering has no way of dealing with this systematically.

My aim, therefore, is to help clarify the conceptual basis used by the minority within the scientific community striving explicitly for a change and, at the same time, to promote a dialogue with those amongst us advocating a more traditional line. I will outline the paradigm change by drawing on some results of my own process of rethinking software engineering. I hope thereby to be able to contribute to the rethinking of others and, thus, to promote a paradigm change in the community at large.

In order to initiate of communication with the reader, I would like to start by explaining what I am trying to do in this paper, and prepare you for some shortcomings which I have not been able to avoid.

The first section is devoted to introducing the concept of **paradigm change** as I will apply it in connection with software engineering. I do this with some care, hoping to avoid misleading analogies with other fields of study. For the purpose of the present discussion, I introduce two perspectives from which we can look at the world we have to deal with. One of them is used as a ruling paradigm in the discipline of software engineering as it stands. The paradigm change I advocate consists in adopting the other perspective as a **primary point of view**.

In keeping with this idea, I had originally used the notions of 'established' vs. 'revised' point of view for naming these perspectives. This

239

Timothy R. Colburn et al. (eds.), Program Verification, 239—259.
© 1993 *Kluwer Academic Publishers.*

has proved unsatisfactory, however, since it confuses two levels of argumentation.

The two perspectives actually coexist in time; they are to some extent complementary, and no individual known to me argues from one of these perspectives only. The change would not imply giving up one of these perspectives in favour of the other but rather improving our understanding of how we should allow them to interact.

In order to make this transparent, I now make use of the notions **'product-oriented'** vs. **'process-oriented'** perspective. I introduce these terms in the first section before applying them in the second section.

The second section is the core of this paper. It serves to portray several aspects relevant to software engineering as they appear from the two perspectives. This section is like a mosaic consisting of individual pieces which together make a whole. Some basic concepts, including 'program', 'quality' and 'system', as well as several more complex issues, such as the role of documents in learning and communication and the relation between software and user competence, are taken up here. Each of these aspects has a paragraph devoted to it, highlighting some important points and commenting on their assumptions and their background as I see them. Each paragraph stands on its own, though they are related, and there is some redundancy in the argumentation. Their sequence in the text is not arbitrary, but there is no attempt to connect the individual paragraphs in a continuous flow of narrative.

While I feel that this approach is useful, since I aim to give a synopsis, it also has its inherent shortcomings: I cannot deal with quotations from the software engineering literature as I ought to, and I cannot give meaningful examples within the individual paragraphs. If I tried to do either of these to a satisfactory degree, this paper would turn into a book. My mosaic pieces, therefore, are like sketches; I have to rely on the willingness of the reader to make them concrete on the background of his or her experience.

The last section serves to place the paradigm change I advocate for software engineering in the wider context of an attempt to gear our development and use of technology more closely to human values. The human concerns at stake which necessitate this attempt are of vital importance to all of us. In spite of its obvious imperfections, I hope that this paper will be a small step towards bringing about such important changes as they apply to our field of work.

PERSPECTIVES AND PARADIGM CHANGE

In what follows, I shall contrast two perspectives on software engineering, a perspective being understood here as *a class of related views* on selected features of an area of concern.

The **product-oriented perspective** regards software as a product standing on its own, consisting of a set of programs and related defining texts. In doing so, the product-oriented perspective abstracts from the characteristics of the given base machine and considers the usage context of the product to be fixed and well understood, thus allowing software requirements to be determined in advance.

The **process-oriented perspective**, on the other hand, views software in connection with human learning, work and communication, taking place in an evolving world with changing needs. Processes of work, learning and communication occur both in software development and use. During development, we find software to be the object of such processes. In use, software acts both as their support and their constraint. From the process-oriented perspective, an actual product is perceived as emerging from the totality of interleaved processes of analysis, design, implementation, evaluation and feedback, carried out by different groups of people involved in system development in various roles. Both the functionality of the product and its quality as experienced by the users are held to be deeply influenced by the way in which these processes are carried out.

I refer to these perspectives as paradigms, in the sense of (Kuhn, 1962), since they provide a frame of reference suggesting questions we ask, quality considerations we aim for, and guidelines for interpreting results in our scientific and technical work. The product-oriented perspective has been adopted as the ruling paradigm in the discipline of software engineering since its founding conference held in Garmisch-Partenkirchen in 1968 (see Naur and Randell, 1969). In this paper, my intention is to criticize the product-oriented perspective in that role, since it does not permit us to treat systematically questions pertaining to the relationship between software and the living human world, which are of paramount importance for the adequacy of software to fulfil human needs. I argue in favour of a paradigm change in which the process-oriented perspective would replace the product-oriented perspective as the ruling paradigm.

We must be careful, of course, when taking up the concept of 'paradigm change', which has been proposed to characterize the development of the natural sciences in the course of several centuries, and applying it to a technical discipline on a very much smaller scale. In order not to draw invalid or misleading analogies, we must, in particular, remind ourselves that, historically, there have been quite different kinds and effects of paradigm changes: Whereas, for example, the Ptolemaic world view was entirely abandoned as a consequence of what is called 'The Copernican Revolution' in (Kuhn, 1979), Newtonian mechanics remained an important scientific discipline within its scope of validity, although its underlying world view had to be given up as a ruling paradigm in physics in our century.

In my opinion, the change of paradigms discussed in this paper is somewhat analogous to the Newtonian case. There can be no question of dropping the product-oriented perspective. Clearly, both product and process aspects have to be taken into account and reconciled in any specific project.

From the product-oriented perspective, software development is portrayed as follows: It starts with a so-called top-level requirement document, which serves to state requirements informally; it subsequently proceeds in several stages, involving successive defining documents and specifications, which eventually are transformed into a program to be executed on a computer. Most leading authors do acknowledge the importance of process aspects, arising in particular in the areas of requirements definition, quality assurance, user acceptance and software modifiability (note how these very terms reflect the underlying point of view!). Taking the product-oriented perspective as a ruling paradigm, however, these problems must be considered as additional aspects outside the realm of systematic treatment. They may influence how we proceed in actual projects, while the product-oriented perspective models what we supposedly should aim for, ideally. We thus find ourselves in a dichotomy between textbook wisdom and the needs of real life, aspects of vital importance being relegated to a place reminiscent of epicycles in pre-Copernican astronomy.

I hold that this situation is inherently unsatisfactory and can only be remedied if we adopt — in research, teaching and professional practice — the richer process-oriented perspective as our primary point of view, providing us with a framework in which to embed the treatment of product aspects. As will be shown at the end of this paper, this shift of emphasis, discussed here in connection with software engineering, can be considered as part of a more general move towards rethinking our basic assumptions and concerns in science and technology with a view to taking better account of human needs.

RELATED VIEWS FROM THE TWO PERSPECTIVES

In order to concretize the product-oriented and the process-oriented perspectives, I shall discuss several basic concepts as they appear from the respective points of view. The concepts are related to one another and have been selected so as to exhibit salient features of software engineering. My aim is to illustrate, by this list of comparisons, a shift of emphasis which can then be generalized to cover other topics of concern not explicitly dealt with in this paper.

Programs

Product-Oriented View
- Programs are formal mathematical objects.
- Programs are derived by formalized procedures starting from an abstract specification.
- Program correctness is established by mathematical proofs with respect to the specification.

Process-Oriented View
- Programs are tools or working environments for people.
- Programs are designed in processes of learning and communication so as to fit human needs.
- Program adequacy is established in processes of controlled use and subsequent revisions.

Clearly, both these views are valid and serve to portray programs in different roles. Before the advent of software engineering, the discussion had focussed on programs in their role as working instructions for the computer, and emphasized efficiency as the primary quality characteristic to be aimed for. The shift to what is now the predominant point of view in software engineering came as a reaction to the software crisis of the late 1960s. With the advent of the powerful third generation computers, their complex operating systems and more sophisticated applications in many fields, the original view of programs proved to be insufficient.

The shift of emphasis to the product-oriented view aimed at coping with the enormous problems of developing complex software in large teams. By insisting on program and specification texts as the basis of human understanding and of proofs for program correctness, it provided a starting-point for the task of dealing with errors in programming. By associating successive development stages with the controlled transition between increasingly formalized defined documents leading eventually to computer-executable code, it offered a means of structuring software production in a way that could be controlled, the defining documents taking the place of intermediate results. By providing guidelines on how to subdivide programs into parts, it supported the division of work in large development teams.

On the other hand, the product-oriented view leaves the relationship between programs and the living human world entirely open:

• There is no way of checking the relevance of the specification.
• Learning and communication between developers and users can only be accommodated in the earliest stages.
• Questions pertaining to the use of programs are deferred until the program is delivered.
• The interface betwen users and the computer as a tool is treated locally (within one program component), and often late (the functionality of the program being determined first, and its subsequently being made more 'user-friendly' to increase its acceptance). This attitude is characterized by the widespread use of the term 'syntactic sugar', referring to the attempt at improving input—output formats by purely syntactic means as a side-issue in program development.

In a historical setting, where computer programs were used to tackle relatively small and well-defined problems (S- and P-programs, according to the classification given in (Lehman, 1980), and where there was a clear-cut distinction between non-formalized work-steps and the use of computer programs, these shortcomings were at first largely ignored by software engineering authors.

Consequently, people and whole organizations had to adapt their working and communication patterns to computer programs.

With the advent of interactive systems closely intertwined with complex work processes, of networks allowing for the combination of small automated components into complex agglomerates leading to different modes of use, and, with the increasing dependence of society on computing resources, the embedment of computer programs in their usage context (E-programs according to the Lehman classification) has become a matter of vital concern whose systematic treatment can no longer be deferred.

In connection with software development, the term 'system' is used by different authors with different connotations. In order to examine these, the definition of 'system' as given by Nygaard is useful:

A *system* is a part of the world, which a *person* (or group of persons) during some time interval and for some reason — *chooses* to regard as a whole consisting of *components*, each component characterised by *properties* which are selected as being relevant and by *actions* which relate to these properties and those of other components. (Nygaard and Håndlykken, 1981)

As software developers, we are mainly concerned with 'systems' in two ways:

- The part of the real world which we take into account when developing programs is considered a system; I will call this the **referent system.**
- The set of programs we produce and their interfaces are considered a system; I will call this the **software system**.

The referent system consists of the microworld (objects, actions, rules) whose counterparts are to be modelled by the programs; of selected aspects of the environment in which the programs are to be embedded; and of rules for the interaction between the program and its environment.

From the points of view being contrasted in this paper, we find important differences pertaining, for example, to the aspects of the environment considered as part of the referent system, the development of these systems in time, and the relationship between the software system and its environment.

Product-Oriented View
- The referent system is chosen with a view to successfully developing the software system (**implementation-oriented reduction**).
- The referent system is described in terms of information processing according to fixed rules (**informationally closed**, 'information' being used in its narrow, technical sense).
- The referent system is considered as being essentially **static**, having

two states: before and after the software system is introduced. The effects of the software system on its environment are considered to be predictable.
- The software developer is **outside** the referent system; development and use are considered separately.
- The software system is considered as being **productionally closed**; it is viewed as **one** set of programs based on a complete and consistent understanding of the referent system.
- The **interaction** between software and its environment is **predefined** as part of the development.

Process-Oriented View
- The referent system is chosen with a view to successfully designing the work processes to be supported by the software system (**embedment-oriented reduction**).
- The referent system is described in terms of interleaving work, learning and communication processes (**informationally open**, 'information' referring to human processing of meaning).
- The referent system is considered as being **dynamic** and **evolving** in time: the software system will contribute to bringing about changes in the environment, which will in turn react on the referent system in an unpredictable manner.
- The software developer **becomes part** of the referent system; processes of development and use are considered to influence one another.
- The software system is considered as being **productionally open**; the actual set of programs at any stage is considered as **a version**, subject to later revisions and embodying limited and possibly conflicting insight into different parts of the referent system.
- The **interaction** between software and its environment is **tailored** by the users **to the actual needs** arising in their work processes.

In delimiting the referent system and in describing its connection to the software system, the product-oriented view relies on basic assumptions borrowed from a mechanistic world view:
1. For the design of high-quality computer programs, it is sufficient to consider (formalized) information-processing aspects of the usage context. They can be considered on their own, other aspects being ignored ('abstracted from'). These other aspects include the importance of social communication and cooperation for meaningful

work; the role of our body and our senses in dealing with real objects rather than with their computer-simulated counterparts; the environmental conditions under which we perform our work.
2. Human information processing and decision-making is largely equivalent to the functioning of computer programs. The division of work between people and computers is therefore arbitrary and should be determined by technical feasibility. There is no guideline as to what should and what should not be automated from a human point of view.
3. The suitability of a program in the context of work processes is determined by the information it produces. There is no consideration of how this information can be meaningfully used in communicative action.

When arguing for the process-oriented view as a basic paradigm, I do not deny the need to take reductionist steps with respect to the richness of reality in order to specify a microworld suitable for modelling in a software system. The chances are, however, that when focussing on the meaningful use of the software system as a tool in its usage environment, we may make different choices in selecting objects, attributes and actions to be included in this microworld, and we may allow for the combination of individual functions of the software system in a more flexible manner. Rather than making a given software system more user-friendly, we will arrive at a software system with different scope, functionality and possibilities for human intervention.

Quality

Product-Oriented View
- Quality is associated with features of the product.
- Quality characteristics refer to the reliability, efficiency etc. of programs; they can be influenced by changing the program.
- Quality is determined by validation (combination of testing and proving).
- Quality is defined by looking from the program to the users (e.g. 'user friendliness', 'acceptability').

Process-Oriented View
- Quality is associated with processes of using the product.

- Quality characteristics refer to the reliability, efficiency etc. of program use; they can be influenced by changing the work situation.
- Quality is determined by evaluation (argumentation, trial use, critical appraisal).
- Quality is defined by looking from the users to the program (e.g. 'relevance', suitability', 'adequacy').

The different notions of quality and their respective roles in judging the merits of software make it particularly clear that both the product-oriented and the process-oriented perspective must be used, so that they complement one another, and so that the human-centered, process-oriented view must have priority.

To determine the quality of computer use in organizations, we need to consider issues concerned with qualification, motivation, training, work organization and traditions of use, together with the features of the product itself. Furthermore, they must be treated as evolving in time. As has been shown in a recent empirical study of routine computer use in the US, processes of computer use are highly creative and expected. To bridge the gap between the functions of the DP system and their actual work-tasks, user communities develop practices that can be described as 'fitting' system functions to actual needs, 'enhancing' system functions by manual practices, 'working around' inadequate system functions in order to achieve human work-goals. By way of an extreme case, instances of reliable use of unreliable software systems were given, where the user community had developed work practices enabling them to work around known program defects.

The relation between quality as experienced by users and features designed into software systems by developers is not, at present, well understood, as is evidenced by increasing concern with quality assurance for software. In consequence, developers act in a situation of uncertainty with respect to the implications of their design decisions. To cope with this, techniques such as reviewing designs in teams, establishing roles in development teams, prototyping, and role-sharing between users and developers have been recommended. (See, for example, several papers in (Budde et al., 1984) and the papers on prototyping in (Agresti, 1986).) In order to make these techniques fruitful, designers must ensure that software systems and underlying design decisions **can be modified** as user needs become better understood. They are then able to enter into processes of communication with users, which, step by step, help to tune DP systems to actual user needs.

Thus, technical considerations concerning product quality, such as transparency and modifiability, must be seen as related to the overall aim of achieving high-quality computer use in organizations. While they cannot guarantee instant success, they are essential prerequisites in processes bringing quality about step by step.

Software development

Product-Oriented View

- Software development aims at producing one software system.
- Software production is followed by software maintenance, comprising the removal of faults and the adaption to new needs.
- Software production is considered on its own.

Process-Oriented View

- Software development aims at a sequence of related versions of a software system.
- 'Maintenance' is rejected as meaningless in connection with software. The tasks associated with maintenance are mapped onto the development of versions.
- Software development is considered part of system development.

The product-oriented view, then, assumes that the desired functionality of the software system is known in advance, and optimizes the path to this product. By contrast, the process-oriented view assumes that there is no such path known in advance, and that the desired functionality of the software system will gradually become unveiled as a result of interleaved processes of development and use.

The product-oriented view describes software development in terms of a linear model, relying on phases, ordered in time, and leading to increasingly formal defining documents. These serve as a starting-point for the next development phase, until the software system is developed as a final step. By contrast, the process-oriented view relies on a cyclic model of (re-)design, (re-)implementation and (re-)evaluation, each cycle leading to a version of the software system which can be evaluated in the context of work processes (see also Floyd, 1981).

For each version, the tasks formerly associated with phases have to be carried out incrementally, starting from the previous version and resulting in its successor. They are, however, no longer assumed to occur sequentially in time. While it is of paramount importance to be able to **isolate** objectives, requirements, system functions, design components and program components as **different universes of discourse**, to separate them in defining documents and to consider their relation to one another, the actual **tasks** of requirements analysis, design, implementation and so on may well occur **interwoven in time**.

The product-oriented view offers **one predefined model** for describing software development. The process-oriented view, on the other hand, offers a cyclical base model, which can be used as a starting-point for **developing a suitable strategy**, taking into account the needs of the actual development setting by choosing the cycles appropriately. Various forms of prototyping allowing for exploration, experimentation and evolution can thus be accommodated and tailored to the situation in hand (Floyd, 1984).

Programs, errors and user competence

Product-Oriented View
- 'Competence' refers to operating the program.
- 'Error' refers to operating the program incorrectly.
- Errors are classified in terms of their **effect** in handling the program (e.g. 'syntax errors').
- Errors are considered to be an expression of incompetence, competence being equated with the ability for correct use in predefined situations.
- Programs should be intelligent; 'intelligence' refers to information processing, symbol manipulation and derivations based on rules.
- Programs should implement decisions according to predefined rules.
- Users are **categorized** according to their competence.
- Programs should monitor users on the basis of **a user model** maintained by the program.

Process-Oriented View
- 'Competence' refers to carrying out work tasks with the help of the program.
- 'Error' refers to one of a class of quite different human situations ranging from simple mistakes to mismatches between user expectation and program functions.
- Errors are classified in terms of their **origin** as human actions (e.g. 'spelling errors').

- Errors are inherently connected with learning; they are therefore a precondition for acquiring the competence to use programs in open situations.
- The use of programs should be intelligent; 'intelligence' refers to competent human dealing with open situations, complex needs and changing goals.
- Programs should support human decision-making according to actual needs and unforeseen events.
- The **competence** of users is considered to be continually **changing**, each use of the program being a learning process.
- Users should be able to **intervene** in the program actions and adapt them to their individual needs.

It should be clear that the point of view adopted primarily in system development will have profound implications for the design decisions to be taken, affecting, in particular, the man-machine interface and the treatment of errors by programs. In order to be able to place errors in the context of their origin, the emphasis must be put on the human tasks to be achieved from requirements analysis onwards (see also the related arguments in the paragraph on systems). This again requires cooperation with the actual users as part of a mutual learning process. Human and political preconditions must be created, which make a partnership between developers and users feasible and meaningful. This implies that, in the course of a paradigm change, our attitudes towards the users of our programs must also change.

Adopting the process-oriented view in designing and installing programs for use may have far-reaching effects on the psychological work conditions of the users. Where product-oriented error treatment prevails, users are confronted with recurring experiences of failure, often without knowing why. Feelings of inferiority with respect to the computer are likely to be increased. Process-oriented error treatment, to the extent that it can be programmed, will help to increase user confidence, as it permits us to understand the context of the error and will help us to avoid it when the situation comes up again. Of course, human guidance in learning how to use programs in open situations is an essential prerequisite here.

From the process-oriented view, programs must be designed so as to facilitate and encourage intelligent use. This affects the terminology used in the interfaces, the choice of functions and the possibilities for combining them and for intervening. It is strongly connected with the notion of adequacy as a software quality characteristic to be aimed for.

The points made above can also be seen as related to the critique of artificial intelligence as given by Dreyfus (Dreyfus, 1979), and they will become more and more important as expert systems spread in organizations. The task will then be to develop and use expert systems in a manner that will allow users to develop and increase their competence, rather than restricting them to one fixed rule system whose working, furthermore, is not transparent to them.

Generally, adopting the process-oriented view will lead to the design of software systems in which the user can intervene so as to enhance the functionality in the context of his work.

Understanding programs, and the role of documents

Because of the abstract nature of software, we can learn and communicate about it in personal discussions with experts (program designers, users), by defining documents and by trial use. The product-oriented and the process-oriented view differ in the respective importance they attach to these three ways of learning about software and how they see them to interact. Since documents play a key role here, further differences come in from the demands made on the content and the form of defining documents. Implicitly, these views draw on differing positions as to how understanding comes about: its relation to theory, to meaning, to form and formalization, to action in the living human world. In this paragraph, I can barely touch on the fundamental issues involved.

Product-Oriented View
- Programs should be understandable from **documents only** (defining documents are part of the product).
- The meaning of a program is defined by its formal semantics as given **in its specification** and derived from semantic atoms in the document by applying semantic rules; the relation of the specification to the real world is open.
- The important part of a specification is the model to be embodied in the program, which should be described using **one formalism**, explanations in natural language are not considered systematically.
- Knowledge is acquired by understanding formal rules, documents must be complete, consistent and unambiguous, this being the only way for making the interpretation reliable.

- Documents should give **one context-free description** of what the program does while abstracting from how it is done (one unified developer's view).
- Documents should describe the **results** of design (and other) processes.

Process-Oriented View

- Personal **discussions** and **trial use** are **indispensable** for understanding programs, documents must faciliate these activities.
- The meaning of a program must be seen in connection with the intentions of the authors, it relates to the totality of possible program uses; the relation of the specification to the real world must be clarified by **argumentations about the specification**.
- The **essential part** of any document is written in **natural language**, introducing concepts, giving definitions and explanations surrounding the model, formalisms should be used as needed but add no new quality.
- Knowledge is acquired to a large extent through **examples and analogies** with existing background knowledge; if well chosen, they will help significantly to determine what is relevant and to increase human ambiguity tolerance.
- Documents should be written from well **defined view-points** explaining what the program does, how and why in a language understandable to the readers.
- Documents should also describe **how the result was obtained**: experiences in learning, decisions and reasons for them, rejected alternatives.

In (Lehman, 1980) there is a complex definition of a program as being "a model of a model of a theory of an abstraction of a model of a part of the real world or a universe of discourse." The definition can be viewed and used both from the product-oriented and from the process-oriented perspective.

For example, in (Turski, 1985) we find a (mathematical) notion of 'theory' such that the specification embodies the theory, the program then being a model of the specification. In (Naur, 1984), by contrast, an entirely different concept of theory is used (going back to Ryle, 1949). Theory, here, resides in the head of the programmer, who is *building a theory* about the part of reality or the universe of discourse relevant as a starting-point for the program as he or she goes through processes of analysis, design and implementation. Any program will only reflect a limited part of such a theory.

In the first use of the word, theory itself is considered a product which can be isolated from the human mind and considered on its own, while in the other, theory is considered to be forever changing and developing and to be inherently connected with human learning processes. The richer notion of theory would go a long way towards explaining why the idea of having documents on software be self-explanatory without the help of the designers cannot be realized.

From the process-oriented view, understanding existing programs by users or by newly arriving programmers can similarly be regarded as theory-building.

The points made above with respect to documents are of special importance when trying to overcome communication problems with users. While the product-oriented view will lead to one model of the program as seen relevant by the developers, the process-oriented view will encourage having different, overlapping views on the program tied in with specific uses in communication processes. The emphasis will shift from issues of form to the content: Which universe of discourse is addressed? What terminology is to be used? Which examples are relevant to the readership? Who is writing for whom to read? Documents, then, have a well-defined role in the communication process (see Floyd and Keil, 1983).

As a specific technique for showing how the result of the design was obtained, the use of diary has been recommended in (Naur, 1983). This technique can also be applied in teams by having a shared project archive as a common reference. In order to make the processes of design transparent, it has to be organized in a diary-like manner, and we have to be able to trace individual design issues as they developed in time. In order to be useful, such a project diary must be made an essential ingredient in discussions and decision-making by the team.

Methods

Product-Oriented View
- Methods are products; proper use of methods will lead to uniform results.
- Methods serve to constrain communication.
- Methods will make software development less dependent on people.

- Methods are generally applicable.
- Methods are static, well-defined and context-independent.
- Methods serve to specify aspects of the product.

Process-Oriented View
- There are no methods, only processes of method development and use; these processes influence the result of using the method.
- Methods serve to support communication.
- Methods are meaningful only as steps to be taken by people.
- Methods are associated with an area of application and a perspective on software development.
- Methods are evolving with experience, need to be tailored to individual needs and situations.
- Methods serve to support processes of cooperation where the product is gradually discovered.

The product-oriented view of methods is to a large extent responsible for a mismatch between the expectations of method-users and the effects obtained after the method has been introduced. Typical complaints include that the results obtained with the use of the method adopted vary enormously depending on *who* uses the method; trying to use a method *literally* will lead to problems in specific situations and even to low-quality results; people will be found to develop modes of working around or even against methods which were supposedly adopted.

Taking the process-oriented view, by contrast, methods as described in books serve as an inspiration for communities developing their own shared systematic ways of working: based on their background knowledge, modified and refined as they gain experience, and tailored to their specific situation. Individual guidelines offered by methods are rejected or adopted on the basis of their usefulness in dealing with the problems in hand and of their compatibility with the attitudes of the organization using the method. Any one method as described by its author will thus lead to several quite different schools of method application. What we have to pay attention to, when introducing a method, is to *develop a suitable and teachable tradition for its use.*

Application area and perspective as inherent aspects of methods have been stressed in (Mathiassen, 1981). As explained in more detail in (Floyd, 1986), the perspective comprises both the value judgements of the method authors and assumptions about which tasks are most important in software development and how they should be connected. The application area must be characterized by taking several aspects of the development situation into account: the application orientation, the size, the expected life-span of the software to be produced and so on.

From the product-oriented perspective, then, methods appear as static *rule systems* governing people's behaviour in a sterotyped, standardized manner. From the process-oriented perspective, they appear as *second-order learning processes* based on an ever-growing class of individual past learning processes in actual projects, using guidelines offered by method authors and tailored to the situation in hand, and preparing us for future learning processes in developing software.

As pointed out in (Floyd, 1986), existing methods overemphasize product aspects in two ways: they embody, as part of their perspective, a product-oriented view of software development *and* they often enforce a product-oriented way of method use. The paradigm change advocated in this paper, as applied to methods, would mean to work on the development of methods *giving priority to the process-oriented view in both ways*: by supporting processes of creative cooperation in teams of developers and users, and by allowing individual method components to be fitted to the individual needs of a given situation in a flexible manner.

Closing Remarks to this Section

As has been said in the introduction, this is a mosaic consisting of interlocking sketches on relevant aspects in software engineering. They are not meant to present results, primarily, but to give examples in order to encourage the reader to consciously adopt the product- and the process-oriented perspectives in given situations. It will then be found that we need new ways of teaching, new strategies in projects, an improved understanding of our relations to one another, but also suitable techniques, tools and forms of organization to make the process-oriented perspective fruitful in our work. Thus, the paradigm change does not mean rejecting technology, but finding ways to a more human-oriented technology.

THE WIDER CONTEXT

The argumentation in the sketches in the previous section brought out several recurrent themes to characterize the product-oriented perspective: linear thinking with no awareness of feedback leading to deeper insights; separation from human cognitive processes and their results; equating living processes with their products; focussing on the solid aspects of the product which remains static, with no regard to the fluid nature of the ever-changing context of its use.

These modes of thinking go along with deep human attitudes: to feel responsible only for what we are doing, but not for the consequences of our actions in their context; to separate technical, formal issues sharply from values and from concerns for living, social processes; to act as if we were solving well-defined problems, rather than taking part and being instrumental in processes of change; to rely on an objective reality which we must uncover as part of system development, rather than to realize that we are constructing this reality by our interwoven processes of learning and communication.

Giving priority to the process-oriented view would imply dealing with conflicts and contradictions which are abstracted from in the product-oriented view. It would require us to change our attitudes towards those who will use our products: to think of them as partners in spite of conflicting interests; to learn to give and take criticism in a supportive and constructive manner; to learn to work in technology keeping in mind human values and changing human needs. It would mean going beyond the mechanistic world-view embodied in the product-oriented perspective.

As a closing remark, I would like to remind the reader that in this paper I have contrasted two perspectives and not two communities of people holding these perspectives. We all argue and act from both of these perspectives. The criticism implied in my argumentation refers to the fact that existing methods and scientific approaches in software engineering embody the product-oriented view almost exclusively. Working towards overcoming this is both highly imperative, considering the role of information technology in the living human world, and also inspiring since it will provide the opportunity for deep insights into the nature of cognitive processes, for richer relations with the people we meet in our work, and thus for personal development and growth while doing high-quality work in technology.

258   CHRISTIANE FLOYD

ACKNOWLEDGEMENTS

The final version of this paper has been deeply influenced by several discussions with Stein Bråten both at and subsequent to the Århus conference. I would also like to thank Heinz Züllighoven for his critical remarks, and Michaela Reisin and Edda Sveinsdottir for their support.

Technische Universität Berliny
Institut für Angewardte Informatick

REFERENCES

Agresti, W. W.: 1986, *New Paradigms for Software Development*, IEEE Computer Society Press, Order Number 707, Washington, D.C.
Budde, R., Kuhlenkamp, K., Mathiassen, L. and Züllighoven, H.: 1984, 'Approaches to Prototyping', *Proceedings of the Working Conference on Prototyping*, Springer-Verlag, Berlin-Heidelberg-New York-Tokyo.
Boehm, B. W.: 1976, 'Software Engineering', *IEEE Transactions on Computers* C-25(12), 1226—1241.
Dreyfus, H. L.: 1972, *What Computers Can't Do — The Limits of Artificial Intelligence*. Harper & Row, New York.
Floyd, C.: 1981, 'A Process-Oriented Approach to Software Development', in: *Systems Architecture, Proceedings of the 6th European ACM Regional Conference*, Westbury House, pp. 285—294.
Floyd, C.: 1984, 'A Systematic Look at Prototyping', in: R. Budde *et al.* (Eds.), *Approaches to Prototyping*, Springer-Verlag, pp. 1—18.
Floyd, C.: 1986, 'A Comparative Evaluation of System Development Methods', in: T. W. Olle, H. G. Sol, and A. A. Verrijn-Stuart (Eds.), *Information Systems Design Methodologies: Improving the Practice*, North-Holland, pp. 19—54.
Floyd, C., and Keil, R.: 1986, 'Adapting Software Development for Systems Design With the User', in: U. Briefs, C. Ciborra, and L. Schneider (Eds.), *System Design For, With and By the Users*, North-Holland, pp. 163—172.
Kuhn, T. S.: 1962, *The Structure of Scientific Revolutions*, University of Chicago.
Kuhn, T. S.: 1979, *The Copernican Revolution — Planetary Astronomy in the Development of Western Thought*, Harvard University Press, Cambridge-London.
Lehman, M. M.: 1980, 'Programs, Life Cycles and Laws of Software Evolution', *Proceedings of the IEEE* 68(9).
Mathiassen, L.: 1981, *Systemudvikling og systemudviklingsmetode*, Datalogisk Afdeling, Aarhus Universitet.
Naur, P.: 1982, 'Formalization in Program Development', *BIT* 22, 437—453.
Naur, P.: 1983, 'Program Development Studies Based on Diaries', in: *Psychology of Computer Use*, Academic Press, London, pp. 159—170.
Naur, P.: 1984, 'Programming as Theory Building', in: *Microprocessing and Microprogramming*, Vol. 15. North-Holland, pp. 253—261.

Naur, P., and Randell, B.: 1969, 'Software Engineering', *Report on a Conference sponsored by the NATO Science Committee, Garmisch 7—11.10.69,* Brussels.
Nygaard, K. and Håndlykken, P.: 1981, 'The System Development Process — Its Setting, Some Problems and Needs for Methods', in: H. Hünke (Ed.), *Software Engineering Environments. Proceedings of the Symposium on Software Engineering Environments,* North-Holland, pp. 157—172.
Ryle, G.: 1949, *The Concept of Mind,* Hutchinson, and Penguin Books, England, 1983.
Turski, W. M.: 1985, *Informatics — A Propaedeutic View. PWN,* Polish Scientific Publishers, Warsaw.

PETER NAUR

THE PLACE OF STRICTLY DEFINED NOTATION IN HUMAN INSIGHT

Strictly defined notation is familiar to anybody from its occurrence in games. More importantly, strictly defined notation is a building element of constructed models, employed widely in science and technology. Constructed models give rise to major issues related to their use and construction, essentially involving informal issues. Minor issues of constructed models are the relations between different models of the same modellee and between different sets of model building elements. Such issues have aspects that relate purely to strictly defined items. The one-sided attention given to purely formal issues in much literature on programming logic is characterized.

1. INTRODUCTION

Strictly defined notation being the bread and butter of any work in programming logic, it may appear either unnecessary or impertinent of me to address this forum with an attempt at a clarification and characterization of its place in human insight. I have, in fact, been in serious doubt over making such an attempt, and have decided on it only after special request from the program committee.

Bertrand Russell, one of the fathers of the modern movement for formalization, has some remarks that indicate the subtlety of the issue. In one passage he says (Russell, 1956):

This suggests a word of advice to such of my readers as may happen to be professors. I am allowed to use plain English because everybody knows that I could use mathematical logic if I chose. Take the statement: 'Some people marry their deceased wives' sisters'. I can express this in a language which only becomes intelligible after years of study, and this gives me freedom. I suggest to young professors that their first work should be written in a jargon only to be understood by the erudite few. With that behind them, they can ever after say what they have to say in a language "understanded of the people". In these days, when our very lives are at the mercy of the professors, I cannot but think that they would deserve our gratitude if they adopted my advice.

Elsewhere Russell says (Russell, 1922):

... for a good notation has a subtlety and suggestiveness which at times make it seem almost like a live teacher.

261

Timothy R. Colburn et al. (eds.), Program Verification, 261–274.
© 1993 *Kluwer Academic Publishers.*

But he also says (Russell, 1901):

It is not easy for the lay mind to realize the importance of symbolism in discussing the foundations of mathematics, and the explanation may perhaps seem strangely para-doxical. The fact is that symbolism is useful because it makes things difficult.

On this background it may not be out of place to try to characterize formalization and strictly defined notation in more detail.

2. STRICTLY DEFINED NOTATION

When talking about strictly defined notation, what I have in mind is first of all notation, that is written symbols employed in a certain context of communicating persons to denote issues that are well established in the context. The written symbols of a notation may have any form whatever. Often in order to make clear that they are indeed items of a notation they are chosen to be unique graphics, icons. Alternatively they may be formed from commonly employed graphics, such as letters and digits.

As a further issue of strictly defined notation as discussed here, the notation has to be introduced into the context by a specific act of definition. Typically this act will employ sentences of ordinary language, such as: 'Let Q denote such and such'.

Moreover, the act of definition is meant to be understood as exhaustive in stating the assumed properties of the symbols. In other words, the defined symbols can be assumed to denote nothing further than what is stated in their definition. This is what I have in mind when saying that the definition is strict.

Strictly defined notation and other issues of strict definition are intimately tied to issues of language that are important to keep clear. First of all, any definition involves language in so far as the act of definition itself is linguistic. Indeed, the act consists of a person saying, in a certain kind of situation and context: I define such and such to denote or mean such and such. Thus definition depends on an already established linguistic practice. This indicates immediately the important point that it is false to assume that linguistic activity generally depends on definition. The members of a human community may continue to talk together indefinitely without ever making use of definitions.

Second, the pronouncement of a definition can only be valid within a certain context and in certain situations. What the relevant context and situations are must be understood by whoever makes use of the definition, but that understanding can only be a matter of a person's direct, intuitive insight.

A prominent place of strict definition is their use in games. All games of cards and such games as chess, checkers, go, and many others, depend entirely on strict definition. In the situation and context of any particular game, the players have to understand the elements of the game, such as the cards or the chessmen, as being endowed with certain particular properties, given to them by acts of definition. A game of chess by definition is a sequence of actions involving chessmen placed on the board in such and such a manner. A bishop, in chess, by definition is a piece that, inter alia, can only move diagonally.

As evidenced by the readiness with which even quite young children will play games involving strict definition, human beings are able to grasp at least certain kinds of strict definition so as to have their actions in a certain context conform to them, or proceed in accordance with them. In such a context the person may be said to view or regard certain items in a particular way. This is best seen as the result of a decision on the part of the person, the decision to play the game in question. Such decisions may be undone at any time, there being nothing compulsive about them. Thus a person playing a game of chess may interrupt the attention to the game at any time, and might for example switch to considering the workmanship displayed in the chessmen, viewed as handicraft.

The special way of regarding certain items that a person displays when acting according to strict definitions is nothing specific, nothing that might justify a claim such that the person thinks in a specific way, formally. Rather, similar special ways of regarding certain items may be found in many other contexts, involving no strict definition. For example, elaborate cooking and serving of food will often involve a concern for visual decoration, in addition to the primary issue of producing edibles. Thus cakes may be produced that may be regarded as works of sculpture, and napkins may be arranged so as to be artificial flowers.

It must further be noted that when a person is engaged in an activity involving strictly defined rules, for example when playing a game of chess, it would be entirely misleading to claim that the person thereby acts as though controlled by strict rules, like an automaton, or that the person thereby displays a special logical or formal mode of thinking.

The point is that the strict rules specify only part of the game. Another part of the game specifies a human intent that must be adopted by the players, the desire to win the game. While the winning situation is defined strictly within the rules of the game, the aspect of human intent associated with the desire to win, and thereby with all such aspects as good and intelligent play, are matters of the player's attitude, and in no manner restricted or controlled by rules. They are on a par with the player's intent to play the game, rather than to engage in some other activity, and in the same manner unrestricted by rules.

Thus a person who masters a set of strictly defined items, in the sense of being able to play a game or argue in terms of them, has not thereby acquired a formal way of thinking, if such a way of speaking makes sense. This follows also from the fact that even highly proficient users of formal items continue to make trivial mistakes when handling them. Even the most trained bookkeeper makes mistakes, and has to apply checks and make balances. Even highly professional mathematicians have to check their formal arguments, and even so mistakes slip through to publication in mathematical journals.

In contrast to playing games in which the strict rules only serve to determine a general frame to activity, acting by following a formal description one rule at a time, for a human being, is troublesome and error prone. For example, a person might be instructed to find a route through a city by following a series of instructions of the form: go ahead until the n'th street on the left/right and then continue along that. Clearly such a form will be a very unreliable guide for a person who wants to reach a particular destination. If at any point the person happens to deviate from the intended behaviour, for example by misinterpreting what in the actual location is to be counted as a side street, or by making a wrong count, the likelihood is that the person will miss the intended destination completely. Similar risks attend to many other kinds of written instruction. The difficulty will probably have been felt acutely by almost anyone who has tried to use computer equipment guided solely by user manuals.

In conclusion, strictly defined items influence a person's behaviour in accordance with their definition when the person decides to regard them in a special way.

3. CONSTRUCTED MODELS BUILT FROM SYMBOLS

While the employment of strictly defined items in games may help to clarify certain sides of the manner in which such items are grasped by human beings, a matter of far greater importance to human affairs is their employment in what I shall call constructed models. Such employment accounts for the application of mathematics to physics, astronomy, biology, and many other fields, and for the use of computers in the same fields and in administration of public affairs and in business.

Very briefly, we are dealing with a constructed model when some aspect of the world, the model, is seen by some persons as having a special relation, that of modelling, to another aspect of the world, the modellee, the modelling relation being such that by manipulating or handling the model the persons may gain useful insight into the modellee. For example, a program running in a computer may be a model of an atomic power reactor, or, more precisely, of certain aspects of such a reactor. The program may for example make it possible to find out what temperatures, pressures, radiation intensities, and fluid flows, would occur in the reactor under various circumstances. It need hardly be mentioned that constructed models in this sense are immensely important to countless activities of industrial societies.

As implied in the designation, constructed models are the result of deliberate design and construction activities. What they are made from is unspecified in the notion, however. This is where strictly defined notation, mathematics, and data and data processes, come in, as useful model building elements.

In order to clarify the issues of human understanding engendered by the use of mathematics and computers in constructed models it is pertinent to distinguish between the actual models and the elements of which they are constructed. We thus have to consider:

(1) The modellee: an aspect of the world. As a simple example taken from classical mechanics, let us take the things that Galileo Galilei let fall from the Leaning Tower of Pisa, and more particularly the heights from which he let them fall and the times required for their fall.

(2) The model: an aspect of the world seen as modelling the modellee. As illustration consider two different models of the things

falling from the Leaning Tower. The first one is as follows:

(M1) $h = 1/2 \, gt^2$

with the explanation that h is the height of fall, t the time of fall, and g the acceleration of gravity. The second model is a computer program:

(M2)   display text ('Give fall time, in seconds:');
        input real (time)
        display text ('Fall distance, in meters:');
        display real (4.905 × time↑2);

(3) Model building elements. We are here interested in models built from strictly defined, symbolic elements. Model M1 is built from elements of mathematical analysis: numbers, functions, etc. Model M2 is built from elements of computing: variables, algorithms, procedures, programming languages, etc.

In terms of this picture it is clear that there are two different major issues of human insight related to constructed models. The one is the relation of model to modellee. This is a matter of grasping a relation between aspects of the world that have no inherent strictly defined properties and the strictly defined items of the model. Let us call this the modelling insight. It involves first of all insight into where each item of the model, such as the values of the quantities h and t, can be found in the world. More refined modelling insight includes concrete awareness of the limitations of the modelling, such as inaccuracy and dependence on situation and context of the thing falling. It may be noted that with this view all the strictly defined items are part of the model, not of the modellee. For example, the height of falling, as an intuitive notion, is part of the modellee, but any strictly definable measure of that height is part of the model. The insight into the relation between the height and the measure of the height is part of the modelling insight.

The second major issue of human insight into constructed models is the relation between the model building elements and the model. This issue is prominent to the person who is engaged on the actual building of the model, and who has to find a way of combining the available model building elements so as to arrive at a model having appropriate properties. When building models in the form of programs this person is the programmer, who must of course at the same time have sufficient insight into the modellee so as to be able to form a theory of the program, in the sense of Ryle, as discussed in (Naur, 1985, in Section 1.4).

In addition to these major issues of human insight there are several minor issues, minor in the sense that the construction and the use of contructed models can proceed, and has often proceeded, without any need for them to be taken up. They all relate to comparisons between several different constructed models or several different sets of model building elements. One issue is the comparison betwen two models. Ma and Mb, of the same modellee, Ma being built from building elements Ea and Mb from building elements Eb. A closely related issue is the construction of a model Mb, from building elements Eb, when model Ma, built from elements Ea, is given. This is often encountered in a form where Ea is a so-called specification language, Ma is a specification, and Eb is a programming language.

A third issue is the comparison of several different sets of model building elements.

With a view to the place of strictly defined notation in these different issues, it must be noted that the major issues of constructed models, viz. the modelling insight and the model building insight, both are concerned with the modellee and thus inherently involve items that are not strictly defined.

By way of contrast, the minor issues are concerned with models and model building elements, i.e. items composed wholly of strictly defined items. Thus among the minor issues there are some that may be called formal, in the sense that they deal only with strictly defined items. The central example is the question of a proof that two models Ma and Mb, built from elements Ea and Eb respectively, are equivalent in some strictly defined sense. When Ea is a specification language and Eb is a programming language this is the issue of proving the problem Mb correct.

By no means all the issues related only to models and model building elements are formal, however. Informal issues arise when the appropriateness of the model building elements to various kinds of modellees, and convenience of both models and their building elements to the persons who deal with them are considered. These are important questions of model building, but since neither appropriateness nor convenience are notions that are, or could be, adequately grasped in terms of strict definitions, they cannot be reduced to formal questions.

The distinction made here between the modelling insight, which is beyond the reach of a formal approach, and the insight into the formal issues of models and model building elements, has been characterized by Einstein (1921) thus:

As far as the propositions of mathematics refer to reality, they are not certain; and as far as they are certain, they do not refer to reality.

In passing it may be mentioned that constructed models have given rise to extensive speculation and controversy, occasioned by what I shall call metaphysical superpositions upon them. Such metaphysical superpositions upon constructed models most prominently are beliefs that the items of certain constructed models are inherent in their modellees, that they are in some sense true to the modellees, and that they account fully for the properties of the modellees. In the nineteenth century a commonly held view of this kind was that the models built upon Newtonian mechanics would account fully for the observable world. Nowadays it is similarly believed widely that models built upon quantum mechanics will serve for that purpose. The significance of constructed models does not depend on such metaphysical superpositions, however.

4. PROGRAMMING LOGIC AND INFORMAL ISSUES

If viewed in the perspective of constructed models, as discussed above, much of the work in programming logic appears to be problematic in manners that will be characterized in what follows. As one entry into these issues, some views of E. W. Dijkstra shall be discussed. These are relevant since their influence on at least the work of Martin-Löf is explicitly acknowledged (Martin-Löf, 1979). Dijkstra has given a brief statement of his view of programs as constructed models in 'A Position Paper on Software Reliability' (Dijkstra, 1977). In this paper Dijkstra says that

a computer program is a tool that one can use by virtue of the knowledge of its explicitly stated properties. Those stated properties are known as 'its functional specification' or 'its specification' for short.

Later he says that

the question whether or not the program meets the specification can, in principle at least, be settled by mathematical means. The rigorous separation of responsibilities did isolate for the program designer a task that is within the realm of applicability of scientific methods.

This is later contrasted with

... the unformalized question whether a tool meeting those specifications is in such-and-such unformalized and ill-understood environment a pleasant tool to use. Correctness is a scientific issue, pleasantness is a non-scientific one....

Dijkstra's argumentation in this passage is problematic first because it depends entirely on the assumption that there is, or has been established, a functional specification in addition to the program itself. This circumstance is, however, by no means essential to the activity of developing programs. It is easy to conceive of programs for which no specifications have been written. For example, program M2 shown above has been written without the benefit of any specification. Indeed, for a programmer who has ordinary familiarity with the law of falling bodies of classical Newtonian physics the insistence on writing a specification for this program would be pointless.

As the second problematic issue, in characterizing the relation of the program to its environment, in other words the relation between the model and the modellee, Dijkstra talks of pleasantness. This seems a somewhat euphemistic word to use about such calamities as aeroplanes colliding in mid air or atomic reactors exploding as a result of errors in the programs written for their control and design. One may perhaps venture the opinion that Dijkstra's view in this matter is too narrow to be worth taking seriously.

Third, Dijkstra suggests that only formal questions can be of scientific interest. With such a view the only scientific questions are those of pure mathematics, while all other fields of study, including physics, astronomy, chemistry, and biology, would be inherently unscientific. With this suggestion Dijkstra is insidiously summoning the prestige of science in the support of formality. In doing so he will gain the support neither of Quine (1960): 'Science is self-conscious common sense', nor of Einstein (1936): 'The whole of science is nothing more than a refinement of everyday thinking'.

In spite of their problematic nature, Dijkstra's views on programming appear to be widely shared by workers in the field of programming logic. However, if one attends closely to the actual work done in the field one will find out only a display of the formal argumentation, but also, much less conspicuously, indications of the insufficiency of that argumentation. These indications have the form of remarks on clearly informal issues. For illustration I will present some observations on the papers given in the Proceedings of the Workshop on Programming

Logic organized by the Göteborg Programming Methodology Group in October 1987, to be denoted WPL 1987 in what follows.

Of the 24 papers given in WPL 1987, 13 have remarks that have a bearing on clearly informal issues. The remarks appear almost exclusively in the introductions and final concluding discussions of the papers, thereby contributing to the placing of the work of the papers in a context of informal, human intents. The remarks can be placed in two groups. The first group has remarks about *informal needs* that the work of the paper intends to satisfy. In this group one finds the following formulations. De Bruijn (p. 1) writes:

... the current formal systems do not adequately describe how people actually think, and, moreover, do not quite match the goals we have in mathematical education. Therefore it is attractive to try to put a substantial part of the mathematical vernacular into the formal system.

Manna and Waldinger (p. 44), when discussing derivations of sample imperative programs, say that

if the derivation requires many gratuitous, unmotivated steps, it may be impossible for a person or system to discover it . . .

Dybjer (p. 100) talks about "what properties a good programming logic should have" after having quoted statements such as that "Domain theory allows reasoning about recursion in a most flexible way, but at a heavy cost of complexity" and "domains and partial objects are not essential even for difficult proofs of termination". Smyth (p. 173) justifies his work with the following remark:

it has long been recognized . . . that quasi-uniformities provide a convenient setting for power space constructions. The main obstacle to integrating this material with computer science has been the lack of a decent account of completeness of quasi-uniformities; hence, again, the work reported in this paper.

Again on page 184: "We seek a decent fixed point theorem . . .". Winskel (p. 442), after discussing formal properties of several models of electrical circuits, asks:

How useful are the results above in the practice of verifying circuits? The model $|G|$ should be used wherever possible because it is so much simpler than that in $|W|$. But how easy is it in practice to show a circuit meets conditions sufficient to ensure the simpler model may be used?

Larsen (p. 445) states:

... this quest for compositionality is an important one to solve if Hennessy-Milner Logic is going to be used in design and verification of large systems. So far, not even small systems have been dealt with using Hennessy—Milner Logic. However, equally important in this respect is the *expressiveness* of the logic: i.e. to what extent does the logic allow us to express 'interesting' properties as formulas.

Another group of remarks in WPL 1987 makes *claims about informal properties of the formal systems being constructed.* Thus de Bruijn writes (p. 2):

Experience with teaching MV was acquired in a course 'Language and Structure of Mathematics', given yearly since 1979 at Eindhoven University of Technology. This course was designed on the claim that MV does not only serve to present mathematics better, but also to understand it better.

On page 3:

... the grammar of the mathematical vernacular is not harder but very much easier than the one of natural language.

Dybjer writes (p. 108):

The theories of partial objects (III) and (IV) allow reasoning about recursion in a most flexible way. ... Partial objects are perhaps not essential even for difficult proofs of termination, but their presence does not need to increase the complexity of the proof.

Backhouse writes (p. 116):

For some time now there has been considerable enthusiasm for understanding the connection between programs and proofs. Indeed, within the theory of types developed by Per Martin-Löf and within the theory of constructions developed by Coquand and Huet it is often claimed that there is an 'identity' between programs and proof objects. The benefits are claimed to be considerable: on the one hand we can utilise our intuitions and experience as theorem-provers when designing computer programs. ... This claimed identity does not, however, always stand up to a detailed scrutiny.

After then discussing some specific problems and a technique for resolving them in terms of so-called hypothetical rules, Backhouse concludes:

We now have considerable experience of the use of hypothetical rules and have found them very attractive to use ... they offer, in our view, a more elegant solution to these problems than that adopted by Nuprl. ...

Paulin-Mohring (p. 134), after presenting some demonstrations, concludes:

The Calculus of Constructions gives a good formalism for developing programs as proof of their specification. But if the derivation of the program follows exactly the algorithm we have in mind, the extracted program is something in some sense too complicated. . . .

Smyth, after substantial formal development, notes (p. 189):

It may be felt that a theory in which we have to distinguish at least three kinds of limits of sequences . . . is too inconvenient for everyday use. This is no doubt true; but in practice the problem is slight. . . .

Abramsky justifies his formal development with, *inter alia*, the following remark (p. 193): "The syntactic presentation of recursive types, powerdomains etc. makes these constructions more 'visible' and easier to calculate with". Hanna and Daeche (p. 301) claim that viewing logic as a partial algebra "allows logic to be specified in a clear, concise and totally rigorous fashion" and on p. 302 claim that their principles lead "to a very clear, natural style of programming".

What these remarks display is, first, that in spite of the generally over-whelming concern for formality and proof, there are many clear indications by which the authors admit the importance of informal concerns to their work. These concerns are predominantly such that serve to guide the formal development in a certain direction, so as to achieve simplicitly, clarity, decency, or expressiveness, of the formal constructions.

Second, in view of the manner in which informal concerns are explicitly stated to be important in setting the direction of the formal development, it is striking, indeed appalling to how slight extent the results of the formal developments are examined with a view to determining whether the informally stated goals of the activity have in fact been reached. Instead of accounts of such examinations one finds the authors' self-confident assurance that the work has been successful. In fact, not a single one of the 24 works under review has even the slightest suggestion that the author is aware that the examination of a formal technique for the way it appears to be simple, or clear, or decent, or expressive, to certain persons employing it, or in application to a particular field, is a difficult matter, raising a host of questions of method and technique. It is curious to observe how the authors in this field, who in the formal aspects of their work require painstaking demonstration and proof, in the informal aspects are satisfied with subjective claims that have not the slightest support, neither in argument nor in verifiable evidence. Surely common sense will indicate that such a manner is scientifically unacceptable.

The deplorable situation of programming logic outlined here is part of a much more widespread pattern of attitudes and manners prevailing in academic computing and mathematics, that tend to accept sales talk in the place of scientifically sound reasoning. These matters go beyond what can be dealt with here. Let it suffice to mention the study of Morris Kline on the disastrous state of academic mathematics and its destructive consequences for education (Kline, 1977). The parallel consequences in computing are as I foresaw them in (Naur, 1975, in Section 1.3).

Datalogisk Institut,
Copenhagen, Denmark.

REFERENCES

Abramsky, S.: 'Domain Theory in Logical Form', pp. 191—208 in *Workshop on Programming Logic*, loc. cit.

Backhouse, R.: 'Overcoming the Mismatch Between Programs and Proofs', pp. 116—122 in *Workshop on Programming Logic*, loc. cit.

Bruijn, N. C. de: 'The Mathematical Vernacular, a Language for Mathematics with Typed Sets', pp. 1—36 in *Workshop Programming Logic*, loc. cit.

Dijkstra, E. W.: 1977, 'A Position Paper on Software Reliability', *ACM SIGSOFT Software Engineering Notes* **2**(5), 3—5.

Dybjer, P.: 'From Type Theory Theory to LCF — A Case Study in Program Verification', pp. 99—109 in *Workshop on Programming Logic*, loc. cit.

Dybjer, P., Nordström, B., Peterson, K., and Smith, J. M. (Eds.): 1987, *Workshop on Programming Logic*, Programming Methodology Group, Report 37. Univ. of Göteborg and Chalmers Univ. of Technology. Göteborg, Sweden, October.

Einstein, A.: 1921, 'Geometry and Experience', pp. 232—246 in *Ideas and Opinions*, C. Selig (Ed.), New York: Crown Publishers, 1954.

Einstein, A.: 1936, 'Physics and Reality', pp. 58—94 in *Out of My Later Years*, New York: Philosophical Library, 1950.

Hanna, K., and Daeche, N.: 'An Algebraic Approach to Computational Logic', pp. 301—326 in *Workshop on Programming Logic*, loc. cit.

Kline, M.: 1977, *Why the Professor Can't Teach — Mathematics and the Dilemma of University Education*, New York: St. Martin's Press.

Larsen, K. G.: 'Proof Systems for Hennessy—Milner Logic with Recursion', pp. 444—481 in *Workshop on Programming Logic*, loc. cit.

Manna, Z., and Waldinger, R.: 'The Deductive Synthesis of Imperative LISP Programs', pp. 39—45 in *Workshop on Programming Logic*, loc. cit.

Martin-Löf, P.: 1979, 'Constructive Mathematics and Computer Programming', pp. 153—175 in *Proc. 6th Int. Congress of Logic, Methodology and Philosophy of Science, Hannover*. Amsterdam: North Holland.

Naur, P.: 1975, 'Programming Languages Natural Languages, and Mathematics', *Comm. ACM* **18**(12), 676—683. This selection Section 1.3.

Naur, P.: 1985, 'Programming as Theory Building', *Microprocessing and Microprogramming* **15**, 253—261. This selection Section 1.4.

Paulin-Mohring, C.: 'An Example of Algorithm Development in the Calculus of Constructions: Binary Search for the Computation of the Lambo Function', pp. 123—135 in *Workshop on Programming Logic*, loc. cit.

Quine, W. O.: 1960, *Word and Object*, Cambridge, Mass.: M.I.T. Press.

Russell, B.: 1901, 'Mathematics and the Metaphysicians', pp. 74—94 in *Mysticism and Logic*, Harmondsworth: Penguin, 1954.

Russell, B.: 1922, 'Introduction', pp. 7—23 in L. Wittgenstein, *Tractatus Logico-Philosophicus*. London: Routledge and Kegan Paul.

Russell, B.: 1956, *Portraits From Memory*.

Smyth, M. B.: 'Quasi-Uniformities: Reconciling Domains with Metric Spaces', pp. 173—190 in *Workshop on Programming Logic*, loc. cit.

Winskel, G.: 'Relating Two Models of Hardware', pp. 427—443 in *Workshop on Programming Logic*, loc. cit.

BRIAN CANTWELL SMITH

LIMITS OF CORRECTNESS IN COMPUTERS

1. INTRODUCTION

On October 5, 1960, the American Ballistic Missile Early-Warning System station at Thule, Greenland, indicated a large contingent of Soviet missiles headed towards the United States [1]. Fortunately, common sense prevailed at the informal threat-assessment conference that was immediately convened: international tensions weren't particularly high at the time, the system had only recently been installed, Kruschev was in New York, and all in all a massive Soviet attack seemed very unlikely. As a result no devastating counter-attack was launched. What was the problem? The moon had risen, and was reflecting radar signals back to earth. Needless to say, this lunar reflection hadn't been predicted by the system's designers.

Over the last ten years, the Defense Department has spent many millions of dollars on a new computer technology called 'program verification' — a branch of computer science whose business, in its own terms, is to 'prove programs correct'. Program verification has been studied in theoretical computer science departments since a few seminal papers in the 1960s [2], but it has only recently started to gain in public visibility, and to be applied to real world problems. General Electric, to consider just one example, has initiated verification projects in their own laboratories: they would like to prove that the programs used in their latest computer-controlled washing machines won't have any 'bugs' (even one serious one can destroy their profit margin) [3]. Although it used to be that only the simplest programs could be 'proven correct' — programs to put simple lists into order, to compute simple arithmetic functions — slow but steady progress has been made in extending the range of verification techniques. Recent papers have reported correctness proofs for somewhat more complex programs, including small operating systems, compilers, and other material of modern system design [4].

What, we do well to ask, does this new technology mean? How good are we at it? For example, if the 1960 warning system had been proven

275

Timothy R. Colburn et al. (eds.), Program Verification, 275—293.
© 1993 *Kluwer Academic Publishers.*

correct (which it was not), could we have avoided the problem with the moon? If it were possible to prove that the programs being written to control automatic launch-on-warning systems were correct, would that mean there could not be a catastrophic accident? In systems now being proposed computers will make launching decisions in a matter of seconds, with no time for any human intervention (let alone for musings about Kruschev's being in New York). Do the techniques of program verification hold enough promise so that, if these new systems could all be proven correct, we could all sleep more easily at night? These are the questions I want to look at in this paper. And my answer, to give away the punch-line, is no. For fundamental reasons — reasons that anyone can understand — there are inherent limitations to what can be proven about computers and computer programs. Although program verification is an important new technology, useful, like so many other things, in its particular time and place, it should definitely not be called verification. Just because a program is 'proven correct', in other words, you cannot be sure that it will do what you intend. First some background.

2. GENERAL ISSUES IN PROGRAM VERIFICATION

Computation is by now the most important enabling technology of nuclear weapons systems: it underlies virtually every aspect of the defense system, from the early warning systems, battle management and simulation systems, and systems for communication and control, to the intricate guidance systems that direct the missiles to their targets. It is difficult, in assessing the chances of an accidental nuclear war, to imagine a more important question to ask than whether these pervasive computer systems will or do work correctly.

Because the subject is so large, however, I want to focus on just one aspect of computers relevant to their correctness: the use of *models* in the construction, use, and analysis of computer systems. I have chosen to look at modelling because I think it exerts the most profound and, in the end, most important influence on the systems we build. But it is only one of an enormous number of important questions. First, there-fore — in order to unsettle you a little — let me just hint at some of the equally important issues I will not address:

1. *Complexity*: At the current state of the art, only very simple programs can be proven correct. Although it is terribly misleading to

assume that either the complexity or power of a computer program is a linear function of length, some rough numbers are illustrative. The simplest possible arithmetic programs are measured in tens of lines; the current state of the verification art extends only to programs of up to several hundred. It is estimated that the systems proposed in the Strategic Defense Initiative (Star Wars), in contrast, will require at least 10,000,000 lines of code [5]. By analogy, compare the difference between resolving a two-person dispute and settling the political problems of the Middle East. There's no *a priori* reason to believe that strategies successful at one level will scale to the other.

2. *Human interaction*: Not much can be 'proven', let alone specified formally, about actual human behaviour. The sorts of programs that have so far been proven correct, therefore, do not include much substantial human interaction. On the other hand, as the moon-rise example indicates, it is often crucial to allow enough human intervention to enable people to over-ride system mistakes. System designers, therefore, are faced with a very real dilemma: should they rule out substantive human intervention, in order to develop more confidence in how their systems will perform, or should they include it, so that costly errors can be avoided or at least repaired? The Three-Mile Island incident is a trenchant example of just how serious this trade-off can get: the system design provided for considerable human intervention, but then the operators failed to act 'appropriately'. Which strategy leads to the more important kind of correctness?

A standard way out of this dilemma is to specify the behaviour of the system *relative to the actions of its operators*. But this, as we will see below, pressures the designers to specify the system totally in terms of internal actions, not external effects. So you end up proving only that the system will *behave in the way that it will behave* (i.e. launch a missile only if the attack is real). Unfortunately, the latter is clearly what is important. Systems comprising computers and people must function properly as integrated systems; nothing is gained by showing that one cog in a misshapen wheel is a very nice cog indeed.

Furthermore, large computer systems are dynamic, constantly changing, embedded in complex social settings. Another famous 'mistake' in the American defense system happened when a human operator mistakenly mounted a training tape, containing a simulation of a full-scale Soviet attack, onto a computer that, just by chance, was automatically pulled into service when the primary machine ran into a problem. For

some tense moments the simulation data were taken to be the real thing
[6]. What does it mean to install a 'correct' module into a complex
social flux?

3. *Levels of Failure*: Complex computer systems must work at many
different levels. It follows that they can fail at many different levels too.
By analogy, consider the many different ways a hospital could fail. First,
the beams used to frame it might collapse. Or they might perform
flawlessly, but the operating room door might be too small to let in a
hospital bed (in which case you would blame the architects, not the
lumber or steel company). Or the operating room might be fine, but the
hospital might be located in the middle of the woods, where no one
could get to it (in which case you would blame the planners). Or, to
take a different example, consider how a letter could fail. It might be so
torn or soiled that it could not be read. Or it might look beautiful, but
be full of spelling mistakes. Or it might have perfect grammar, but
disastrous contents.

Computer systems are the same: they can be 'correct' at one level —
say, in terms of hardware — but fail at another (i.e. the systems built on
top of the hardware can do the wrong thing even if the chips are fine).
Sometimes, when people talk about computers failing, they seem to
think only the hardware needs to work. And hardware does from time
to time fail, causing the machine to come to a halt, or yielding errant
behaviour (as for example when a faulty chip in another American early
warning system sputtered random digits into a signal of how many
Soviet missiles had been sighted, again causing a false alert [7]). And
the connections between the computers and the world can break: when
the moon-rise problem was first recognized, an attempt to override it
failed because an iceberg had accidentally cut an undersea telephone
cable [8]. But the more important point is that, in order to be reliable, a
system has to be correct *at every relevant level*: the hardware is just the
starting place (and by far the easiest, at that). Unfortunately, however,
we don't even know what all the relevant levels are. So-called 'fault
tolerant' computers, for example, are particularly good at coping with
hardware failures, but the software that runs on them is not thereby
improved [9].

4. *Correctness and Intention*: What does *correct* mean, anyway?
Suppose the people want peace, and the President thinks that means
having a strong defense, and the Defense department thinks that means

having nuclear weapons systems, and the weapons designers request control systems to monitor radar signals, and the computer companies are asked to respond to six particular kinds of radar pattern, and the engineers are told to build signal amplifiers with certain circuit characteristics, and the technician is told to write a program to respond to the difference between a two-volt and a four-volt signal on a particular incoming wire. If being correct means *doing what was intended*, whose intent matters? The technician's? Or what, with twenty years of historical detachment, we would say *should have been intended*?

With a little thought any of you could extend this list yourself. And none of these issues even touch on the intricate technical problems that arise in actually building the mathematical models of software and systems used in the so-called 'correctness' proofs. But, as I said, I want to focus on what I take to be the most important issue underlying all of these concerns: the pervasive use of models. Models are ubiquitous not only in computer science but also in human thinking and language; their very familiarity makes them hard to appreciate. So we'll start simply, looking at modelling on its own, and come back to correctness in a moment.

3. THE PERMEATING USE OF MODELS

When you design and build a computer system, you first formulate a model of the problem you want it to solve, and then construct the computer program in its terms. For example, if you were to design a medical system to administer drug therapy, you would need to model a variety of things: the patient, the drug, the absorption rate, the desired balance between therapy and toxicity, and so on and so forth. The absorption rate might be modelled as a number proportional to the patient's weight, or proportional to body surface area, or as some more complex function of weight, age, and sex.

Similarly, computers that control traffic lights are based on some model of traffic — of how long it takes to drive across the intersection, of how much metal cars contain (the signal change mechanisms are triggered by metal detectors buried under each street). Bicyclists, as it happens, often have problems with automatic traffic lights, because bicycles don't exactly fit the model: they don't contain enough iron to trigger the metal detectors. I also once saw a tractor get into trouble because it couldn't move as fast as the system 'thought' it would: the cross-light went green when the tractor was only half-way through the intersection.

To build a model is to conceive of the world in a certain delimited way. To some extent you must build models before building any artifact at all, including televisions and toasters, but computers have a special dependence on these models: *you write an explicit description of the model down inside the computer*, in the form of a set of rules or what are called *representations* — essentially linguistic formulae encoding, in the terms of the model, the facts and data thought to be relevant to the system's behaviour. It is with respect to these representations that computer systems work. In fact that's really what computers are (and how they differ from other machines): they run by manipulating representations, and representations are always formulated in terms of models. This can all be summarized in a slogan: no computation without representation.

The models, on which the representations are based, come in all shapes and sizes. Balsa models of cars and airplanes, for example, are used to study air friction and lift. Blueprints can be viewed as models of buildings; musical scores as models of a symphony. But models can also be abstract. Mathematical models, in particular, are so widely used that it is hard to think of anything that they haven't been used for: from whole social and economic systems, to personality traits in teenagers, to genetic structures, to the mass and charge of subatomic particles. These models, furthermore, permeate all discussion and communication. Every expression of language can be viewed as resting implicitly on some model of the world.

What is important, for our purposes, is that every model deals with its subject matter *at some particular level of abstraction*, paying attention to certain details, throwing away others, grouping together similar aspects into common categories, and so forth. So the drug model mentioned above would probably pay attention to the patients' weights, but ignore their tastes in music. Mathematical models of traffic typically ignore the temperaments of taxi drivers. Sometimes what is ignored is at too 'low' a level; sometimes too 'high': it depends on the purposes for which the model is being used. So a hospital blueprint would pay attention to the structure and connection of its beams, but not to the arrangements of proteins in the wood the beams are made of, nor to the efficacy of the resulting operating room.

Models *have* to ignore things exactly because they view the world at a level of abstraction ('abstraction' is from the Latin *abstrahere*, 'to pull or draw away'). And it is good that they do: otherwise they would drown in the infinite richness of the embedding world. Though this isn't the place for metaphysics, it would not be too much to say that every act of conceptualization, analysis, categorization, does a certain amount of violence to its subject matter, in order to get at the underlying regularities that group things together. If you don't commit that act of violence — don't ignore some of what's going on — you would become so hypersensitive and so overcome with complexity that you would be unable to act.

To capture all this in a word, we will say that models are inherently *partial*. All thinking, and all computation, are similarly partial. Furthermore — and this is the important point — thinking and computation *have* to be partial: that's how they are able to work.

4. FULL-BLOODED ACTION

Something that is not partial, however, is action. When you reach out your hand and grasp a plow, it is the real field you are digging up, not your model of it. Models, in other words, may be abstract, and thinking may be abstract, and some aspects of computation may be abstract, but action is not. To actually build a hospital, to clench the steering wheel and drive through the intersection, or to inject a drug into a person's body, is to act in the full-blooded world, not in a partial or distilled model of it.

This difference between action and modelling is extraordinarily important. Even if your every thought is formulated in the terms of some model, to act is to take leave of the model and participate in the whole, rich, infinitely variegated world. For this reason, among others, action plays a crucial role, especially in the human case, in grounding the more abstract processes of modelling or conceptualization. One form that grounding can take, which computer systems can already take advantage of, is to provide feedback on how well the modelling is going. For example, if an industrial robot develops an internal three-dimensional representation of a wheel assembly passing by on a conveyor belt, and then guides its arm towards that object and tries to pick it up, it can use video systems or force sensors to see how well the model corresponded to what was actually the case. The world doesn't care about the model: the claws will settle on the wheel just in case the actualities mesh.

Feedback is a special case of a very general phenomenon: you often learn, when you do act, just how good or bad your conceptual model was. You learn, that is, if you have adequate sensory apparatus, the capacity to assess the sensed experience, the inner resources to revise and reconceptualize, and the luxury of recovering from minor mistakes and failures.

5. COMPUTERS AND MODELS

What does all this have to do with computers, and with correctness? The point is that computers, like us, participate in the real world: they take real actions. One of the most important facts about computers, to put this another way, is that we plug them in. They are not, as some theoreticians seem to suppose, pure mathematical abstractions, living in a pure detached heaven. They land real planes at real airports; administer real drugs; and — as you know all too well — control real radars, missiles, and command systems. Like us, in other words, although they base their actions on models, they have consequence in a world that inevitably transcends the partiality of those enabling models. Like us, in other words, and unlike the objects of mathematics, they are challenged by the inexorable conflict between the partial but tractable model, and the actual but infinite world.

And, to make the only too obvious point: we in general have no guarantee that the models are right — indeed we have no *guarantee* about much of anything about the relationship between model and world. As we will see, current notions of 'correctness' don't even address this fundamental question.

In philosophy and logic, as it happens, there is a very precise mathematical theory called 'model theory'. You might think that it would be a theory about what models are, what they are good for, how they correspond to the worlds they are models of, and so forth. You might even hope this was true, for the following reason: a great deal of theoretical computer science, and all of the work in program verification and correctness, historically derives from this model-theoretic tradition, and depends on its techniques. Unfortunately, however, model theory doesn't address the model—world relationship at all.

Rather, what model theory does is to tell you how your descriptions, representation, and programs *correspond to your model*.

The situation, in other words, is roughly as depicted in Figure 1. You are to imagine a description, program, computer system (or even a thought — they are all similar in this regard) in the left hand box, and the very real world in the right. Mediating between the two is the inevitable model, serving as an idealized or preconceptualized simulacrum of the world, in terms of which the description or program or whatever can be understood. One way to understand the model is as the glasses through which the program or computer looks at the world: it is the world, that is, as the system sees it (though not, of course, as it necessarily is).

The technical subject of 'model theory', as I have already said, is a study of the relationship on the left. What about the relationship on the right? The answer, and one of the main points I hope you will take away from this discussion, is that, at this point in intellectual history, we have no theory of this right-hand side relationship.

There are lots of reasons for this, some very complex. For one thing, most of our currently accepted formal techniques were developed, during the first half of this century, to deal with mathematics and

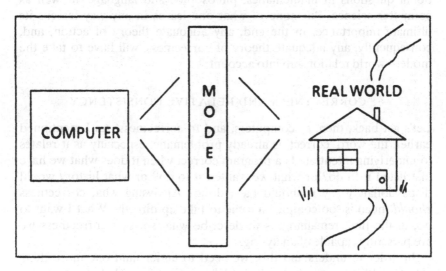

Fig. 1. Computers, models, and the embedding world.

physics. Mathematics is unique, with respect to models, because (at least to a first level of approximation) its subject matter *is* the world of models and abstract structures, and therefore the model—world relationship is relatively unproblematic. The situation in physics is more complex, of course, as is the relationship between mathematics and physics. How apparently pure mathematical structures could be used to model the material substrate of the universe is a question that has exercised physical scientists for centuries. But the point is that, whether or not one believes that the best physical models do more justice and therefore less violence to the world than do models in so-called 'higher-level' disciplines like sociology or economics, formal techniques don't themselves address the question of adequacy.

Another reason we don't have a theory of the right-hand side is that there is very little agreement on what such a theory would look like. In fact all kinds of questions arise, when one studies the model—world relationship explicitly, about whether it can be treated formally at all, about whether it can be treated rigorously, even if not formally (and what the relationship is between those two), about whether any theory will be more than usually infected with prejudices and preconceptions of the theorists, and so forth. The investigation quickly leads to foundational questions in mathematics, philosophy, and language, as well as computer science. But none of what one learns in any way lessens its ultimate importance. In the end, any adequate theory of action, and, consequently, any adequate theory of correctness, will have to take the model—world relationship into account.

6. CORRECTNESS AND RELATIVE CONSISTENCY

Let's get back, then, to computers, and to correctness. As I mentioned earlier, the word 'correct' is already problematic, especially as it relates to underlying intention. Is a program correct when it does what we have instructed it to do? or what we wanted it to do? or what history would dispassionately say it should have done? Analysing what correctness *should* mean is too complex a topic to take up directly. What I want to do, in the time remaining, is to describe what sorts of correctness we are presently capable of analysing.

In order to understand this, we need to understand one more thing about building computer systems. I have already said that, when you design a computer system, you first develop a model of the world, as indicated in the diagram. But you don't, in general, ever get to hold the model in your hand: computer systems, in general, are based on models that are purely abstract. Rather, if you are interested in proving your program 'correct', you develop two concrete things, structured in terms of the abstract underlying model (although these are listed here in logical order, the program is very often written first):

1. A *specification*: a formal description in some standard formal language, specified in terms of the model, in which the desired behaviour is described; and

2. The *program*: a set of instructions and representations, also formulated in the terms of the model, which the computer uses as the basis for its actions.

How do these two differ? In various ways, of which one is particularly important. The program has to say *how the behaviour is to be achieved*, typically in a step-by-step fashion (and often in excruciating detail). The specification, however, is less constrained: all it has to do is to specify *what proper behaviour would be*, independent of how it is accomplished. For example, a specification for a milk-delivery system might simply be: 'Make one milk delivery at each store, driving the shortest possible distance in total'. That's just a description of what has to happen. The program, on the other hand, would have the much more difficult job of saying how this was to be accomplished. It might be phrased as follows: 'drive four blocks north, turn right, stop at Gregory's Grocery Store on the corner, drop off the milk, then drive 17 blocks north-east, . . .'. Specifications, to use some of the jargon of the field, are essentially *declarative*; they are like indicative sentences or claims. Programs, on the other hand, are *procedural*: they must contain instructions that lead to a determinate sequence of actions.

What, then, is a proof of correctness? It is a proof that any system that *obeys the program* will *satisfy the specification.*

There are, as is probably quite evident, two kinds of problems here. The first, often acknowledged, is that the correctness proof is in reality only a proof that two characterizations of something are compatible. When the two differ — i.e. when you try to prove correctness and fail — there is no more reason to believe that the first (the specification) is any more correct than the second (the program). As a matter of technical practice, specifications tend to be extraordinarily complex formal descriptions, just as subject to bugs and design errors and so forth as programs. In fact they are very much like programs, as this introduction should suggest. So what almost always happens, when you write a specification and a program, and try to show that they are compatible, is that you have to adjust both of them in order to get them to converge.

For example, suppose you write a program to factor a number C, producing two answers A and B. Your specification might be:

> *Given a number C, produce numbers A and B such that $A \times B = C$.*

This is a specification, not a program, because it doesn't tell you *how* to come up with A and B. All it tells you is what properties A and B should have. In particular, suppose I say: ok, C is 5,332,114; what are A and B? Staring at the specification just given won't help you to come up with an answer. Suppose, on the other hand, given this specification, that you then write a program — say, by successively trying pairs of numbers until you find two that work. Suppose further that you then set out to prove that your program meets your specification. And, finally, suppose that this proof can be constructed (I won't go into details here; I hope you can imagine that such a proof could be constructed). With all three things in hand — program, specification, and proof — you might think you were done.

In fact, however, things are rarely that simple, as even this simple example can show. In particular, suppose, after doing all this work that you try your program out, confident that it must work because you have a proof of its correctness. You randomly give it 14 as an input, expecting 2 and 7. But in fact it gives you the answers $A = 1$ and $B = 14$. In fact, you realise upon further examination, it will *always* give back $A = 1$ and $B = C$. It does this, *even though you have a proof of its being correct*, because you didn't make your specification meet your intentions. You wanted both A and B to be *different* from C (and also different from 1), but you forgot to say that. In this case you have to modify both the program and the specification. A plausible new version of the latter would be:

> *Given a number C, produce numbers A and B such that $A \neq 1$ and $B \neq 1$ and $A \times B = C$.*

And so on and so forth: the point, I take it, is obvious. If the next version of the program, given 14, produces $A = -1$ and $B = -14$, you would similarly have met your new specification, but still failed to meet your intention. Writing 'good' specifications — which is to say, writing specifications that capture your intention — is hard.

It should be apparent, nonetheless, that developing even straightforward proofs of 'correctness' is nonetheless very useful. It typically forces you to delineate, very explicitly and completely, the model on which both program and specification are based. A great many of the simple bugs that occur in programs, of which the problem of producing 1 and 14 was an example, arise from sloppiness and unclarity about the model. Such bugs are not identified by the proof, but they are often unearthed in the attempt to prove. And of course there is nothing wrong with this practice; anything that helps to erradicate errors and increase confidence is to be applauded. The point, rather, is to show exactly what these proofs consist in.

In particular, as the discussion has shown, when you show that a program meets its specifications, all you have done is to show that two formal descriptions, slightly different in character, are compatible. This is why I think it is somewhere between misleading and immoral for computer scientists to call this 'correctness'. What is called a proof of correctness is really a proof of the compatibility or consistency between two formal objects of an extremely similar sort: program and specification. As a community, we computer scientists should call this *relative consistency*, and drop the word '*correctness*' completely.

What proofs of relative consistency ignore is the second problem intimated earlier. Nothing in the so-called program verification process *per se* deals with the right-hand side relationship: the relationship between the model and the world. But, as is clear, it is over inadequacies on the right-hand side — inadequacies, that is, in the models in terms of which the programs and specifications are written — that systems so commonly fail.

The problem with the moon-rise, for example, was a problem of this second sort. The difficulty was not that the program failed, in terms of the model. The problem, rather, was that the model was overly simplistic; *it didn't correspond to what was the case in the world*. Or, to put it more carefully, since all models fail to correspond to the world in indefinitely many ways, as we have already said, it didn't correspond to what was the case *in a crucial and relevant way*. In other words, to answer one of our original questions, even if a formal specification had been written for the 1960 warning system, and a proof of correctness generated, there is no reason to believe that potential difficulties with the moon would have emerged.

You might think that the designers were sloppy; that they would have thought of the moon if they had been more careful. But it turns out to be extremely difficult to develop realistic models of any but the most artificial situations, and to assess how adequate these models are. To see just how hard it can be, think back on the case of General Electric, and imagine writing appliance specifications, this time for a refrigerator. To give the example some force, imagine that you are contracting the refrigerator out to be built by an independent supplier, and that you want to put a specification into the contract that is sufficiently precise to guarantee that you will be happy with anything that the supplier delivers that meets the contract.

Your first version might be quite simple — say, that it should maintain an internal temperature of between 3 and 6 degrees Centigrade; not use more than 200 Watts of electricity; cost less than $100 to manufacture; have an internal volume of half a cubic meter; and so on and so forth. But of course there are hundreds of other properties that you implicitly rely on: it should, presumably, be structurally sound: you wouldn't be happy with a deliciously cool plastic bag. It shouldn't weigh more than a ton, or emit loud noises. And it shouldn't fling projectiles out at high speed when the door is opened. In general, it is impossible, when writing specifications, to include *everything* that you want: legal contracts, and other humanly interpretable specifications, are always stated within a background of common sense, to cover the myriad unstated and unstatable assumptions assumed to hold in force. (Current computer programs, alas, have no common sense, as the cartoonists know so well.)

So it is hard to make sure that everything that meets your specification will really be a refrigerator; it is also hard to make sure that your requirements don't rule out perfectly good refrigerators. Suppose for example a customer plugs a toaster in, puts it inside the refrigerator, and complains that the object he received doesn't meet the temperature specification and must therefore not be a refrigerator. Or suppose he tries to run it upside down. Or complains that it doesn't work in outer space, even though you didn't explicitly specify that it would only work within the earth's atmosphere. Or spins it at 10,000 rpm. Or even just unplugs it. In each case you would say that the problem lies not with the refrigerator but with the use. But how is *use* to be specified? The point is that, as well as modelling the artifact itself, you have to model the relevant part of the world in which it will be embedded. It follows that the model of a refrigerator as a device that *always* maintains an internal temperature of between 3 and 6 degrees is too strict to cover all possible situations. One could try to model what appropriate use would be, though specifications don't, ordinarily, even try to identify all the relevant circumstantial factors. As well as there being a background set of constraints with respect to which a model is formulated, in other words, there is also a background set of assumptions on which a specification is allowed at any point to rely.

7. THE LIMITS OF CORRECTNESS

It's time to summarize what we've said so far. The first challenge to developing a perfectly 'correct' computer system stems from the sheer complexity of real-world tasks. We mentioned at the outset various factors that contribute to this complexity: human interaction, unpredictable factors of setting, hardware problems, difficulties in identifying salient levels of abstraction, etc. Nor is this complexity of only theoretical concern. A December 1984 report of the American Defense Science Board Task Force on 'Military Applications of New-Generation Computing Technologies' identifies the following gap between current laboratory demonstrations and what will be required for successful military applications — applications they call 'Real World; Life or Death'. In their estimation the military now needs (and, so far as one can tell, expects to produce) an increase in the power of computer systems of nine orders of magnitude, accounting for both speed and amount of information to be processed. That is a 1,000,000,000-fold increase over current research systems, equivalent to the difference between a full century of the entire New York metropolitan area, compared to one day in the life of a hamlet of one hundred people. And remember that even current systems are already several orders of magnitude more complex than those for which we can currently develop proofs of relative consistency.

But sheer complexity has not been our primary subject matter. The second challenge to computational correctness, more serious, comes from the problem of formulating or specifying an appropriate model. Except in the most highly artificial or constrained domains, modelling the embedding situation is an approximate, not a complete, endeavour. It has the best hopes of even partial success in what Winograd has called 'systematic domains': areas where the relevant stock of objects, properties, and relationships are most clearly and regularly pre-defined. Thus bacteremia, or warehouse inventories, or even flight paths of airplanes coming into airports, are relatively systematic domains, at least compared to conflict negotiations, any situations involving intentional human agency, learning and instruction, and so forth. The systems that land airplanes are hybrids — combinations of computer and people — exactly because the unforeseeable happens, and because what happens is in part the result of human action, requiring human interpretation. Although it is impressive how well the phone companies can model telephone connections, lines, and even develop statistical models of telephone use, at a certain level of abstraction, it would nevertheless be impossible to model the content of the telephone conversations themselves.

Third, and finally, is the question of what one does about these first two facts. It is because of the answer to this last question that I have talked, so far, somewhat interchangeably about people and computers. With respect to the ultimate limits of models and conceptualization, both people and computers are restrained by the same truths. If the world is infinitely rich and variegated, no prior conceptualization of it, nor any abstraction, will ever do it full justice. That's ok — or at least we might as well say that it's ok, since that's the world we've got. What matters is that we not forget about that richness — that we not think, with misplaced optimism, that machines might magically have access to a kind of 'correctness' to which people cannot even aspire.

It is time, to put this another way, that we change the traditional terms of the debate. The question is not whether machines can do things, as if, in the background, lies the implicit assumption that the object of comparison is people. Plans to build automated systems capable of making a 'decision', in a matter of seconds, to annihilate Europe, say, should make you uneasy; requiring a person to make the same decision in a matter of the same few seconds should make you uneasy too, and for very similar reasons. The problem is that there is simply no way that reasoning of any sort can do justice to the inevitable complexity of the situation, because of what reasoning is. Reasoning is based on partial models. Which means it cannot be guaranteed to be correct. Which means, to suggest just one possible strategy for action, that we might try, in our treaty negotiations, to find mechanisms to slow our weapons systems down.

It is striking to realise, once the comparison between machines and people is raised explicitly, that we don't typically expect 'correctness' for people in anything like the form that that we presume it for computers. In fact quite the opposite, and in a revealing way. Imagine, in some bygone era, sending a soldier off to war, and giving him (it would surely have been a 'him') final instructions. "Obey your commander, help your fellow-soldier", you might say, "and above all do your country honour." What is striking about this is that it is considered not just a weakness, but a punishable weakness — a breach of morality — to obey instructions *blindly* (in fact, and for relevant reasons, you generally *can't* follow instructions blindly; they have to be interpreted to the situation at hand). You are subject to court-martial, for example, if you violate fundamental moral principles, such as murdering women and children, even if following strict orders.

In the human case, in other words, our social and moral systems seem to have built in an acceptance of the uncertainties and limitations inherent in the model—world relationship. We *know* that the assumptions and preconceptions built into instructions will sometimes fail, and we know that instructions are always incomplete; we exactly rely on judgment, responsibility, consciousness, and so forth, to carry someone through those situations — all situations, in fact — where model and world part company. In fact we never talk about people, in terms of their overall personality, being *correct*; we talk about people being *reliable*, a much more substantive term. It is individual actions, fully situated in a particular setting, that are correct or incorrect, not people in general, or systems. What leads to the highest number of correct human actions is a person's being reliable, experienced, capable of good judgment, etc.

There are two possible morals here, for computers. The first has to do with the notion of experience. In point of fact, program verification is not the only, or even the most common, method of obtaining assurance that a computer system will do the right thing. Programs are usually judged acceptable and are typically accepted into use, not because we prove them 'correct', but because they have shown themselves relatively reliable in their destined situations for some substantial period of time. And, as part of this experience, we expect them to fail: there always has to be room for failure. Certainly no one would ever accept a program without this *in situ* testing: a proof of correctness is at best added insurance, not a replacement, for real life experience. Unfortunately, for the ten million lines of code that is supposed to control and coordinate the Star Wars Defense System, there will never, God willing, be an *in situ* test.

One answer, of course, if genuine testing is impossible, is to run a *simulation* of the real situation. But simulation, as our diagram should make clear, *tests only the left-hand side relationship.* Simulations are defined in terms of models; they don't test the relationship between the model and the world. That is exactly why simulations and tests can never replace embedding a program in the real world. All the wargames we hear about, and hypothetical military scenarios, and electronic battlefield simulators, and so forth, are all based on exactly the kinds of models we have been talking about all along. In fact the subject of simulation, worthy of a whole analysis on its own, is really just our whole subject welling up all over again.

I said earlier that there were two morals to be drawn, for the computer, from the fact that we ask people to be reliable, not correct. The second moral is for those who, when confronted with the fact that genuine or adequate experience cannot be had, would say "oh, well, let's build responsibility and morality into the computers — if people can have it, there's no reason why machines can't have it too." Now I will not argue that this is inherently impossible, in a metaphysical or ultimate philosophical sense, but a few short comments are in order. First, from the fact that humans sometimes *are* responsible, it does not follow that we know what responsibility is: from tacit skills no explicit model is necessarily forthcoming. We simply do not know what aspects of the human condition underlie the modest levels of responsibility to which we sometimes rise. And second, with respect to the goal of building computers with even human levels of full reliability and responsibility, I can state with surety that the present state of artificial intelligence is about as far from this as mosquitos are from flying to the moon.

But there are deeper morals even than these. The point is that even if we could make computers reliable, they still wouldn't necessarily always do the correct thing. *People* aren't provably 'correct', either: that's why we hope they are responsible, and surely one of the major ethical facts is that correctness and responsibility don't coincide. Even if, in another 1000 years, someone were to devise a genuinely responsible computer system, there is no reason to suppose that it would achieve 'perfect correctness' either, in the sense of never doing anything wrong. This isn't a failure, in the sense of a performance limitation; it stems from the deeper fact that models must abstract, in order to be useful. The lesson to be learned from the violence inherent in the model—world relationship, in other words, is that there is an *inherent* conflict between the power of analysis and conceptualization, on the one hand, and sensitivity to the infinite richness, on the other.

But perhaps this is an overly abstract way to put it. Perhaps, instead, we should just remember that there will always be another moon-rise.

Intelligent Systems Laboratory,
Xerox Palo Alto Research Center, U.S.A.

REFERENCES

1. Berkeley, Edmund: 1962, *The Computer Revolution*, Doubleday, pp. 175—177, citing newspaper stories in the *Manchester Guardian Weekly* of Dec. 1, 1960, a UPI dispatch published in the *Boston Traveller* of Dec. 13, 1960, and an AP dispatch published in the *New York Times* on Dec 23, 1960.
2. McCarthy, John: 1963, 'A Basis for a Mathematical Theory of Computation', in: P. Braffort and D. Hirschberg (Eds.), *Computer Programming and Formal Systems*, Amsterdam: North-Holland, 1967, pp. 33—70. Floyd, Robert: 1967, 'Assigning Meaning to Programs', *Proceedings of Symposia in Applied Mathematics* 19 (also in F. T. Schwartz (Ed.) *Mathematical Aspects of Computer Science*, Providence: American Mathematical Society, 1967). Naur, P.: 1966, 'Proof of Algorithms by General Snapshots', *BIT* 6(4), 310—316.
3. Stevens, AL, BBN Inc.: personal communication.
4. See, for example, Boyer, R. S., and J S. Moore, (Eds.): 1981, *The Correctness Problem in Computer Science*, London: Academic Press.
5. Fletcher, James, study chairman, and McMillan, Brockway, panel chairman: 1984, *Report of the Study on Eliminating the Threat Posed by Nuclear Ballistic Missiles (U)*, Vol. 5, *Battle Management, Communications, and Data Processing (U)*, U.S. Department of Defense, February.
6. See, for example, the Hart—Goldwater report to the Committee on Armed Services of the U.S. Senate: 'Recent False Alerts from the Nation's Missile Attack Warning System' (Washington, D.C.: U.S. Government Printing Office, Oct. 9, 1980); Physicians for Social Responsibility, *Newsletter*, 'Accidental Nuclear War', Winter, 1982, p. 1.
7. *Ibid.*
8. Berkeley, *op. cit.* See also Daniel Ford's two-part article 'The Button', *New Yorker*, April 1, 1985, p. 43, and April 8, 1985, p. 49, excerpted from Ford, Daniel, *The Button*, New York: Simon and Schuster, 1985.
9. Developing software for fault-tolerant systems is in fact an extremely tricky business.

PART IV
FOCUS ON FORMAL VERIFICATION

RICHARD A. DE MILLO, RICHARD J. LIPTON, AND
ALAN J. PERLIS

SOCIAL PROCESSES AND PROOFS OF
THEOREMS AND PROGRAMS

> I should like to ask the same question that
> Descartes asked. You are proposing to give a
> precise definition of logical correctness which
> is to be the same as my vague intuitive feeling
> for logical correctness. How do you intend to
> show that they are the same? ... The average
> mathematician should not forget that intuition
> is the final authority.
>
> J. Barkley Rosser

Many people have argued that computer programming should strive to
become more like mathematics. Maybe so, but not in the way they seem
to think. The aim of program verification, an attempt to make programming more mathematics-like, is to increase dramatically one's
confidence in the correct functioning of a piece of software, and the
device that verifiers use to achieve this goal is a long chain of formal,
deductive logic. In mathematics, the aim is to increase one's confidence
in the correctness of a theorem, and it's true that one of the devices
mathematicians *could* in theory use to achieve this goal is a long chain
of formal logic. But in fact they don't. What they use is a proof, a very
different animal. Nor does the proof settle the matter; contrary to what
its name suggests, a proof is only one step in the direction of confidence. We believe that, in the end, it is a social process that determines whether mathematicians feel confident about a theorem — and
we believe that, because no comparable social process can take place
among program verifiers, program verification is bound to fail. We can't
see how it's going to be able to affect anyone's confidence about
programs.

Outsiders see mathematics as a cold, formal, logical, mechanical,
monolithic process of sheer intellection; we argue that insofar as it is
successful, mathematics is a social, informal, intuitive, organic, human
process, a community project. Within the mathematical community, the
view of mathematics as logical and formal was elaborated by Bertrand

297

Timothy R. Colburn et al. (eds.), Program Verification, 297—319.
© 1993 *Kluwer Academic Publishers.*

Russell and David Hilbert in the first years of this century. They saw mathematics as proceeding in principle from axioms or hypotheses to theorems by steps, each step easily justifiable from its predecessors by a strict rule of transformation, the rules of transformation being few and fixed. The *Principia Mathematica* was the crowning achievement of the formalists. It was also the deathblow for the formalist view. There is no contradiction here: Russell did succeed in showing that ordinary working proofs can be reduced to formal, symbolic deductions. But he failed, in three enormous, taxing volumes, to get beyond the elementary facts of arithmetic. He showed what can be done in principle and what cannot be done in practice. If the mathematical process were really one of strict, logical progression, we would still be counting on our fingers.

BELIEVING THEOREMS AND PROOFS

> Indeed every mathematician knows that proof has not been 'understood' if one has done nothing more than verify step by step the correctness of the deductions of which it is composed and has not tried to gain a clear insight into the ideas which have led to the construction of this particular chain of deductions in preference to every other one.
>
> N. Bourbaki

> Agree with me if I seem to speak the truth.
>
> Socrates

Stanislaw Ulam estimates that mathematicians publish 200,000 theorems every year [20]. A number of these are subsequently contradicted or otherwise disallowed, others are thrown into doubt, and most are ignored. Only a tiny fraction come to be understood and believed by any sizable group of mathematicians.

The theorems that get ignored or discredited are seldom the work of crackpots or incompetents. In 1879, Kempe [11] published a proof of the four-color conjecture that stood for eleven years before Heawood [8] uncovered a fatal flaw in the reasoning. The first collaboration between Hardy and Littlewood resulted in a paper they delivered at the June 1911 meeting of the London Mathematical Society; the paper was never published because they subsequently discovered that their proof was wrong [4]. Cauchy, Lamé, and Kummer all thought at one time or another that they have proved Fermat's Last Theorem [3]. In 1945, Rademacher thought he had solved the Riemann Hypothesis; his results not only circulated in the mathematical world but were announced in *Time* magazine [3].

Recently we found the following group of footnotes appended to a brief historical sketch of some independence results in set theory [10]:

(1) The result of Problem 11 contradicts the results announced by Levy [1963b]. Unfortunately, the construction presented there cannot be completed.

(2) The transfer to *ZF* was also claimed by Marek [1966] but the outlined method appears to be unsatisfactory and has not been published.

(3) A contradicting result was announced and later withdrawn by Truss [1970].

(4) The example in Problem 22 is a counterexample to another condition of Mostowski, who conjectured its sufficiency and singled out this example as a test case.

(5) The independence result contradicts the claim of Felgner [1969] that the Cofinality Principle implies the Axiom of Choice. An error has been found by Morris (see Felgner's corrections to [1969]).

The author has no axe to grind; he has probably never even heard of the current controversy in programming; and it is clearly no part of his concern to hold his friends and colleagues up to scorn. There is simply no way to describe the history of mathematical ideas without describing the successive social processes at work in proofs. The point is not that mathematicians make mistakes; that goes without saying. The point is that mathematicians' errors are corrected, not by formal symbolic logic, but by other mathematicians.

Just increasing the number of mathematicians working on a given problem does not necessarily insure believable proofs. Recently, two independent groups of topologists, one American, the other Japanese, independently announced results concerning the same kind of topological object, a thing called a homotopy group. The results turned out to be contradictory, and since both proofs involved complex symbolic and numerical calculation, it was not at all evident who had goofed. But the stakes were sufficiently high to justify pressing the issue, so the Japanese and American proofs were exchanged. Obviously, each group was highly motivated to discover an error in the other's proof; obviously, one proof or the other was incorrect. But neither the

Japanese nor the American proof could be discredited. Subsequently, a third group of researchers obtained yet another proof, this time supporting the American result. The weight of the evidence now being against their proof, the Japanese have retired to consider the matter further.

There are actually two morals to this story. First, a proof does not in itself significantly raise our confidence in the probable truth of the theorem it purports to prove. Indeed, for the theorem about the homotopy group, the horribleness of all the proffered proofs suggests that the theorem itself requires rethinking. A second point to be made is that proofs consisting entirely of calculations are not necessarily correct.

Even simplicity, clarity, and ease provide no guarantee that a proof is correct. The history of attempts to prove the Parallel Postulate is a particularly rich source of lovely, trim proofs that turned out to be false. From Ptolemy to Legendre (who tried time and time again), the greatest geometricians of every age kept ramming their heads against Euclid's fifth postulate. What's worse, even though we now know that the postulate is indemonstrable, many of the faulty proofs are still so beguiling that in Heath's definitive commentary on Euclid [7] they are not allowed to stand alone; Heath marks them up with italics, footnotes, and explanatory marginalia, lest some young mathematician, thumbing through the volume, be misled.

The idea that a proof can, at best, only probably express truth makes an interesting connection with a recent mathematical controversy. In a recent issue of *Science* [12], Gina Bari Kolata suggested that the apparently secure notion of mathematical proof may be due for revision. Here the central question is not "How do theorems get believed?" but "What is it that we believe when we believe a theorem?" There are two relevant views, which can be roughly labeled classical and probabilistic.

The classicists say that when one believes mathematical statement A, one believes that *in principle* there is a correct, formal, valid, step-by-step, syntactically checkable deduction leading to A in a suitable logical calculus such as Zermelo—Fraenkel set theory or Peano arithmetic, a deduction of A à la the *Principia*, a deduction that completely formalizes the truth of A in the binary, Aristotelian notion of truth: "A proposition is true if it says of what is, that it is, and if it says of what is not, that it is not." This formal chain of reasoning is by no means the same thing as an everyday, ordinary mathematical proof. The classical view does not require that an ordinary proof be accompanied by its formal counterpart; on the contrary, there are mathematically sound reasons for allowing the gods to formalize most of our arguments. One theoretician estimates, for instance, that a formal demonstration of one of Ramanujan's conjectures assuming set theory and elementary analysis would take about two thousand pages; the length of a deduction from first principles is nearly inconceivable [14]. But the classicist believes that the formalization is in principle a possibility and that the truth it expresses is binary, either so or not so.

The probabilists argue that since any very long proof can at best be viewed as only probably correct, why not state theorems probabilistically and give probabilistic proofs? The probabilistic proof may have the dual advantage of being technically easier than the classical, bivalent one, and may allow mathematicians to isolate the critical ideas that give rise to uncertainty in traditional, binary proofs. This process may even lead to a more plausible classical proof. An illustration of the probabilist approach is Michael Rabin's algorithm for testing probable primality [17]. For very large integers N, all of the classical techniques for determining whether N is composite become unworkable. Using even the most clever programming, the calculations required to determine whether numbers larger than 10^{104} are prime require staggering amounts of computing time. Rabin's insight was that if you are willing to settle for a very good probability that N is prime (or not prime), then you can get it within a reasonable amount of time — and with vanishingly small probability of error.

In view of these uncertainties over what constitutes an acceptable proof, which is after all a fairly basic element of the mathematical process, how is it that mathematics has survived and been so successful? If proofs bear little resemblance to formal deductive reasoning, if they can stand for generations and then fall, if they can contain flaws that defy detection, if they can express only the probability of truth within certain error bounds — if they are, in fact, not able to *prove* theorems in the sense of guaranteeing them beyond probability and, if necessary, beyond insight, well, then, how does mathematics work? How does it succeed in developing theorems that are significant and that compel belief?

First of all, the proof of a theorem is a message. A proof is not a beautiful abstract object with an independent existence. No mathematician grasps a proof, sits back, and sighs happily at the knowledge that he can now be certain of the truth of his theorem. He runs out into the hall and looks for someome to listen to it. He bursts into a colleague's office and commandeers the blackboard. He throws aside his scheduled topic and regales a seminar with his new idea. He drags his graduate students away from their dissertations to listen. He gets onto the phone and tells his colleagues in Texas and Toronto. In its first incarnation, a proof is a spoken message, or at most a sketch on a chalkboard or a paper napkin.

That spoken stage is the first filter for a proof. If it generates no excitement or belief among his friends, the wise mathematician reconsiders it. But if they find it tolerably interesting and believable, he writes it up. After it has circulated in draft for a while, if it still seems plausible, he does a polished version and submits it for publication. If the referees also find it attractive and convincing, it gets published so that it can be read by a wider audience. If enough members of that larger audience believe it and like it, then after a suitable cooling-off period the reviewing publications take a more leisurely look, to see whether the proof is really as pleasing as it first appeared and whether, on calm consideration, they really believe it.

And what happens to a proof when it is believed? The most immediate process is probably an internalization of the result. That is, the mathematician who reads and believes a proof will attempt to paraphrase it, to put it in his own terms, to fit it into his own personal view of mathematical knowledge. No two mathematicians are likely to internalize a mathematical concept in exactly the same way, so this process leads usually to multiple versions of the same theorem, each reinforcing belief, each adding to the feeling of the mathematical community that the original statement is likely to be true. Gauss, for example, obtained at least half a dozen independent proofs of his 'law of quadratic reciprocity'; to date over fifty proofs of this law are known. Imre Lakatos gives, in his *Proofs and Refutations* [13], historically accurate discussions of the transformations that several famous theorems underwent from initial conception to general acceptance. Lakatos demonstrates that Euler's formula $V - E + F = 2$ was reformulated again and again for almost two hundred years after its first statement, until it finally reached its current stable form. The most compelling transformation that can take place is generalization. If, by the same social process that works on the original theorem, the generalized theorem comes to be believed, then the original statement gains greatly in plausibility.

A believable theorem gets used. It may appear as a lemma in larger proofs; if it does not lead to contradictions, then we are all the more inclined to believe it. Or engineers may use it by plugging physical values into it. We have fairly high confidence in classical stress equations because we see bridges that stand; we have some confidence in the basic theorems of fluid mechanics because we see airplanes that fly.

Believable results sometimes make contact with other areas of mathematics — important ones invariably do. The successful transfer of a theorem or a proof technique from one branch of mathematics to another increases our feeling of confidence in it. In 1964, for example, Paul Cohen used a technique called forcing to prove a theorem in set theory [2]; at that time, his notions were so radical that the proof was hardly understood. But subsequently other investigators interpreted the notion of forcing in an algebraic context, connected it with more familiar ideas in logic, generalized the concepts, and found the generalizations useful. All of these connections (along with the other normal social processes that lead to acceptance) made the idea of forcing a good deal more compelling, and today forcing is routinely studied by graduate students in set theory.

After enough internalization, enough transformation, enough generalization, enough use, and enough connection, the mathematical community eventually decides that the central concepts in the original theorem, now perhaps greatly changed, have an ultimate stability. If the various proofs feel right and the results are examined from enough angles, then the truth of the theorem is eventually considered to be established. The theorem is thought to be true in the classical sense — that is, in the sense that it *could* be demonstrated by formal, deductive logic, although for almost all theorems no such deduction ever took place or ever will.

THE ROLE OF SIMPLICITY

> For what is clear and easily comprehended attracts; the complicated repels.
>
> David Hilbert

> Sometimes one has to say difficult things, but one ought to say them as simply as one knows how.
>
> G. H. Hardy

As a rule, the most important mathematical problems are clean and easy to state. An imporant theorem is much more likely to take form A than form B.

A: Every ——— is a ———.
B: If ——— and ——— and ——— and ——— and ——— except for special cases
    a) ———
    b) ———
    c) ———,
then unless
    i) ——— or
    ii) ——— or
    iii) ———,
every ——— that satisfies ——— is a ———.

The problems that have most fascinated and tormented and delighted mathematicians over the centuries have been the simplest ones to state. Einstein held that the maturity of a scientific theory could be judged by how well it could be explained to the man on the street. The four-color theorem rests on such slender foundations that it can be stated with complete precision to a child. If the child has learned his multiplication tables, he can understand the problem of the location and distribution of the prime numbers. And the deep fascination of the problem of defining the concept of 'number' might turn him into a mathematician.

The correlation between importance and simplicity is no accident. Simple, attractive theorems are the ones most likely to be heard, read, internalized, and used. Mathematicians use simplicity as the first test for a proof. Only if it looks interesting at first glance will they consider it in detail. Mathematicians are not altruistic masochists. On the contrary, the history of mathematics is one long search for ease and pleasure and elegance — in the realm of symbols, of course.

Even if they didn't want to, mathematicians would have to use the criterion of simplicity; it is a psychological impossibility to choose any but the simplest and most attractive of 200,000 candidates for one's attention. If there are important, fundamental concepts in mathematics that are not simple, mathematicians will probably never discover them.

Messy, ugly mathematical propositions that apply only to paltry classes of structures, idiosyncratic propositions, propositions that rely on inordinately expensive mathematical machinery, propositions that require five blackboards or a roll of paper towels to sketch — these are unlikely ever to be assimilated into the body of mathematics. And yet it is only by such assimilation that proofs gain believability. The proof by itself is nothing; only when it has been subjected to the social processes of the mathematical community does it become believable.

In this paper, we have tended to stress simplicity above all else because that is the first filter for any proof. But we do not wish to paint ourselves and our fellow mathematicians as philistines or brutes. Once an idea has met the criterion of simplicity, other standards help determine its place among the ideas that make mathematicians gaze off abstractedly into the distance. Yuri Manin [14] has put it best: A good proof is one that makes us wiser.

DISBELIEVING VERIFICATIONS

> On the contrary, I find nothing in logistic for the discoverer but shackles. It does not help us at all in the direction of conciseness, far from it; and if it requires twenty-seven equations to establish that 1 is a number, how many will it require to demonstrate a real theorem?
>
> Henri Poincaré

> One of the chief duties of the mathematician in acting as an advisor to scientists ... is to discourage them from expecting too much from mathematics.
>
> Norbert Weiner

Mathematical proofs increase our confidence in the truth of mathematical statements only after they have been subjected to the social mechanisms of the mathematical community. These same mechanisms doom the so-called proofs of software, the long formal verifications that correspond, not to the working mathematical proof, but to the imaginary logical structure that the mathematician conjures up to describe his feeling of belief. Verifications are not messages; a person who ran out into the hall to communicate his latest verification would rapidly find himself a social pariah. Verifications cannot really be read; a reader can flay himself through one of the shorter ones by dint of heroic effort, but that's not reading. Being unreadable and — literally — unspeakable, verifications cannot be internalized, transformed, generalized, used, connected to other disciplines, and eventually incorporated into a community consciousness. They cannot acquire credibility gradually, as a mathematical theorem does; one either believes them blindly, as a pure act of faith, or not at all.

At this point, some adherents of verification admit that the analogy to mathematics fails. Having argued that A, programming, resembles B, mathematics, and having subsequently learned that B is nothing like what they imagined, they wish to argue instead that A is like B', their mystical version of B. We then find ourselves in the peculiar position of putting across the argument that was originally theirs, asserting that yes, indeed, A does resemble B; our argument, however, matches the terms up differently from theirs. (See Figures 1 and 2.) Verifiers who wish to abandon the simile and substitute B' should as an aid to understanding abandon the language of B as well — in particular, it would help if they did not call their verifications 'proof'. As for ourselves, we will continue to argue that programming is like mathematics, and that the same social processes that work in mathematical proofs doom verifications.

There is a fundamental logical objection to verification, an objection on its own ground of formalistic rigor. Since the requirement for a program is informal and the program is formal, there must be a transition, and the transition itself must necessarily be informal. We have been distressed to learn that this proposition, which seems self-evident to us, is controversial. So we should emphasize that as anti-

Mathematics		*Programming*
theorem	...	program
proof	...	verification

Fig. 1. The verifiers' original analogy.

Mathematics		*Programming*
theorem	...	specification
proof	...	program
imaginary		
formal		
demonstration	...	verification

Fig. 2. Our analogy.

formalists, we would not object to verification on these grounds; we only wonder how this inherently informal step fits into the formalist view. Have the adherents of verification lost sight of the informal origins of the formal objects they deal with? Is is their assertion that their formalizations are somehow incontrovertible? We must confess our confusion and dismay.

Then there is another logical difficulty, nearly a basic, and by no means so hair-splitting as the one above. The formal demonstration that a program is consistent with its specifications has value only if the specification and the program are independently derived. In the toy program atmosphere of experimental verification, this criterion is easily met. But in real life, if during the design process a program fails, it is changed, and the changes are based on knowledge of its specifications; if the specifications are changed, and those changes are based on knowledge of the program gained through the failure. In either case, the requirement of having independent criterion to check against each other is no longer met. Again, we hope that no one would suggest that programs and specifications should not be repeatedly modified during the design process. That would be a position of incredible poverty — the sort of poverty that does, we fear, result from infatuation with formal logic.

Back in the real world, the kinds of input/output specifications that accompany production software are seldom simple. They tend to be long and complex and peculiar. To cite an extreme case, computing the payroll for the French National Railroad requires more than 3000 pay rates (one uphill, one downhill, and so on). The specifications for any reasonable compiler or operating system fill volumes — and no one believes that they are complete. There are even some cases of black-box code, numerical algorithms that can be shown to work in the sense that they are used to build real airplanes or drill real oil wells, but work for no reason that anyone knows; the input assertions for these algorithms are not even formulable, let alone formalizable. To take just one example, an important algorithm with the rather jaunty name of Reverse Cuthill—McKee was known for years to be far better than plain Cuthill—McKee, known empirically, in laboratory tests and field trials and in production. Only recently, however, has its superiority been theoretically demonstrable [6], and even then only with the usual informal mathematical proof, not with a formal deduction. During all of the years when Reverse Cuthill—McKee was unproved, even though it automatically made any program in which it appeared unverifiable, programmers perversely went on using it.

It might be countered that while real-life specifications are lengthy and complicated, they are not deep. Their verifications are, in fact, nothing more than extremely long chains of substitutions to be checked with the aid of simple algebraic identities.

All we can say in response to this is: Precisely. Verifications are long and involved but shallow; that's what's wrong with them. The verification of even a puny program can run into dozens of pages, and there's not a light moment or a spark of wit on any of those pages. Nobody is going to run into a friend's office with a program verification. Nobody is going to sketch a verification out on a paper napkin. Nobody is going to buttonhole a colleague into listening to a verification. Nobody is ever going to read it. One can feel one's eyes glaze over at the very thought.

It has been suggested that very high level languages, which can deal directly with a broad range of mathematical objects or functional languages, which it is said can be concisely axiomatized, might be used to insure that a verification would be interesting and therefore responsive to a social process like the social process of mathematics.

In theory this idea sounds hopeful; in practice, it doesn't work out. For example, the following verification condition arises in the proof of a fast Fourier transform written in MADCAP, a very high level language [18]:

If $S \in \{1, -1\}$, $b = \exp(2\pi i S/N)$, r is an integer, $N = 2^r$,

(1) $C = \{2j : 0 \leqslant j < N/4\}$ and

(2) $a = \langle a_r : a_r = b^{r \bmod(N/2)}, 0 \leqslant r < N/2 \rangle$ and

(3) $A = \{j : j \bmod N < N/2, 0 \leqslant j < N\}$ and

(4) $A^* = \{j : 0 \leqslant j < N\} - A$ and

(5) $F = \langle f_r : f_r = \sum_{k_1 \in R_n} k_1(b^{k_1 \lfloor r/2^{r-1} \rfloor \bmod N}), R_r = \{j : (j - r)$

         $\bmod(N/2) = 0\} \rangle$ and $k \leqslant r$

then

(1) $A \cap (A + 2^{r-k-1}) = \{x : x \bmod 2^{r-k} < 2^{r-k-1}, 0 \leqslant x < N\}$

(2) $\qquad \langle \triangleright a_c \triangleright a_c \rangle = \langle a_r \colon a_r = b^{r2^k \bmod(N/2)}, 0 \leqslant r < N/2 \rangle$

(3) $\qquad \langle \triangleright (F_{A \cap (A+2^{r-k-1})} + F_{\{j \colon 0 \leqslant j < N\} - A \cap (A+2^{r-k-1})})$

$\qquad \triangleright (\langle \triangleright a_c \triangleright a_c \rangle$

$\qquad *(F_{A \cap (A+2^{r-k-1})} + F_{\{j \colon 0 \leqslant j < N\} - A \cap (A+2^{r-k-1})})))\rangle$

$\qquad = \langle f_r \colon f_r = \sum_{k_1 \in R_r} k_1(b^{\lfloor r/2^{r-k-1} \rfloor j \bmod N}),$

$\qquad R_r = \{j \colon (j-r) \bmod 2^{r-k-1} = 0\}\rangle$

(4) $\qquad \langle \triangleright (F_A + F_A \cdot) \triangleright a^*(F_A - F_A \cdot)\rangle = \langle f_r \colon f_r = \sum_{k_1 \in R_r}$

$\qquad k_1(b^{k_1 \lfloor r/2^{r-1} \rfloor \bmod N}), R_r = \{j \colon (j-r) \bmod(N/2) = 0\}\rangle$

This is not what we would call pleasant reading.

Some verifiers will concede that verification is simply unworkable for the vast majority of programs but argue that for a few crucial applications the agony is worthwhile. They point to air-traffic control, missile systems, and the exploration of space as areas in which the risks are so high that any expenditure of time and effort can be justified.

Even if this were so, we would still insist that verification renounce its claim on all other areas of programming; to teach students in introductory programming courses how to do verification, for instance, ought to be as farfetched as teaching students in introductory biology how to do open-heart surgery. But the stakes do not affect our belief in the basic impossibility of verifying any system large enough and flexible enough to do any real-world task. No matter how high the payoff, no one will ever be able to force himself to read the incredibly long, tedious verifications of real-life systems, and unless they can be read, understood, and refined, the verifications are worthless.

Now, it might be argued that all these references to readability and internalization are irrelevant, that the aim of verification is eventually to construct an automatic verifying system.

Unfortunately, there is a wealth of evidence that fully automated verifying systems are out of the question. The lower bounds on the length of formal demonstrations for mathematical theorems are immense [19], and there is no reason to believe that such demonstrations for programs would be any shorter or cleaner — quite the contrary. In fact, even the strong adherents of program verification do not take seriously the possibility of totally automated verifiers. Ralph London, a proponent of verification, speaks of an out-to-lunch system, one that could be left unsupervised to grind out verifications; but he doubts that such a system can be built to work with reasonable reliability. One group, despairing of automation in the foreseeable future, has proposed that verifications should be performed by teams of 'grunt mathematicians', low level mathematical teams who will check verification conditions. The sensibilities of people who could make such a proposal seem odd, but they do serve to indicate how remote the possibility of automated verification must be.

Suppose, however, that an automatic verifier could somehow be built. Suppose further that programmers did somehow come to have faith in its verifications. In the absence of any real-world basis for such belief, it would have to be blind faith, but no matter. Suppose that the philosopher's stone had been found, that lead could be changed to gold, and that programmers were convinced of the merits of feeding their programs into the gaping jaws of a verifier. It seems to us that the scenario envisioned by the proponents of verification goes something like this: The programmer inserts his 300-line input/output package into the verifier. Several hours later, he returns. There is his 20,000-line verification and the message 'VERIFIED'.

There is a tendency, as we begin to feel that a structure is logically, provably right, to remove from it whatever redundancies we originally built in because of lack of understanding. Taken to its extreme, this tendency brings on the so-called Titanic effect; when failure does occur, it is massive and uncontrolled. To put it another way, the severity with which a system fails is directly proportional to the intensity of the designer's belief that it cannot fail. Programs designed to be clean and tidy merely so that they can be verified will be particularly susceptible to the Titanic effect. Already we see signs of this phenomenon. In their notes on Euclid [16], a language designed for program verification, several of the foremost verification adherents say, "Because we expect all Euclid programs to be verified, we have not made special provisions for exception handling . . . Runtime software errors should not occur in verified programs." Errors should not occur? Shades of the ship that shouldn't be sunk.

So, having for the moment suspended all rational disbelief, let us suppose that the programmer gets the message 'VERIFIED'. And let us suppose further that the message does not result from a failure on the part of the verifying system. What does the programmer know? He knows that this program is formally, logically, provably, certifiably correct. He does not know, however, to what extent it is reliable, dependable, trustworthy, safe; he does not know within what limits it will work; he does not know what happens when it exceeds those limits. And yet he has that mystical stamp of approval: 'VERIFIED'. We can almost see the iceberg looming in the background over the unsinkable ship.

Luckily, there is little reason to fear such a future. Picture the same programmer returning to find the same 20,000 lines. What message would he really find, supposing that an automatic verifier could really be built? Of course, the message would be 'NOT VERIFIED'. The programmer would make a change, feed the program in again, return again. 'NOT VERIFIED'. Again he would make a change, again he would feed the program to the verifier, again 'NOT VERIFIED'. A program is a human artifact; a real-life program is a complex human artifact; and any human artifact of sufficient size and complexity is imperfect. The message will never read 'VERIFIED'.

THE ROLE OF CONTINUITY

> We may say, roughly, that a mathematical idea is "significant" if it can be connected, in a natural and illuminating way, with a large complex of other mathematical ideas.
>
> G. H. Hardy

The only really fetching defense ever offered for verification is the scaling-up argument. As best we can reproduce it, here is how it goes:

(1) Verification is now in its infancy. At the moment, the largest tasks it can handle are verifications of algorithms like FIND and model programs like GCD. It will in time be able to tackle more and more complicated algorithms and trickier and trickier model programs. These verifications are comparable to mathematical proofs. They are read. They generate the same kinds of interest and excitement that theorems do. They are subject to the ordinary social processes that work on mathematical reasoning, or on reasoning in any other discipline, for that matter.

(2) Big production systems are made up of nothing more than

algorithms and model programs. Once verified, algorithms and model programs can make up large, workaday production systems, and the (admittedly unreadable) verification of a big system will be the sum of the many small, attractive, interesting verifications of its components.

With (1) we have no quarrel. Actually, algorithms were proved and the proofs read and discussed and assimilated long before the invention of computers — and with a striking lack of formal machinery. Our guess is that the study of algorithms and model programs will develop like any other mathematical activity, chiefly by informal, social mechanisms, very little if at all by formal mechanisms.

It is with (2) that we have our fundamental disagreement. We argue that there is no continuity between the world of FIND or GCD and the world of production software, billing systems that write real bills, scheduling systems that schedule real events, ticketing systems that issue real tickets. And we argue that the world of production software is itself discontinuous.

No programmer would agree that large production systems are composed of nothing more than algorithms and small programs. Patches, *ad hoc* constructions, bandaids and tourniquets, bells and whistles, glue, spit and polish, signature code, blood-sweat-and-tears, and, of course, the kitchen sink — the colorful jargon of the practicing programmer seems to be saying something about the nature of the structures he works with; maybe theoreticians ought to be listening to him. It has been estimated that more than half the code in any real production system consists of user interfaces and error messages — *ad hoc*, informal structures that are by definition unverifiable. Even the verifiers themselves sometimes seem to realize the unverifiable nature of most real software. C. A. R. Hoare has been quoted [9] as saying, "In many applications, algorithm plays almost no role, and certainly presents almost no problem." (We wish we could report that he thereupon threw up his hands and abandoned verification, but no such luck.)

Or look at the difference between the world of GCD and the world of production software in another way: The specifications for algorithms are concise and tidy, while the specifications for real-world systems are immense, frequently of the same order of magnitude as the systems themselves. The specifications for algorithms are highly stable, stable over decades or even centuries; the specifications for real systems vary daily or hourly (as any programmer can testify). The specifications for algorithms are exportable, general; the specifications for real systems are idiosyncratic and *ad hoc*. These are not differences in degree. They are differences in kind. Babysitting for a sleeping child for one hour does not scale up to raising a family of ten — the problems are essentially, fundamentally different.

And within the world of real production software there is no continuity either. The scaling-up argument seems to be based on the fuzzy notion that the world of programming is like the world of Newtonian physics — made up of smooth, continuous functions. But, in fact, programs are jagged and full of holes and caverns. Every programmer knows that altering a line or sometimes even a bit can utterly destroy a program or mutilate it in ways that we do not understand and cannot predict. And yet at other times fairly substantial changes seem to alter nothing; the folklore is filled with stories of pranks and acts of vandalism that frustrated the perpetrators by remaining forever undetected.

There is a classic science-fiction story about a time traveler who goes back to the primeval jungles to watch dinosaurs and then returns to find his own time altered almost beyond recognition. Politics, architecture, language — even the plants and animals seem wrong, distorted. Only when he removes his time-travel suit does he understand what has happened. On the heel of his boot, carried away from the past and therefore unable to perform its function in the evolution of the world, is crushed the wing of a butterfly. Every programmer knows the sensation: A trivial, minute change wreaks havoc in a massive system. Until we know more about programming, we had better for all practical purposes think of systems as composed, not of sturdy structures like algorithms and smaller programs, but of butterflies' wings.

The discontinuous nature of programming sounds the death knell for verification. A sufficiently fanatical researcher might be willing to devote two or three years to verifying a significant piece of software if he could be assured that the software would remain stable. But real-life programs need to be maintained and modified. There is no reason to believe that verifying a modified program is any easier than verifying the original the first time around. There is no reason to believe that a big verification can be the sum of many small verifications. There is no reason to believe that a verification can transfer to any other program — not even to a program only one single line different from the original.

And it is this discontinuity that obviates the possibility of refining verifications by the sorts of social processes that refine mathematical proofs. The lone fanatic might construct his own verification, but he would never have any reason to read anyone else's, nor would anyone else ever be willing to read his. No community could develop. Even the most zealous verifier could be induced to read a verification only if he thought he might be able to use or borrow or swipe something from it. Nothing could force him to read someone else's verification once he had grasped the point that no verification bears any necessary connection to any other verification.

BELIEVING SOFTWARE

> The program itself is the only complete description of what the program will do.
>
> P. J. Davis

Since computers can write symbols and move them about with negligible expenditure of energy, it is tempting to leap to the conclusion that anything is possible in the symbolic realm. But reality does not yield so easily; physics does not suddenly break down. It is no more possible to construct symbolic structures without using resources than it is to construct material structures without using them. For even the most trivial mathematical theories, there are simple statements whose formal demonstrations would be impossibly long. Albert Meyer's outstanding lecture on the history of such research [15] concludes with a striking interpretation of how hard it may be to deduce even fairly simple mathematical statements. Suppose that we encode logical formulas as binary strings and set out to build a computer that will decide the truth of a simple set of formulas of length, say, at most a thousand bits. Suppose that we even allow ourselves the luxury of a technology that will produce proton-size electronic components connected by infinitely thin wires. Even so, the computer we design must densely fill the entire observable universe. This precise observation about the length of formal deductions agrees with our intuition about the amount of detail embedded in ordinary, workaday mathematical proofs. We often use "Let us assume, without loss of generality ..." or "Therefore, by renumbering, if necessary ..." to replace enormous amounts of formal detail. To insist on the formal detail would be a silly waste of resources. Both symbolic and material structures must be engineered with a very cautious eye. Resources are limited; time is limited; energy is limited. Not even the computer can change the finite nature of the universe.

We assume that these constraints have prevented the adherents of verification from offering what might be fairly convincing evidence in support of their methods. The lack at this late date of even a single verification of a working system has sometimes been attributed to the youth of the field. The verifiers argue, for instance, that they are only now beginning to understand loop invariants. At first blush, this sounds like another variant of the scaling-up argument. But in fact there are large classes of real-life systems with virtually no loops — they scarcely ever occur in commercial programming applications. And yet there has never been a verification of, say, a Cobol system that prints real checks; lacking even one makes it seem doubtful that there could at some time in the future be many. Resources, and time, and energy are just as limited for verifiers as they are for all the rest of us.

We must therefore come to grips with two problems that have occupied engineers for many generations: First, people must plunge into activities that they do not understand. Second, people cannot create perfect mechanisms.

How then do engineers manage to create reliable structures? First, they use social processes very like the social processes of mathematics to achieve successive approximations at understanding. Second, they have a mature and realistic view of what 'reliable' means: in particular, the one thing it never means is 'perfect'. There is no way to deduce logically that bridges stand, or that airplanes fly, or that power stations deliver electricity. True, no bridges would fall, no airplanes would crash, no electrical systems black out if engineers would first demonstrate their perfection before building them — true because they would never be built at all.

The analogy in programming is any functioning, useful, real-world system. Take for instance an organic-chemical synthesizer called SYN-CHEM [5]. For this program, the criterion of reliability is particularly straightforward — if it synthesizes a chemical, it works; if it doesn't, it doesn't work. No amount of correctness could ever hope to improve on this standard; indeed, it is not at all clear how one could even begin to formalize such a standard in a way that would lend itself to verification. But it is a useful and continuing enterprise to try to increase the number of chemicals the program can synthesize.

It is nothing but symbol chauvinism that makes computer scientists think that our structures are so much more important than material structures that (a) they should be perfect, and (b) the energy necessary to make them perfect should be expended. We argue rather that (a) they cannot be perfect, and (b) energy should not be wasted in the futile attempt to make them perfect. It is no accident that the probabilistic view of mathematical truth is closely allied to the engineering notion of reliability. Perhaps we should make a sharp distinction between program reliability and program perfection — and concentrate our efforts on reliability.

The desire to make programs correct is constructive and valuable. But the monolithic view of verification is blind to the benefits that could result from accepting a standard of correctness like the standard of correctness for real mathematical proofs, or a standard of reliability like the standard for real engineering structures. The quest for workability within economic limits, the willingness to channel innovation by recycling successful design, the trust in the functioning of a community of peers — all the mechanisms that make engineering and mathematics really work are obscured in the fruitless search for perfect verifiability.

What elements could contribute to making programming more like engineering and mathematics? One mechanism that can be exploited is the creation of general structures whose specific instances become more reliable as the reliability of the general structure increases. (This process has recently come to be called 'abstraction', but we feel that for a variety of reasons 'abstraction' is a bad term. It is easily confused with the totally different notion of abstraction in mathematics, and often what has passed for abstraction in the computer science literature is simply the removal of implementation details.) This notion has appeared in several incarnations, of which Knuth's insistence on creating and understanding generally useful algorithms is one of the most important and encouraging. Baker's team-programming methodology [1] is an explicit attempt to expose software to social processes. If reusability becomes a criterion for effective design, a wider and wider community will examine the most common programming tools.

The concept of verifiable software has been with us too long to be easily displaced. For the practice of programming, however, verifiability must not be allowed to overshadow reliability. Scientists should not confuse mathematical models with reality — and verification is nothing but a model of believability. Verifiability is not and cannot be a dominating concern in software design. Economics, deadlines, cost—

benefit ratios, personal and group style, the limits of acceptable error —
all these carry immensely much more weight in design than verifiability
or nonverifiability.

So far, there has been little philosophical discussion of making
software reliable rather than verifiable. If verification adherents could
redefine their efforts and reorient themselves to this goal, or if another
view of software could arise that would draw on the social processes of
mathematics and the modest expectations of engineering, the interests
of real-life programming and theoretical computer science might both
be better served.

Even if, for some reason that we are not now able to understand, we
should be proved wholly wrong and the verifiers wholly right, this is not
the moment to restrict research on programming. We know too little
now to sense what directions will be most fruitful. If our reasoning
convinces no one, if verification still seems an avenue worth exploring,
so be it; we three can only try to argue against verification, not blast it
off the face of the earth. But we implore our friends and colleagues not
to narrow their vision to this one view no matter how promising it may
seem. Let it not be the only view, the only avenue. Jacob Bronowski has
an important insight about a time in the history of another discipline
that may be similar to our own time in the development of computing:

A science which orders its thought too early is stifled ... The hope of the medieval
alchemists that the elements might be changed was not as fanciful as we once thought.
But it was merely damaging to a chemistry which did not yet understand the composition of water and common salt.

ACKNOWLEDGMENTS

We especially wish to thank those who gave us public forums — the 4th
POPL program committee for giving us our first chance; Bob Taylor
and Jim Morris for letting us express our views in a discussion at Xerox
PARC; L. Zadeh and Larry Rowe for doing the same at the Computer
Science Department of the University of California at Berkeley; Marvin
Dennicoff and Peter Wegner for allowing us to address the DOD
conference on research directions in software technology.

We also wish to thank Larry Landweber for allowing us to visit for a
summer the University of Wisconsin at Madison. The environment and
the support of Ben Noble and his staff at the Mathematics Research
Center was instrumental in letting us work effectively.

318   RICHARD A. DE MILLO ET AL.

The seeds of these ideas were formed out of discussions held at the DOD Conference on Software Technology in 1976 at Durham, North Carolina. We wish to thank in particular J. R. Suttle, who organized this conference and has been of continuing encouragement in our work.

We also wish to thank our many friends who have discussed these issues with us. They include: Al Aho, Jon Barwise, Manuel Blum. Tim Budd, Lucio Chiaraviglio, Philip Davis, Peter Denning, Bernie Elspas, Mike Fischer, Ralph Griswold, Leo Guibas, David Hansen, Mike Harrison, Steve Johnson, Jerome Kiesler, Kenneth Kunen, Nancy Lynch, Albert Meyer, Barkley Rosser, Fred Sayward, Tim Standish, Larry Travis, Tony Wasserman, and Ann Yasuhara.

We also wish to thank both Bob Grafton and Marvin Dennicoff of ONR for their comments and encouragement.

Only those who have seen earlier drafts of this paper can appreciate the contribution made by our editor, Mary-Claire van Leunen. Were it the custom in computer science to list a credit line 'As told to . . .', that might be a better description of the service she performed.

Georgia Institute of Technology and
Yale University.

REFERENCES

1. Baker, F. T.: 1972, 'Chief Programmer Team Management of Production Programming', *IBM Syst. J.* **11**(1), 56—73.
2. Cohen, P. J.: 1963, 'The Independence of the Continuum Hypothesis', *Proc. Nat. Acad. Sci., USA.* Part I **50**, 1143—1148; Part II, **51**, 105—110.
3. Davis, P. J.: 1972, 'Fidelity in Mathematical Discourse: Is One and One Really Two?', *The Amer. Math. Monthly* **79**(3), 252—263.
4. Bateman, P. and Diamond, H.: 1978, 'John E. Littlewood (1885—1977): An Informal Obituary', *The Math. Intelligencer* **1**(1), 28—33.
5. Gelerenter, H. *et al.*: 1973, 'The Discovery of Organic Synthetic Roots by Computer', *Topics in Current Chemistry* **41**, Springer-Verlag, pp. 113—150.
6. George, J. Alan: 1971, 'Computer Implementation of the Finite Element Method', Ph.D. Thesis, Stanford U., Stanford, Calif.
7. Heath, Thomas L.: 1956, *The Thirteen Books of Euclid's Elements*, Dover, New York, pp. 204—219.
8. Heawood, P. J.: 1890, 'Map Colouring Theorems', *Quarterly J. Math., Oxford Series* **24**, 322—339.
9. Hoare, C. A. R.: 1978, Quoted in *Software Management*, C. McGowan and R. McHenry (Eds.), *Research Directions in Software Technology*, M.I.T. Press, Cambridge, Mass.

10. Jech, Thomas J.: 1973, *The Axiom of Choice*. North-Holland Pub. Co., Amsterdam, p. 118.
11. Kempe, A. B.: 1879, 'On the Geographical Problem of the Four Colors', *Amer. J. Math.* **2**, 193—200.
12. Kolata, G. Bari: 1976, 'Mathematical Proof: The Genesis of Reasonable Doubt', *Science* **192**, 989—990.
13. Lakatos, Imre: 1976, *Proofs and Refutations: The Logic of Mathematical Discovery*, Cambridge University Press, England.
14. Manin, Yu I.: 1977, *A Course in Mathematical Logic*, Springer-Verlag, pp. 48—51.
15. Meyer, A.: 1974, 'The Inherent Computational Complexity of Theories of Ordered Sets: A Brief Survey', *Int. Cong. of Mathematicians*, Aug.
16. Popek, G. *et al.*: 1977, 'Notes on the Design of Euclid', *Proc. Conf. Language Design for Reliable Software, SIGPLAN Notices* (ACM) **12**(3), pp. 11—18.
17. Rabin, M. O.: 1976, 'Probabilistic Algorithms', in: J. F. Traub (Ed.), *Algorithms and Complexity: New Directions and Recent Results*, Academic Press, New York.
18. Schwartz, J.: 1973, 'On Programming', *Courant Rep.*, New York U., New York.
19. Stockmeyer, L.: 1974, 'The Complexity of Decision Problems in Automata Theory and Logic'. Ph.D. Thesis, M.I.T., Cambridge, Mass.
20. Ulam, S. M.: 1976, *Adventures of a Mathematician*, Scribner's, New York, p. 288.

JAMES H. FETZER

PROGRAM VERIFICATION: THE VERY IDEA

> I hold the opinion that the construction of
> computer programs is a mathematical activity
> like the solution of differential equations, that
> programs can be derived from their specifica-
> tions through mathematical insight, calculation,
> and proof, using algebraic laws as simple and
> elegant as those of elementary arithmetic.
>
> C. A. R. Hoare

There are those, such as Hoare [20], who maintain that computer programming should strive to become more like mathematics. Others, such as DeMillo, Lipton and Perlis [8], contend this suggestion is mistaken because it rests upon a misconception. Their position emphasizes the crucial role of social processes in coming to accept the validity of a proof or the truth of a theorem, no matter whether within purely mathematical contexts or without: "We believe that, in the end, it is a social process that determines whether mathematicians feel confident about a theorem" [8, p. 271]. As they perceive it, the situation with respect to program verification is worse insofar as no similar social process occurs between program verifiers. The use of verification to guarantee the performance of a program is therefore bound to fail. Although Hoare's work receives scant attention in their paper, there should be no doubt that his approach — and that of others, such as E. W. Dijkstra [10], who share a similar point of view — is the intended object of their criticism.

Their presentation has aroused enormous interest and considerable controversy, ranging from unqualified agreement [expressed, for example, by Glazer [13]: "Such an article makes me delight in being . . . a member of the human race"] to unqualified disagreement [expressed, for example, by Maurer [28]: "The catalog of criticisms of the idea of proving a program correct . . . deserves a catalog of responses . . ."]. Indeed, some of the most interesting reactions have come from those whose position lies somewhere in between, such as van den Bos [37], who maintains that,

321

Timothy R. Colburn et al. (eds.), Program Verification, 321—358.
© 1993 *Kluwer Academic Publishers.*

Once one accepts the quasi-empiricism in mathematics, and by analogy in computer science, one can either become an adherent of the Popperian school of conjectures (theories) and refutations [32], or one may believe Kuhn [23], who claims that the fate of scientific theories is decided by a social forum . . .[1]

Perhaps better than any other commentor, van den Bos seems to have put his finger on what may well be the crucial issue raised by [8], namely: if program verification, like mathematical validation, could only occur as the result of a fallible social process, if it could occur at all, then what would distinguish programming procedures from other expert activities, such as judges deciding cases at law and referees reviewing articles for journals? If it is naive to presume that mathematical demonstrations, program verifications and the like are fundamentally distinct from these activities, on what basis can they be differentiated?

The purpose of this article is to investigate the arguments that DeMillo, Lipton and Perlis have presented in an effort to disentangle several issues that seem to have become intricately intertwined. In particular, their position, in part, rests upon a difference in social practice that could change if program verifiers were to modify their behavior. It also depends, in part, upon problems that arise from the complexity of the programs to be verified. There appear to be two quite different kinds of 'program complexity', however, only one of which succumbs to their arguments. Moreover, if program verifiers were to commence collaborating in their endeavors, the principal rationale underlying their position would tend to disappear. Indeed, while social processes are crucial in determining what theorems the mathematical community takes to be true and what proofs it takes to be valid, they do not thereby make them true or valid. The absence of similar social processes in determining which programs are correct, accordingly, does not affect which programs are correct. Nevertheless, there are reasons for doubting whether program verification can succeed as a generally applicable and completely reliable method for guaranteeing the performance of a program. Therefore, it looks as though DeMillo, Lipton and Perlis have offered some bad arguments for some positions that need further elaboration and deserve better support.

MATHEMATICS AS A FALLIBLE SOCIAL PROCESS

> Outsiders see mathematics as a cold, formal, logical, mechanical, monolithic process of sheer intellection; we argue that insofar as it is successful, mathematics is a social, informal, intuitive, organic, human process, a community project.
>
> DeMillo, Lipton and Perlis

The conception of mathematical procedure portrayed by DeMillo, Lipton and Perlis initially drew a distinction between *proofs* and *demonstrations*, where demonstrations are supposed to be long chains of formal logic, while proofs are not. The difference intended here, strictly speaking, appears to be between proofs and what are typically referred to as 'proof sketches', where proof sketches are incomplete (or 'partial') proofs. Indeed, proofs are normally defined in terms of demonstrations, where a proof of theorem *T*, say, occurs just in case theorem *T* can be shown to be the last member of a sequence of formulae where every member of that sequence is either given (as an axiom or as an assumption) or else derived from preceding members of that sequence (by relying upon the members of a specified set of rules of inference) [6, p. 182]. In fact, what is known as *mathematical induction* is a special case of the application of demonstrative procedures to infinite sequences, where these processes, which tend to rely upon recursive techniques, are completely deductive [2, p. 169].

Moreover, when DeMillo, Lipton and Perlis offer 'proof sketches' as the objects of mathematicians' attention rather than proofs, it becomes possible to make (good) sense of otherwise puzzling statements such as:

In mathematics, the aim is to increase one's confidence in the correctness of a theorem, and it's true that one of the devices mathematicians could in theory use to achieve this goal is a long chain of formal logic. But in fact they don't. What they use is a proof, a very different animal. Nor does the proof settle the matter; contrary to what its name suggests, a proof is only one step in the direction of confidence. [8, p. 271]

Thus, while a proof, strictly speaking, is a (not necessarily long) chain of formal logic that is no different than a demonstration, a proof sketch is "a very different animal", where a proof sketch, unlike a proof, may often be "only one step in the direction of confidence." Nevertheless, although these reflections offer an interpretation under which their statements appear to be true, it leaves open a larger question, namely: whether the aim of proofs in mathematics can be adequately characterized as that of "increasing one's confidence in the correctness of theorems" rather than as formal demonstrations.

In support of their depiction, DeMillo, Lipton and Perlis emphasize the tentative and fallible character of mathematical progress, where out of some 200,000 theorems said to be published each year, "A number of these are subsequently contradicted or otherwise disallowed, others are thrown into doubt and most are ignored" [8, p. 272]. Since numerous purported proofs are unable to withstand critical scrutiny, they suggest, the acceptability or believability of a specific mathematical result depends upon its reception and ultimate evaluation by the mathematical community. In this spirit, they describe what appears to be a typical sequence of activity within this arena, where, say, a proof begins as an idea in someone's mind, receives translation into a sketch, is discussed with colleagues and, if no substantial objections arise, is developed and submitted for publication, where, if it survives the criticism of other mathematicians, then it tends to be accepted [8, p. 273].

In this sense, the behavior of typical members of the mathematical community in the discovery and promotion of specific findings certainly assumes the dimensions of a social process involving more than one person interacting together to bring about a certain outcome. Although it may be difficult to imagine a Bertrand Russell or a David Hilbert rushing to his colleages for their approval of his findings, there would appear to be no good reasons to doubt that average mathematicians frequently behave in the manner described. Therefore, I would tend to agree that mathematicians' mistakes are typically discovered or corrected through casual interaction with other mathematicians. The restraints imposed by symbolic logic, after all, exert their influence only through their assimilation as habits of thought and as patterns of reasoning by specific members of a community of this kind: discoveries and corrections of mistakes usually occur when one mathematician gently nudges another "in the right direction." (Relevant discussions can be found in [4] and [24].)

To the extent to which DeMillo, Lipton and Perlis should be regarded as endorsing the view that review procedures exercised by colleagues and peers tend to improve the quality of papers that appear in mathematics journals, there seems to be little grounds for disagreement. For potential proofs are often strengthened, theorems altered to correspond to what is provable, and various arguments discovered to be deeply flawed through social interaction. Nevertheless, a community of mathematicians who are fast and sloppy referees is not especially difficult to imagine: where is the university whose faculty do not occasionally compose shoddy and inaccurate reviews — even for very good journals? After all, what makes (what we call) a *proof* a proof is its validity rather than its acceptance (by us) as valid, just as what makes a sentence true is what it asserts to be the case is the case, nor merely that it is believed (by us) and therefore referred to as *true*. Social processing, therefore, is neither necessary nor sufficient for a proof to be valid, as DeMillo, Lipton and Perlis implicitly concede [8, p. 272].

DEDUCTIVE VALIDITY AND PSYCHOLOGICAL CERTAINTY

> ... a theorem either can or cannot be derived from a set of axioms. I don't believe that the correctness of a theorem is to be decided by a general election.
>
> L. Lamport

Confidence in the truth of a theorem (or in the validity of an argument), of course, appears to be a psychological property of a person-at-a-time: one and the same person at two different times can vary greatly in his confidence over the truth of the same theorem or the validity of the same argument, just as two different persons at the same time might vary greatly in their confidence that that same theorem is true or that that same argument is valid. Indeed, there is nothing inconsistent about scenarios in which, say, someone is completely confident that a specific formula is a theorem (when it happens to be false) or else completely uncertain whether a particular argument is valid (when it happens to be valid). No doubt, mathematicians are sometimes driven to discover demonstrations of theorems after they are already completely convinced of their truth. Demonstrations, in such cases, cannot increase the degree of confidence when that degree is already maximally strong. But that is not to deny they can still fulfill other — non-psychological — functions, such as providing objective evidence of the truth of one's subjective belief.

From the point of view of a traditional theory of knowledge, the role of demonstration becomes readily apparent; for the classical conception of knowledge characterizes 'knowledge' in terms of three necessary and sufficient conditions as warranted, true belief (for example [7, ch. 2]). Hence, an individual z who is in a state of belief with respect to a certain formula f, where z believes that f is a theorem, say, cannot be properly qualified as possessing knowledge that f is a theorem unless his belief can be supported by means of reasons, evidence, or warrants, which might be one or another of three different kinds, depending upon the nature of the objects that might be known. For results in logic and mathematics fall within the domain of deductive methodology and require demonstrations. Lawful and causal claims fall within the domain of empirical inquiries and require inductive warrants. Observational and experimental findings fall within the domain of perceptual investigations and acquire support on the basis of direct sense experience.[2]

With respect to deductions, the term *verification* can be used in two rather different senses. One of these occurs in pure mathematics and in pure logic, in which theorems of mathematics and of logic are subject to demonstration. These theorems characterize claims that are always true as a function of the meanings assigned to the specific symbols by means of which they are expressed. Theorem-schemata and theorems in this sense are subject to verfication by deriving them from no premises at all (within systems of natural deduction) or from primitive axioms (within axiomatic formal systems).[3] The other occurs in ordinary reasoning and in scientific contexts in general, whenever *conclusions* are shown to follow from specific sets of *premises*, where there is no presumption that these conclusions might be derived from no premises at all or that those premises should be true as a function of their meaning. Thus, within a system of natural deduction or an axiomatic formal system, the members of the class of consequences that can be derived from no premises at all or that follow from primitive axioms alone may be said to be *absolutely* verifiable. By contrast those members of the class of consequences that can only be derived relative to specific sets of premises whose truth is not absolutely verifiable, may be said to be *relatively* verifiable.[4]

The difference between absolute and relative verifiability, moreover, is extremely important for the theory of knowledge. For theorems that can be verified in the absolute sense cannot be false, so long as the rules are not changed or the axioms are not altered. But conclusions that are verified in the relative sense can still be false (even when the premises and the rules remain the same). The absolute verification of a theorem thus satisfies both necessary and sufficient conditions for its warranted acceptance as true, but the relative verification of a conclusion does not. Indeed, as an epistemic policy, the degree of confidence that anyone should invest in the conclusion of an argument should never exceed the degree of confidence that ought to be invested in its premises — even when it is valid! Unless the premises of an argument cannot be false — unless those premises themselves are absolutely verifiable — it is a mistake to assume the conclusion of a valid argument cannot be false. No more can appropriately be claimed than that its conclusion must be true if all its premises are true, which is the defining property of a valid demonstration.

CONSTRUCTING PROOFS AND VERIFYING PROGRAMS

> Formal proofs carry with them a certain objectivity. That a proof is formalizable, that the formal proofs have the structural properties that they do, explains in part why proofs are convincing to mathematicians.
>
> T. Tymoczko

The truth of the conclusion of a valid deductive argument, therefore, can never be more certain than the truth of its premises — unless its truth can be established on other grounds. While deductive reasoning preserves the truth (insofar as the conclusion of a valid argument cannot be false if its premises are true), the truth of those premises can be guaranteed, in general, only under those special circumstances that arise when they themselves are verifiable in the absolute sense. Otherwise, the truth of the premises of any argument has to be established on independent grounds, which might be deductive, inductive or perceptual. Yet none of these types of warrants provides an infallible foundation for any inference to the truth of the conclusions they support. Perhaps few of us would be inclined to think that our senses are infallible, i.e. that things must always be the way they appear to be. The occurrence of illusions, hallucinations and delusions disabuses us of that particular fantasy. The mistakes we make about reasoning are far more likely to occur concerning inductive and deductive arguments, whose features are frequently not clearly understood. We should not overlook that, apart from imagination and conjecture, which serve as sources of ideas but do not establish their truth, all of our states of knowledge — other than those of pure mathematics and logic — are ultimately dependent for their support upon direct and indirect connections to experience.

The features that distinguish (good) deductive arguments are the following:

(a) they are *demonstrative*, i.e. if their premises were true, their conclusions could not be false (without contradiction);

(b) they are *non-ampliative*, i.e. there is no information or content in their conclusions that is not already contained in their premises; and,

(c) they are *additive*, i.e. the addition of further information in the form of additional premises can neither strengthen nor weaken these arguments, which are already maximally strong.

Thus, the non-ampliative property of (good) deductive arguments can serve to explain both their demonstrative and additive characteristics. Demonstrative arguments, of course, are said to be 'valid', while valid arguments with true premises are said to be 'sound' (and cannot possibly have false conclusions).[5]

Compared to deductive arguments, (good) inductive arguments are (a) *non-demonstrative*, (b) *ampliative*, and (c) *non-additive*. Arguments satisfying appropriate inductive standards likewise should be said to be 'proper', while proper arguments with true premises may be said to be 'correct' (but can have conclusions that are false even when their premises are true). It should come as no surprise, therefore, that the purposes served by such different types of arguments are quite distinct, indeed. For inductive arguments are meant to be *knowledge-expanding*, while deductive arguments are meant to be *truth-preserving*. The ways in which inductive arguments expand our knowledge assume various forms. Reasoning from samples to populations, from the observed to the unobserved and from the past to the future always involves drawing inferences to conclusions that contain more information or content than do their premises. The most familiar instances of inductive reasoning we all employ in our daily lives concern the behavior of ordinary physical things — things that may or may not work right, fit properly, or function smoothly (such as electrical appliances, including microwave ovens and personal computers). When we interact with systems such as these, we invariably base our expectations upon our experience: we draw conclusions concerning their behavior in the future from their behavior in the past. All such reasoning is ampliative and — as we all too often discover to our dismay — is both non-demonstrative and non-additive.

The function of a program, of course, is to convey instructions to a computer. Most programs today are written in high-level programming languages, such as Pascal, LISP and Prolog, which simulate 'abstract machines', whose instructions can be more readily composed than can those of the machines that ultimately execute them. Thus, a source program written in a high-level language is translated into an equivalent low-level 'object program' written in machine language either directly, by an interpreter; or indirectly, by a compiler. The 'target machine' then executes the object program when instructed to do so. If programs are verifiable, therefore, then they must be subject to deductive procedures. Indeed, precisely this conception is advanced by [19, p. 576]:

Computer programming is an exact science in that all the properties of a program and all the consequences of executing it can, in principle, be found out from the text of the program itself by means of purely deductive reasoning.

Thus, if programs are absolutely verifiable, then there must exist some *program rules of inference* or *primitive program axioms* permitting inferences to be drawn concerning the performance that a machine will display when such a program is executed. If they are relatively verifiable, then there must be sets of premises concerning the text of such a program from which it is possible to derive conclusions concerning the performance that that machine will display when that program is executed. If these conditions cannot be satisfied, however, then the very idea of program verification will have been misconceived.

SOCIAL PROCESSES AND PROGRAM VERIFICATIONS

> We do not argue that strict logical deduction should be the only way that mathematics should be done, or even that it should come first; rather, it should come last, after the theorems to be proved, and their proofs, are well understood.
>
> W. D. Maurer

The critical dimension of deduction and induction, furthermore, is the justification of corresponding classes of rules as acceptable principles of inference. For, in their absence, it would be impossible to ascertain which, among all those arguments that — rightly or wrongly — are supposed to be valid or proper, actually are. Within this domain, "thinking doesn't make it so." The fact that a community of mathematicians happens to agree upon the validity of an argument or the truth of a theorem, alas, no more guarantees the validity of that argument or the truth of that theorem than agreement within a society of observers that the Earth is flat (for which a variety of mutually convincing arguments are advanced) could guarantee that that belief is true. In the absence of classes of rules of inference whose acceptability can be justified, in other words, validity and propriety are merely subjective properties of arguments insofar as specific persons happen to hold them in high esteem, where their standing may vary from person to person and from time to time.

Since computer programs, like mathematical proofs, are syntactical entities consisting of sequences of lines (strings of signs and the like), they both appear to be completely formalized entities for which completely formal procedures appear to be appropriate. Yet programs differ from theorems, at least to the extent to which programs are supposed to possess a semantic significance that theorems seem to lack. For the sequences of lines that compose a program are intended to stand for operations and procedures that can be performed by a machine, whereas the sequences of lines that constitute a proof do not.[6] Even if the social acceptability of a mathematical proof is neither necessary nor sufficient for its validity, the suggestion might be made that the existence of social processes of program verification may actually be even more important for the success of this endeavor than it is for validating proofs. The reason that could be advanced in support of this position is the opportunity that such practices would provide for more than one programmer to inspect a program for a suitable relationship between its syntax and its intended semantics, i.e. the behavior expected of the machine.

There are no reasons for believing that DeMillo, Lipton and Perlis are mistaken in their observation that the verification of programs is not a popular pastime [8]. The argument we are considering thus serves to reinforce the importance of this disparity in behavior between the members of the mathematical community and the members of the programming fraternity. Their position implicitly presumes that programmers are inherently unlikely to collaborate on the verification of programs, if only because it is such a tedious and complex activity. Suppose, however, that conditions within society were to change in certain direct and obvious ways, say that substantial financial rewards were offered for the best team efforts in verifying programs, with prizes of up to $10,000,000 awarded by Ed McMahon and guest appearances on *The Tonight Show*. Surely under circumstances of this kind, the past tendency of program verifiers to engage in solitary enterprise might be readily overcome, resulting in the emergence of a new wave in program verification, "the collaborative verification group," dedicated to mutual efforts by more than one programmer to verify particular programs. Surely under these conditions — which are not completely far-fetched in the context of modern times — a social process for the verification of programs could emerge within the computer science community that would be the counterpart to the social process for the validation of proofs in mathematics. Under these circumstances, there would not be any difference of this kind.

If this were to come about, then the primary assumption underlying the position of DeMillo, Lipton and Perlis would no longer apply. Regardless of what other differences might distinguish them, in this respect their social processing of theorems and of programs would be the same. If we refer to differences between subjects or activities that could not be overcome, no matter what efforts we might undertake, as differences 'in principle', and to those that could be overcome, by making appropriate efforts, as differences 'in practice', then it should be obvious that DeMillo, Lipton and Perlis have identified a difference in practice that is not also a difference in principle. What this means is that to the extent to which their position depends upon this difference, it represents no more than a contingent, *de facto* state-of-affairs that might be a mere transient stage in the development of program verification within the computer science community. If there is an 'in principle' difference between them, it must lie elsewhere, because divergence in social practice is a difference that could be overcome.

THE CONCEPTION OF PROBABILISTIC PROOFS

> If proofs bear little resemblance to formal deductive reasoning, if they can stand for generations and then fall, if they can contain flaws that defy detection, if they can express

only the probability of truth within certain
error bounds — if they are, in fact, not able to
prove theorems in the sense of guaranteeing
them beyond probability and, if necessary,
beyond insight, well, then, how does mathematics work?

DeMillo, Lipton and Perlis

When DeMillo, Lipton and Perlis [8, p. 273] advance the position that
"a proof can, at best, only probably express truth," therefore, it is
important to discover exactly what they mean, since proofs are deductive and accordingly enjoy the virtues of demonstrations. There are
several alternatives. In the first place, this claim might reflect the differences that obtain between proofs and proof sketches, insofar as incomplete or partial proofs do not satisfy the same objective standards —
and therefore need not convey the same subjective certainty — as do
complete proofs. In the second place, it might reflect the fallibility of
acceptance of the premises of such an argument — which could be valid
yet have at least one false premise and therefore be unsound — because
of which acceptance of its conclusion would be tempered with uncertainty as well. While both of these ideas find expression in their article,
however, their principal contention appears to be of a rather different
character altogether.

Thus, DeMillo, Lipton and Perlis distinguish between the 'classicists'
and the 'probabilists', where classicists maintain that:

... when one believes mathematical statement A, one believes that in principle there is
a correct, formal, valid, step by step, syntactically checkable deduction leading to A in a
suitable logical calculus ...

which is a complete proof lying behind a proof sketch as the object of
acceptance or of belief. Probabilists, by comparison, instead maintain
that:

... since any very long proof can at best be viewed as only probably correct, why not
state theorems probabilistically and give probabilistic proofs? The probabilistic proof
may have the dual advantage of being technically easier than the classical, bivalent one,
and may allow mathematicians to isolate the critical ideas that give rise to uncertainty in
traditional, binary proofs.

Thus, in application to proofs or to proof sketches, there appear to be
three elements to this position: first, that long proofs are difficult to
follow; second, that probabilistic proofs are easier to follow; and, third,

that probabilistic proofs may disclose the problematic aspects of ordinary proofs. Rabin's algorithm for testing probable prime numbers is offered as an illustration, where, "if you are willing to settle for a very good probability that N is prime (or not prime), then you can get it within a reasonable amount of time — and with (a) vanishingly small probability of error."

Their reference to "traditional, binary proofs" is important insofar as traditional proofs (in the classical sense) are supposed to be either valid or invalid: there is nothing 'probable' about them. Therefore, I take DeMillo, Lipton and Perlis to be endorsing an alternative conception, according to which arguments are amenable to various measures of strength (or corresponding 'degrees of conviction'), which might be represented by, say, some number between zero and one inclusively, with some of the properties associated with probabilities, likelihoods, or whatever [11, Part III]. A hypothetical scale could be constructed accordingly, such that degrees of partial support of measures zero and one for instance are distinguished from worthless fallacies (whose premises, for example, might be completely irrelevant to their conclusions), on the one hand, and from demonstrative arguments (the truth of whose premises guarantees the truth of their conclusions), on the other, as extreme cases not representing *partial* support (Figure 1).

From this perspective, the existence of even a 'vanishingly small' probability of error is essential to a probabilistic proof. If there were no probability of error at all a proof could not be probabilistic. Thus, if the commission of an error were either a necessity (as in the case of a fallacy of irrelevance) or an impossibility (as in the case of valid demon-

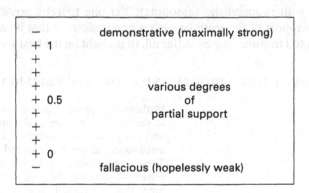

Fig. 1. A measure of evidential support.

stration), then an argument would have to be other than probabilistic. Indeed, the existence of fallacies of irrelevance exemplifies an important point discussed above: fallacious arguments, though logically flawed, can nevertheless exert enormous persuasive appeal — otherwise, we would not have to learn how to detect and avoid them [29, Ch 10]. But, if this is the case, then DeMillo, Lipton and Perlis, when appropriately understood, are advocating a conception of mathematics according to which (a) classical proofs are practically impossible (so that demonstrations are not ordinarily available), yet (b) worthless fallacies are still unacceptable (so that our conclusions are nevertheless supported). This position strongly suggests the possibility that DeMillo, Lipton and Perlis should be viewed as advocating the conception of mathematics as a domain of inductive procedure.

When consideration is given to the distinguishing characteristics of inductive arguments, this interpretation seems to fit their position quite well. As stated earlier, inductive arguments are (a) nondemonstrative, (b) ampliative, and (c) nonadditive. This means (a) that their conclusions can be false even when their premises are true (permitting the possibility of error); (b) that their conclusions contain some information or content not already contained in their premises (otherwise they would not be nondemonstrative); and, (c) the addition of further evidence in the form of additional premises can either strengthen or weaken these arguments (whether that evidence is discovered days, months, years or centuries later). Thus, in accepting the primality of some very large number N probabilistically, for example, one goes beyond the content contained in the premises (which do not guarantee the truth of that conclusion) and runs a risk of error (which cannot be avoided with probabilistic reasoning). Yet one thereby possesses evidence in support of the truth of such a conclusion, so that its acceptance is warranted to some degree. After all, that might be the best we can do.

MATHEMATICAL PROOFS AND PROBABLE VERIFICATIONS

> Mathematical proofs increase our confidence in the truth of mathematical statements only after they have been subjected to the social mechanisms of the mathematical community. These same mechanisms doom the so-called proofs of software, the long formal verifications that correspond, not to the working

> mathematical proof, but to the imaginary
> logical structure that the mathematician con-
> jures up to describe his feeling of belief.
>
> DeMillo, Lipton and Perlis

Indeed, whether or not we can do better appears to be at the heart of the controversy surrounding this position. DeMillo, Lipton and Perlis, after all, do not explicitly deny the existence of classical bivalent proofs (although they recommend probabilistic proofs as more appropriate to their subject matter). Moreover, they implicitly concede the existence of classical bivalent proofs (insofar as the pursuit of probabilistic proofs may even lead to their discovery). They even go so far as to suggest the social processing of a probabilistic proof, like that of a traditional proof, can involve "enough internalization, enough transformation, enough generalization, enough use, . . ." that the mathematical community accepts it as correct: "The theorem is thought to be true in the classical sense — that is, in the sense that it could be demonstrated by formal, deductive logic, although for almost all theorems no such deduction ever took place or ever will" [8, pp. 273–274].

The force of their position appears to derive from the complexity that confronts those who would attempt to undertake mathematical proofs and program verifications. They report a formal demonstration of a conjecture by Ramanujan would require 2000 pages to formalize. They lament that Russell and Whitehead "in three enormous, taxing volumes, (failed) to get beyond the elementary facts of arithmetic." These specific examples may be subject to dispute: for Ramanujan's conjecture, precisely how is such a fanciful estimate supposed to be derived and verified? For Russell and Whitehead, is *Principia Mathematica* therefore supposed to be a failure? And yet their basic point ("The lower bounds on the length of formal demonstrations for mathematical theorems are immense and there is no reason to believe that such demonstrations for programs would be any shorter or clearer — quite the contrary") nevertheless merits serious consideration.

One of the most important ambiguities to arise within this context emerges at this juncture. While DeMillo, Lipton and Perlis [8, p. 278] suggest that the *scaling-up argument* (the contention that very complex programs and proofs can be broken down into much simpler programs and proofs for the purposes of verification and of demonstration) is the best the other side can produce, they want to deny it should be taken seriously insofar as it is supposed to depend upon an untenable assumption:

The scaling-up argument seems to be based on the fuzzy notion that the world of programming is like the world of Newtonian physics — made up of smooth, continuous functions. But, in fact, programs are jagged and full of holes and caverns. Every programmer knows that altering a line or sometimes even a bit can utterly destroy a program or mutilate it in ways that we do not understand and cannot predict.

Indeed, since this argument is supposed to be the best argument in defense of the verificationist position, what they call "the discontinuous nature of programming" is said to sound "the death knell for verification."

Maurer has strenuously objected to their complaints that the verification of one program can never be transferred to any other program ("even to a program only one single line different from the original") and that there are no grounds to suppose that the verification of a large program could be broken down into smaller parts ("there is no reason to believe that a big verification can be the sum of many small verifications") [28, p. 278]. In response, he has observed that, while the modification of a correct program can produce an incorrect one for which "no amount of verification can prove it correct" and while minute changes in correct programs can produce "wildly erratic behavior . . . if only a single bit is changed," that does not affect the crucial result, namely: that proofs of the correctness of a program can be transferred to other programs, when those other programs are carefuly controlled modifications of the original program, if not in whole then at least in part. Indeed, quite frequently, "if a program is broken up into a main program and n subroutines, we have $n + 1$ verifications to do, and that is all we have to do in proving program correctness."

CUMULATIVE COMPLEXITY AND PATCH-WORK COMPLEXITY

> . . . verifications cannot be internalized, transformed, generalized, used, connected to other disciplines, and eventually incorporated into a community consciousness. They cannot acquire gradual credibility, as a mathematical theorem does; one either believes them blindly, as a pure act of faith, or not at all.
>
> DeMillo, Lipton and Perlis

DeMillo, Lipton and Perlis, however, cannot be quite so readily dismissed, for their contentions on behalf of their position are quite intriguing:

> No programmer would agree that large production systems are composed of nothing more than algorithms and small programs. Patches, *ad hoc* constructions. Band-Aids and tourniquets, bells and whistles, glue, spit and polish, signature code, blood-sweat-and-tears, and, of course, the kitchen sink — the colorful jargon of the practicing programmer seems to be saying something about the nature of the structures he works with; maybe theoreticians ought to be listening to him. [8, p. 277]

Thus, it appears to be because most real software tends to consist of a lot of error messages and user interfaces — *ad hoc*, informal structures that are by definition 'unverifiable' — that they want to view the verificationist position as so far removed from the realities of programming life. But I think the strong arguments advanced on both sides of this issue suggest that two different concepts may be intimately intertwined that should be unraveled, where these involve differing dimensions of the nature of program complexity.

Maurer's position, for example, tends to support the view of complexity according to which more complex programs consist of less complex programs interacting according to some specific arrangement, a conception that is quite compatible with the scaling-up argument that DeMillo, Lipton and Perlis are inclined to disparage. This conception of complexity might be described as:

(C1) Cumulative Complexity = $_{df}$
the complexity of larger programs arising when they consist of (relatively straightforward) arrangements of smaller programs;

as opposed to an alternative conception of complexity that is quite different:

(C2) Patch-Work Complexity = $_{df}$
the complexity of larger programs arising when they consist of complicated, *ad hoc*, peculiar arrangements of smaller programs;

where the verificationist attitude appears to be appropriate to cumulative complexity, but far less adequate (perhaps hopelessly inappropriate) for cases of patch-work complexity, while the anti-verificationist attitude appears to have the opposite virtues in relation to cases of both of these kinds, respectively. Thus, the differences that distinguish large from small programs in the case of cumulative complexity tend to be differences of degree, whereas the differences that distinguish cumulative from patch-work complexity are differences in kind.[7]

If this reconstruction of the situation is approximately correct, then the verificationist approach, in principle, would appear to apply to small programs and to large programs when they exemplify cumulative complexity. However, it would not apply — or, at best, only in part — to those that exhibit patch-work complexity. Moreover, the *in principle* qualification is important here, because DeMillo, Lipton and Perlis offer reasons to doubt that large programs are or ever will be subject to verification, even when they are not 'patch-work programs':

> The verification of even a puny program can run into dozens of pages, and there's not a light moment or a spark of wit on any of those pages. Nobody is going to run into a friend's office with a program verification. Nobody is going to sketch a verification out on a paper napkin. Nobody is going to buttonhole a colleague into listening to a verification. Nobody is ever going to read it. One can feel one's eyes glaze over at the very thought. [8, p. 276]

This enchanting passage is almost enough to beguile one into the belief that a program verification is among the most boring and insipid of all the world's objects. Yet even if that were true — even if it destroys all our prospects for a social process of program verification parallel to that found in mathematics — it would not establish the potential for their production as pointless and without value, if its purpose is properly understood.

Indeed, one of the most important insights that can be gleaned from reviewing this debate is an appreciation of the role of formal demonstrations in both mathematical and in programming contexts. For while it is perfectly appropriate for DeMillo, Lipton and Perlis to accentuate the historical truth that the vast majority of mathematical theorems and computer programs will never be subjected to the exquisite pleasures of formal validation or of program verification, it remains enormously important, as a theoretical possibility, that those theorems and programs could have been or could still be subjected to the critical scrutiny that such a thorough examination would provide. Indeed, from this point of view, the theoretical possibility of subjecting them to rigorous appraisal ought to be regarded as more important than its actual exercise. For it is this potentiality, whether or not it is actually exercised, that affords an objective foundation for the intersubjective evaluation of knowledge claims: z knows that something is the case only when z's belief that that is the case can be supported deductively, inductively, or whatever, not as something that has been done but as something that could be done were it required. This may be called 'an examiner's view' of knowledge [14, p. 319].[8]

THEOREMS, ALGORITHMS, AND PROGRAMS

> Lamport and Maurer display an amazing inability to distinguish between algorithms and programs.
>
> DeMillo, Lipton and Perlis

The argument that has gone before, however, depends upon an assumption that DeMillo, Lipton and Perlis seem to be unwilling to grant: the presumption that programs are like theorems from a certain point of view. That some analogy exists between theorems and programs is a tempting inference, not least of all given their own analogy with mathematics. In Table I we begin with the most seemingly obvious comparison one might make. Thus, from this perspective, proofs are to theorems as verifications are to programs — demonstrations of their truth or correctness.[9] (To avoid ambiguity, we will assume the correctness of a program means 'program correctness' rather than 'specification correctness', even though, in other contexts, specification correctness is important as well. Our concern here is whether or not a program satisfies a certain set of specifications instead of whether those specifications are something a user may want to change at any point.)

This analogy, however, might not be as satisfactory as it initially appears. If we consider that programs can be viewed as functions from

TABLE I
A plausible analogy

	Mathematics	Programming
Objects of Inquiry:	Theorems	Programs
Methods of Inquiry:	Proofs	Verifications

inputs to outputs, the features of mathematical proofs that serve as counterparts to inputs, outputs, and programs, respectively, are premises, conclusions, and rules of inference. Thus, in lieu of the plausible analogy with which we began, a more adequate conception emerges in Table II, where the acceptability of this analogy itself depends upon interpreting programs as well as rules of inference as functions (from a domain into a range). This analogy is still compatible with the first, however, so long as it is understood that rules of inference are used to derive theorems from premises, whereas programs are used to derive outputs from inputs. Thus, to the extent to which establishing that a certain theorem follows from certain premises (using certain rules of inference) is like establishing a certain output is generated by a certain input (by a certain program), proving a theorem is indeed like verifying a program.

TABLE II
A more likely analogy

	Mathematics	Programming
Domain:	Premises	Input
Function:	Rules of inference	Programs
Range:	Theorems	Output

In order to discern the deeper disanalogy between these activities, therefore, it is necessary to realize that while algorithms satisfy the conditions previously specified (of qualifying as functions from a domain to a range), programs need not do so. One reason, of course, emerges from considerations of complexity, especially when programs are patch-work complex programs, because programs of this kind have idiosyncratic features that make a difference to the performance of those programs, yet are not readily amenable to deductive procedures. The principal claim advanced by DeMillo, Lipton and Perlis (with respect to the issue here), is that patch-work complex programs have *ad hoc*, informal aspects such as error messages and user interfaces, that are unverifiable 'by definition' [8, p. 227]. In cases of this kind, presumably, no axioms relating these special features of specific programs to their performance when executed are available, making verification impossible.

A more general reason, however, arises from features of other programs that are obviously intended to bring about the performance of special tasks by the machines that execute them. Illustrations of such tasks include the input and output behavior that is supposed to result as a causal consequence of their execution. Thus, the IBM PC manual for Microsoft BASIC defines various 'I/O Statements' in the following fashion:[10]

(A)	Statement:	Action:
	BEEP	Beeps the speaker.
	CIRCLE (x, y) r	Draws a circle with center x, y and radius r.
	COLOR b, p	In graphics mode, sets background color and pallette of foreground colors.
	LOCATE row, col	Positions the cursor.
	PLAY string	Plays music as specified by string.

Although these 'statement' commands are expected to produce their corresponding 'action' effects, it should not be especially surprising that prospects for their verification are problematic.

Advocates of program verification, however, might argue that commands like these are special cases. They are amenable to verification procedures, not in the sense of absolute verifiability, but rather in the sense of relative verifiability. The programs in which these commands appear could be subject to verification, provided special 'causal axioms' are available relating the execution of these commands to the performance of the corresponding tasks. The third reason, therefore, ought to be even more disturbing for advocates of program verification. For it should be evident that even the simplest and most commonplace program implicitly possesses precisely the same causal character. Consider, for example, the following program, simple, written in Pascal:

```
(B)     program simple (output);
        begin
            writeln ('2 + 2 =', 2 + 2);
        end
```

This program, of course, instructs the machine to write '2 + 2 =' on a line followed by its solution '4' on the same line. For either of these outcomes to occur, however, obviously depends upon various different causal factors, including the characteristics of the compiler, the processor, the printer, the paper and every other component whose specific features influence the execution of such a program.

Taking all together, these considerations support the theoretical necessity to distinguish programs as encodings of algorithms from the logical structures that they represent. A program, after all, is a particular implementation of an algorithm in a form that is suitable for execution by a machine. In this sense, program, unlike an algorithm, qualifies as a causal model of a logical structure of which a specific algorithm may be a specific instance. The consequences of this realization are enormous, insofar as causal models of logical structures need not have the same properties that characterize those logical structures themselves. Algorithms, rather than programs, thus appear to be the appropriate candidates for analogies with *pure mathematics*, while programs bear comparison with *applied mathematics*. Propositions in applied mathematics, unlike those in pure mathematics, run the risk of observational and experimental disconfirmation.

From this point of view, it becomes possible to appreciate why De-Millo, Lipton and Perlis accentuate the role of program testing as follows:

> It seems to us that the only potential virtue of program proving [verification] lies in the hope of obtaining perfection. If one now claims that a proof of correctness can raise confidence, even though it is not perfect or that an incomplete proof can help one locate errors, that that claim must be verified! There is absolutely no objective evidence that program verification is as effective as, say, *ad hoc* theory testing in this regard.

If not for the presumption that programs are not algorithms in some fundamental respects, these remarks would be very difficult — if not impossible — to understand. But when the assumption is made that,

(D1) algorithm = $_{df}$
— a logical structure of the type function suitable for the derivation of outputs when given inputs;

(D2) program = $_{df}$
— a causal model of an algorithm obtained by implementing that function in a form that is suitable for execution by a machine;

it is no longer difficult to understand why they should object to the conception of program verification as an inappropriate and unjustifiable exportation of a deductive procedure applicable to theorems and to algorithms for the purpose of evaluating causal models that are executed by machine.[11]

ABSTRACT MACHINES VERSUS TARGET MACHINES

> I find digital computers of the present day to be very complicated and rather poorly defined. As a result, it is usually impractical to reason logically about their behavior. Sometimes, the only way of finding out what they will do is by experiment. Such experiments are certainly not mathematics. Unfortunately, they are not even science, because it is impossible to generalize from their results or to publish them for the benefit of other scientists.
>
> C. A. R. Hoare

The conception of computer programs as causal models and the difference between programs and algorithms deserve elaboration, especially insofar as there are various senses in which something might or might not qualify as a program or as a causal model. The concept of a program is highly ambiguous, since the term 'program' may be used to refer to (i) algorithms, (ii) encodings of algorithms, (iii) encodings of algorithms that can be compiled, or (iv) encodings of algorithms that can be compiled and executed by a machine. There are other program senses as well.[12] As an effective decision procedure, an algorithm is more abstract than is a program, insofar as the same algorithm might be implemented in different forms suitable for execution by various machines by using different languages. From this perspective, the senses of program defined by (ii), (iii) and (iv) provide conceptual benefits that the sense defined by (i) does not. Indeed, were 'program' defined by sense (i), programs could not fail to be verifiable.

The second sense is of special importance within this context, especially in view of the distinction between 'abstract machines' and 'target machines'. As noted earlier, source programs are written in high-level languages that simulate abstract machines, whose instructions can be more readily composed than can those of the target machines that ultimately execute them. It is entirely possible, in sense (ii), to envision the composition of a program as involving no more than the encoding of an algorithm in a programming language, no matter whether that program is now or ever will be executed by a machine or not. Indeed, it might be said that the composition of a program involves no more than the encoding of an algorithm in a programming language, even if that language cannot be executed by any machine at all. An instance of this state-of-affairs, moreover, is illustrated by the mini-language CORE, introduced by Marcotty and Ledgard [27] as a means for explaining the features characteristic of programming languages in general, without encountering the complexities involved in discussions of Pascal, LISP and so on. In cases of this kind, these languages may reflect properties of abstract machines for which there exist no actual target machine counterparts.

The crucial difference between programs in senses (i) and (ii) and programs in senses (iii) and (iv), therefore, is that (i) and (ii) can be satisfied merely by reference to abstract machines, whereas (iii) and (iv) require the existence of target machines. In the case of a mini-language like CORE, it might be argued that the abstract machine is the target machine. But this contention overlooks the difference at stake here because an abstract machine no more qualifies as a machine than an artificial flower qualifies as a flower. Compilers, interpreters, processors and the like are properly characterized as *physical things*, i.e. as systems in space/time for which causal relations obtain. Abstract machines are properly characterized as *abstract entities*, i.e. as systems not in space/time for which only logical relations obtain. It follows that, in sense (i) and (ii), the intended interpretations of programs are abstract machines that are not supposed to have physical machine counterparts. But in senses (iii) and (iv), the intended interpretations of programs are abstract machines that are supposed to have physical machine counterparts. And this difference is crucial: it corresponds to the intended difference between definitions (D1) and (D2).

On the basis of these distinctions, it should be evident that algorithms — and programs in senses (i) and (ii) — are subject to absolute verification by means of deductive procedures. This possibility occurs because the properties of abstract machines that have no physical machine counterparts can be established by definition, i.e. through stipulations or conventions, which might be formalized either by means of program rules of interference or by means of primitive program axioms. In this sense, the abstract machine under consideration simply *is* the abstract entities and relations thereby specified. By comparison, programs in senses (iii) and (iv) are merely subject to relative verification, at best, by means of deductive procedures. Their differences from algorithms arise precisely because, in these cases, the properties of the abstract machines they represent, in turn, stand for physical machines whose properties can only be established inductively. With programs, unlike algorithms, there are no 'program rules of inference' or 'primitive program axioms' whose truth is ascertainable by definition.

In either case, however, all these rules and axioms relate the occurrence of an input I to the occurrence of an output O, which can be written in the form of claims that input I causes output O more or less as follows:

(C)　　$I \ c \ O$;

thus, given this rule or axiom, the occurrence of output O may be inferred from the occurrence of input I together with a rule or axiom of that form:

(D)　　From '$I \ c \ O$' and 'I' infer to 'O;'

thus, from axiom or rule '$I \ c \ O$' and input 'I', output 'O' validly follows. The difference between algorithms and programs, from this point of view, is that patterns of reasoning of form (D) are absolutely verifiable in the case of algorithms but are only relatively verifiable in the case of programs — a difference that reflects the fact that claims of form (C) can be established by deductive procedures as definitional stipulations with respect to algorithms, but can only be ascertained by inductive procedures, as lawful or causal generalizations, in the case of programs, thus understood.

THE VERY IDEA OF PROGRAM VERIFICATION

> When the correctness of a program, its compiler, and the hardware of the computer have all been established with mathematical certainty, it will be possible to place great reliance on the results of the program, and predict their properties with a confidence limited only by the reliability of the electronics.
>
> C. A. R. Hoare

When entertained from this point of view, the fundamental difficulty encountered in attempting to apply deductive methodology to the verification of programs does not appear to arise from either the idiosyncracy of various features of those programs or from the inclusion of instructions for special tasks to be performed. Both types of cases can be envisioned as matters that can be dealt with by introducing special rules and special axioms that correspond to either the *ad hoc* features of those patch-work complex programs or to the special behavior that is supposed to be exhibited in response to special commands. The specific inputs '*I*' and the specific outputs '*O*', after all, can be taken to cover these special kinds of cases. Let us therefore take this for granted in order to provide the strongest case possible for the verificationist position and thereby avoid any chance of being charged with having attacked a straw man.

As we discovered, the crucial problem confronting program verification is establishing the truth of claims of form (C) above, which might be done in two possible ways. The first is to interpret rules and axioms of form (C) as definitional truths concerning the abstract machine thereby defined. The other is to interpret rules and axioms of form (C) as empirical claims concerning the possible behavior of the target machine thereby described. Interpreted in the first way, the performance of an abstract machine can be conclusively verified, but it possesses no significance at all for the performance of any physical system. Interpreted in the second way, the performance of an abstract machine possesses significance for the performance of a physical system, but it cannot be conclusively verified. And the reason, by now, should be obvious; for programs are subject to 'relative' rather than 'absolute' verification, in relation to 'rules and axioms' in the form of lawful and causal generalizations as premises — empirical claims whose truth can never be established with certainty!

The very idea of program verification trades upon an equivocation. Interpreted in senses (i) and (ii), there is no special difficulty that arises in 'verifying' that output O follows from input I as a logical consequence of axioms of the form, $I \ c \ O$. Under such an interpretation, however, nothing follows from the verification of a 'program' concerning the performance of any physical machine. In this case, the absolute verification of an abstract machine is theoretically possible and is not particularly problematic. Interpreted in senses (iii) and (iv), however, that output O follows from input I as a logical consequence of axioms of the form, $I \ c \ O$, cannot be subject to absolute verification, precisely because the truth of these axioms depends upon the causal properties of physical systems, whose presence or absence is only ascertainable by means of inductive procedures. In this case, the absolute verification of an abstract machine is logically impossible because its intended interpretation is a target machine whose behavior might not be described by those axioms, whose truth can only be established by induction.

This conclusion strongly suggests the conception of programming as a mathematical activity requires qualification in order to be justified. For while it follows from the axioms for the theory of natural numbers that, say,

(E) $2 + 2 = 4$;

the application of that proposition — which may be true of the abstract domain to which its intended interpretation refers — for the purpose of describing the causal behavior of physical things like alcohol and water need not remain true:

(F)   2 units of water + 2 units or alcohol = 4 units of mixture.

For while the abstract proposition (E) is true, the empirical proposition (F) is false. The difference involved here is precisely that between *pure* mathematics and *applied* mathematics.[13] When the function of a program is merely to satisfy the constraints imposed by an abstract machine for which there is no intended interpretation with respect to any physical system, then the behavior of that system can be subject to conclusive absolute verification. This scenario makes Hoare's [20] four basic principles true because then:
(1) computers are mathematical machines;
(2) computer programs are mathematical expressions;
(3) a programming language is a mathematical theory; and,
(4) programming is a mathematical activity.
But if the function of a program is to satisfy the constraints imposed by an abstract machine for which there is an intended interpretation with respect to a physical system, then the behavior of that system cannot be subject to conclusive absolute verification but requires instead empirical inductive investigation to support inconclusive relative verifications. In cases of this kind, Hoare's four principles are false and require displacement as follows:
(1) computers are applied mathematical machines;
(2) computer programs are applied mathematical expressions;
(3) a programming language is an applied mathematical theory; and
(4) programming is an applied mathematical activity;

where propositions in applied mathematics, unlike those in pure mathematics, run the risk of observational and experimental disconfirmation.

COMPUTER PROGRAMS AS APPLIED MATHEMATICS

> A geometrical theory in physical interpretation can never be validated with mathematical certainty, no matter how extensive the experimental tests to which it is subjected; like any other theory of empirical science, it can acquire only a more or less high degree of confirmation.
>
> C. G. Hempel

The differences between pure and applied mathematics are very great, indeed. As Einstein remarked, insofar as the laws of mathematics refer to reality, they are not certain; and insofar as they are certain, they do not refer to reality. DeMillo, Lipton and Perlis likewise want to maintain that, to the extent to which verification has a place in programming practice, it applies to the evaluation of algorithms; and to the extent to which programming practice goes beyond the evaluation of algorithms, it cannot rely upon verification. Indeed, from the perspective of the classical theory of knowledge, their position makes excellent sense; for the investigation of the properties of programs (thus understood) falls within the domain of inductive methodology, while the investigation of the properties of algorithms falls within the domain of deductive methodology. As we have discovered, these are not the same.

Since the behavior of algorithms can be known with certainty (within the limitations of deductive procedure), but the behavior of programs can only be known with uncertainty (within the limitations of inductive procedure), the degree of belief (or 'strength of conviction') to which specific algorithms and specific programs are entitled can vary greatly. In particular, a hypothetical scale once again may be constructed, where, to the extent to which a person can properly claim to be rational with respect to his beliefs, there should be an appropriate correspondence (which need not necessarily be an identity) between his degree of subjective belief that something is the case (as displayed by Figure 2) and the measure of objective evidence in its support (as displayed by Figure 1). Otherwise, such a person does not distribute those degrees of

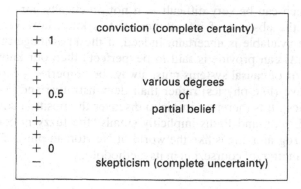

```
  −              conviction (complete certainty)
+ 1
+
+
+
+              various degrees
+ 0.5                of
+              partial belief
+
+
+ 0
  −              skepticism (complete uncertainty)
```

Fig. 2. A measure of subjective belief.

belief in accordance with the available evidence and is to that extent irrational [11, Ch. 10].

The conception of a program as a causal model suitable for execution by a machine reflects the interpretation of programs as causal factors that interact with other causal factors to bring about a specific output as an effect of the introduction of specific input. As everyone appears willing to admit, the execution of a program qualifies as causally complex, insofar as even a correct program can produce "wildly erratic behavior . . . if only a single bit is changed." The reason the results of executing a program cannot provide deductive support for the evaluation of a program [as Hoare 20, p. 116 acknowledges], moreover is that the behavior displayed by a causal system is an effect of the complete set of relevant factors whose presence or absence made a difference to its production. Indeed, this reflection has an analog with respect to inductive procedure, since it is a fundamental principle of inductive methodology that measures of evidential support should be based upon the complete set of relevant evidence that is currently available, where any finding whose truth or falsity increases or decreases the evidential support for a conclusion is evidentially relevant.[14]

The principle of maximal specificity for causal systems and the principle of total evidence for inductive methodology are mutually reinforcing: reasoning about programs tends to be (a) non-demonstrative, (b) ampliative, and (c) non-additive precisely because the truth of a generalization about a causal system depends upon *its complete specifi-*

cation, which can be very difficult — if not practically impossible — to obtain. In the absence of information of this kind, however, the best knowledge available is uncertain. Indeed, if the knowledge that deductive warrants can provide is said to be 'perfect', then our knowledge of the behavior of causal systems must always be 'imperfect', experimental and tentative (like physics) rather than demonstrative and certain (like mathematics). It is therefore ironic to discover the position advanced by DeMillo, Lipton and Perlis implicitly entails "the fuzzy notion that the world of programming is like the world of Newtonian physics" — not in its subject matter, of course, but in its methodology.

INDUCTIVE TESTING AND COMPUTER PROGRAMS

> Could the God that plays dice trigger a nuclear holocaust by a random error in a military computer?
>
> H. A. Pagels

At least two lines of defense might be advanced against this conclusion, one of which depends upon the possibility of an *ideal* programmer, the other upon the prospects for verification by *machine*. The idea of an ideal programmer is that of a programmer who knows as much about algorithms, programming and computers as there is to know. When this programmer is satisfied with a program, by hypothesis that program is 'correct'. The catch is two-fold. First, how could any programmer possess the knowledge required to be an 'ideal programmer'? Unless this person were God, we may safely assume the knowledge he possesses has been ascertained by means of the usual methods, including (fallible) inductive reasoning about the future behavior of complex systems based upon evidence about their past behavior. Second, even a correct program is but one feature of a complex causal system. The performance of a computer while executing a program depends not only upon the software but also upon the firmware, the hardware, the input/output devices and all the rest. While it would not be mistaken to suggest that, *ceteris paribus*, if these other components perform as they should, the system will perform as it should, such claims are not testable.

The emphasis here is on the word 'should'. Since the outputs that result from various inputs are complex effects of an interacting arrangement of software, firmware, hardware, and so on, the determination that some specific component of such a system functions properly ('as it should') depends upon assumptions concerning the specific states of each of the other, in the absence of which, strictly speaking, no program as such can be subject to test. Even when the specific states of the relevant components have been explicitly specified, the production of output O given input I on one occasion provides no guarantee that output O would be produced by input I on another occasion. Indeed, the type of system created by the interaction of these component parts could be probabilistic rather than deterministic. Repeated tests of any such system can provide only inductive evidence of reliability. Taken together, these considerations suggest that even the idea of an ideal programmer cannot improve the prospects for program verification.[15]

A fascinating example of the kind of difficulty that can be encountered is illustrated by the distinction between 'hard' and 'soft' errors, which occur when quantum phenomena are implicated in situations like these:

In the 1980s a new generation of high-speed computers will appear with switching devices in the electronic components which are so small they are approaching the molecular microworld in size. Old computers were subject to 'hard errors' — a malfunction of a part, like a circuit burning out or a broken wire, which had to be replaced before the computer could work properly. But the new computers are subject to a qualitatively different kind of malfunction called 'soft errors' in which a tiny switch fails during only one operation — the next time it works fine again. Engineers cannot repair computers for this kind of malfunction because nothing is actually broken. [31, p. 125]

It should be clear by now that the difference between proving a theorem and verifying a program does not depend upon the presence or absence of any social process during their production, but rather upon the presence or absence of causal significance. A summary of their parallels is illustrated in Table III. The chart shows the difference in social processes is merely a difference in practice, but the difference in causal significance is a real difference in principle.

The suggestion can also be made that program verification might be performed by higher-level machines that have the capacity to validate proofs that are many orders of magnitude more complex than those that can be mechanically verified today. This possibility implicitly raises issues of mentality concerning whether or not computers have the capacity for semantic interpretation as well as for syntactic manipulation [12]. Even assuming that their powers are limited to string pro-

TABLE III
A final comparison

	Proving theorems	Verifying programs
Syntactic objects of inquiry:	Yes	Yes
Social process of production:	Yes	No
Physical counterparts:	No	Yes
Causal significance:	No	Yes

cessing, this is an intriguing prospect, especially since proofs and programs alike are syntactic entities. There are no special difficulties so long as their intended interpretations are abstract machines. When their intended interpretations are target machines, then we encounter the problem of determining the reliability of the verifying programs themselves ("How do we verify the verifiers?"), which invites a regress of relative verifications of relative verifications. As long as our reasoning is valid, causally significant conclusions can be derived only from causally significant premises.

Tymoczko [36], however, suggests that proof procedures in mathematics have three distinctive features: they are convincing, they are formalizable and they are surveyable. His sense of 'surveyability' can be adequately represented by 'replicability'. Proofs are convincing to mathematicians because they can be formalized and replicated. Insofar as the machine verification of programs has the capacity to satisfy the desiderata of formalizability and of replicability, their successfully replicated results ought to constitute evidence for their correctness.[16] This, of course, does not alter the inductive character of the support thereby attained nor does it increase the range of the potential application of program verification, but it would be foolish to doubt the importance of computers in extending our reasoning capabilities, just as telescopes, microscopes, and all the rest have extended our sensible capacities: verifying programs, after all, could be published (just as proofs are published) in order to be subjected to the criticism of the community — even by means of computer trials! Even machine verifications, however, cannot guarantee the performance that will result from executing a program, which is yet one more form of our dilemma.[17]

COMPLEXITY AND RELIABILITY

> We must therefore come to grips with two problems that have occupied engineers for many generations: First, people must plunge into activities that they do not understand. Second, people cannot create perfect mechanisms.
>
> DeMillo, Lipton and Perlis

From this perspective, the admonitions advanced by DeMillo, Lipton and Perlis against the pursuit of perfection when perfection cannot be realized are clearly telling in this era of dependence upon technology. There is little to be gained and much to be lost through fruitless efforts to guarantee the reliability of programs when no guarantees are to be had. When they assess the situation with respect to critical cases (such as air-traffic control, missile systems, and the like) in which human lives are at risk, the ominous significance of their position appears to be overwhelming:

... the stakes do not affect our belief in the basic impossibility of verifying any system large enough and flexible enough to do any real-world task. No matter how high the payoff, no one will ever be able to force himself to read the incredibly long, tedious verifications of real-life systems, and, unless they can be read, understood, and refined, the verifications are worthless.

Thus, even when allowance is made for the possibility of group collaboration, the mistaken assumption that program performance can be guaranteed could easily engender an untenable conception of the situation encountered when human lives are placed in jeopardy as in the case of SDI. If one were to assume the execution of a program could be anticipated with the mathematical precision that is characteristic of demonstrative domains, then one might more readily succumb to the temptation to conclude that decisions can be made with complete confidence in the (possibly unpredictable) operational performance of a complex causal system.

Complex systems like SDI are heterogeneous arrangements of complicated components, many of which combine hardware and software. They depend upon sensors providing real-time streams of data where, to attain rapid and compact processing, their avionics portions are programmed in assembly language — a type of programming that does not lend itself to the construction of program verifications. Systems like these differ in important respects from, say, brief programs that read an ASCII file in order to count the number of characters or words per line. Any analysis of computer-system reliability that collapses these diverse types of systems and programs into a single catch-all category would be indefensibly oversimplified. Therefore it should be emphasized that, in addition to the difference between absolute and relative verifiability with respect to programs themselves, the reliability of inductive testing — not to mention observation techniques — declines as the complexity of the system increases. As a rule of thumb, the more complex the system, the less likely it is to perform as desired [22]. The operational performance of these complex systems should never be taken for granted and cannot be guaranteed.

The blunder that would be involved in thinking otherwise does not result from the presence or the absence of social processes within this field that might foster the criticism of a community, but emanates from the very nature of these objects of inquiry themselves. The fact that one or more persons of saintly disposition might sacrifice themselves to the tedium of eternal verification of tens of millions of lines of code for the benefit of the human race is beside the point. The limitations involved here are not merely practical: *they are rooted in the very character of causal systems themselves.* From a methodological point of view, it might be said that programs are conjectures, while executions are attempted — and all too frequently successful — refutations (in the spirit of Popper [32, 33]). Indeed, the more serious the consequences that would attend the making of a mistake, the greater the obligation to insure that it is not made. In maintaining that program verification cannot succeed as a generally applicable and completely reliable method for guaranteeing the performance of a program, DeMillo, Lipton and Perlis thus arrived at the right general conclusion for the wrong specific reasons. Still, we are all indebted to them for their efforts to clarify a confusion whose potential consequences — not only for the community of computer science, but for the human race — cannot be overstated and had best be understood.

ACKNOWLEDGEMENTS

The original version of this paper was composed while a Post-Doctoral Fellow in computer science at Wright State University. Special thanks to Henry Davis, David Hemmendinger, Jack Kulas and especially Al Sanders for encouraging a philosopher to poke around in their backyard. I am also indebted to two very conscientious but anonymous referees for this magazine and to Rob Kling and Chuck Dunlop for their stimulating criticism and penetrating inquiries, which forced me to clarify the basis for my position.

University of Minnesota,
Duluth, U.S.A.

NOTES

[1] Popper advocates the conception of science as an objective process of 'trial and error' whose results are always fallible, while Kuhn emphasizes the social dimension of scientific communities in accepting and rejecting what he calls 'paradigms'. See, for example, [23], [32—33]. A fascinating collection of papers discussing their similarities and differences is presented in [25].

[2] The point is that there appear to be three kinds of evidence that can be advanced in support of the truth of a sentence, two of which occur in the form of other sentences (whose truth may require their own establishment). While descriptions of the data of more-or-less direct experience presupposes causal interactions between language-users and the world, the construction of arguments does not. [11 esp. pp. 18—24]

[3] For an exceptionally lucid discussion of the differences between so-called natural deduction systems and axiomatic formal systems, where the former operate with sets of inference rules in lieu of formal axioms but the latter operates with sets of axioms and as few as a single rule, see [3].

[4] The claim that a proposition has been 'verified', therefore, can mean more than one thing, since some derivations establish that a claim is a theorem (which cannot possibly be false), while other derivations establish that a claim is a conclusion (which cannot be false, so long as its premises are true). Derivations of both types are equally 'valid', of course, but only the former show that their conclusions cannot be false.

[5] This point differs from the previous note. That the valid conclusion of an argument from true premises cannot be false does not mean that we can know, in general, whether or not a valid argument does or does not have true premises. Thus, the difficulty we confront, from an epistemic point of view, is establishing the truth of our premises as well as the validity of our arguments. The importance of this difference is great.

[6] It could be argued, of course, that even these purely syntactical entities may be viewed as having semantical relations to some domain of abstract entities. In the case of pure mathematics, these entities might be variously entertained as Platonic, as Intuitive, or as Conventional. See, for example, the discussions found in [1] The key issue, however, is not whether there might be some such abstract domain of interpretation but whether there is any such physical domain.

[7] Moreover, differences in degree with respect to measurable magnitudes — such as

heights, weights, and so forth — are not precluded from qualifying as differences in kind. Indeed, to suppose that there cannot be any 'real' difference between two things merely because there are innumerable intermediate degrees of difference between them is known as the fallacy of the continuum. The distinction recommended here is appropriate either way.

8   A fascinating illustration of this attitude may be found in Holt [21, pp. 24—25], who discusses what he takes to be "the three C's" of formal program specification: clarity, completeness, and consistency. He suggests that the formal verification of program correctness represents an idealization, "(much as certain laws of physics are idealizations of the motion of bodies of mass). This idealization encourages careful, proficient reasoning by programmers, *even if they choose not to carry out the actual mathematical steps in detail*." Careful, proficient reasoning by programmers, however, is presumably attainable by methods other than program verification procedures.

9   There are other aspects of specific programs, of course, including heuristics, especially, where the role of heuristics as 'rules of thumb' or as guidelines that allow exceptions but are useful, nonetheless, could be the subject of further inquiry (emphasizing, for example, that discovering heuristics is an inductive enterprise). Here, however, the focus is upon the difference between programs and algorithms as objects of verification.

10   I am grateful to my colleague, David Cole, for proposing these examples.

11   The phrase 'causal model', of course, has been bestowed upon entities as diverse as *scientific theories* (such as classical mechanics and relativity theory), *physical apparatus* (such as arrangements of ropes and pulleys), and even *operational definitions* (such as that IQs are what IQ tests test). Different disciplines tend to generate their own special senses. For discussion, see [15, 18, 34].

12   A different set of five distinctive senses of 'program' is suggested by [30]. Those cited here, however, appear appropriate for our purposes.

13   On the difference between pure and applied mathematics, see especially Hempel [16, 17] which are as relevant today as they were then.

14   That every property whose presence or absence makes a difference to the performance of a causal system must be taken into account in order to understand its behavior is known as *the requirement of maximal specificity*; that inductive reasoning must be based upon all the available relevant evidence is known as *the requirement of total evidence* [11].

15   An alternative position could emphasize the social processing whereby more than one 'very good programmer' might come to accept or to believe in the correctness of a program. As we have already discovered, this will not do. After all, the claim, "This program reflects what programmer z believes are the right commands," could be true, when the sentence, "This program reflects the right commands," happens to be false. The first is a claim about z's *beliefs*, the later about a *program* — even when z is a programming group.

16   For an intriguing discussion of these questions in relation to the proof-by-computer of the four-color problem, see [9, 35 and 36].

17   As Cerutti and Davis [5, pp. 903—904], have observed, "For machine proofs, we can (a) run the program several times, (b) inspect the program, (c) invite other people to inspect the program or to write and run similar programs. In this way, if a common

result is repeatedly obtained, one's degree of belief in the theorm goes up". An interesting discussion of their position may be found in [9, esp., pp. 805—807]; but Detlefsen and Luker beg the question in assuming that people are simply a different kind of computer, a crucial question on which they shed no light.

REFERENCES

1. Benacerraf, P. and Putnam, H., (Eds.): 1964, *Philosophy of Mathematics: Selected Readings*. Prentice-Hall, Englewood Cliffs, N.J.
2. Black, M.: 1967, 'Induction', *The Encylopedia of Philosophy* 4, Edwards, P., Editor-in-Chief. Macmillan, New York, pp. 169—181.
3. Blumberg, A.: 1967, 'Logic, modern', *The Encyclopedia of Philosophy* 5, Edwards, P., Editor-in-Chief, Macmillian, New York, pp. 12—34.
4. Bochner, S.: 1966, *The Role of Mathematics in the Rise of Science*. Princeton Univ. Press, Princeton, N.J.
5. Cerutti, E. and Davis, P.: 1969, 'Formac meets Pappus', *Am. Math. Monthly* 76, 895—904.
6. Church, A.: 1959, 'Logistic system', *Dictionary of Philosophy*. Runes, D., (Ed.) Littlefield, Adam & Co., Ames, Iowa, pp. 182—183.
7. Dancy, J.: 1985, *An Introduction to Contemporary Epistemology*. Blackwell, Oxford.
8. DeMillo, R., Lipton, R. and Perlis, A.: 1979, 'Social processes and proofs of theorems and programs', *Commun. ACM 22*, 5, 271—280.
9. Detlefsen, M. and Luker, M.: 1980, 'The four-color theorem and mathematical proof', *J. Philos.* 77, 12, 803—820.
10. Dijkstra, E. W.: 1976, *A Discipline of Programming*. Prentice-Hall, Englewood Cliffs, N.J.
11. Fetzer, J. H.: 1981, *Scientific Knowledge*. Reidel, Dordrecht, Holland.
12. Fetzer, J. H.: 1988, 'Signs and minds: An introduction to the theory of semiotic systems', in *Aspects of Artificial Intelligence*, Fetzer, J., (Ed.) Kluwer, Dordrecht/Boston/London/Tokyo, pp. 133—161.
13. Glazer, D.: 1979, 'Letter to the editor', *Commun. ACM 22*, 11, 621.
14. Hacking, I.: 1967, 'Slightly more realistic personal probabilities. *Philos. Sci. 34*, 4, 311—325.
15. Heise, D. R.: 1975, *Causal Analysis*. Wiley, New York.
16. Hempel, C. G.: 1949, 'On the nature of mathematical truth', in *Readings in Philosophical Analysis*, Feigl, N. and Sellars, W., (Eds.) Appleton-Century-Crofts, New York, pp. 222—237.
17. Hempel, C. G: 1949, 'Geometry and empirical science', in *Readings in Philosophical Analysis*, Feigl, H. and Sellars, W., (Eds.) Appleton-Century-Crofts, New York, pp. 238—249.
18. Hesse, M.: 1966, *Models and Analogies in Science*, Univ. of Notre Dame Press, Notre Dame, Ind.
19. Hoare, C. A. R.: 1969, 'An axiomatic basis for computer programming', *Commun. ACM 12*, 576—580, 583.

20. Hoare, C. A. R.: 1986, 'Mathematics of programming', *BYTE*, 115—149.
21. Holt, R.: 1986, 'Design goals for the Turing programming language. Technical Report CSRI-187', Computer Systems Research Institute, Univ. of Toronto.
22. Kling, R.: 1987, 'Defining the boundaries of computing across complex organization', in *Critical Issues in Information Systems*, Boland, R. and Hirschheim, R. (Eds.). Wiley, New York.
23. Kuhn, T. S.: 1970, *The Structure of Scientific Revolutions*, 2d ed. Univ. of Chicago Press, Chicago.
24. Lakatos, I.: 1976, *Proofs and Refutations*, Cambridge Univ. Press, Cambridge, U.K.
25. Lakatos, I., and Musgrave, A. (Eds.): 1970, *Criticism and the Growth of Knowledge*, Cambridge Univ. Press, Cambridge, U.K.
26. Lamport, L.: 1979, 'Letter to the editor', *Commun. ACM 22*, 11, 624.
27. Marcotty, M. and Ledgard, H.: 1986, *Programming Language Landscape: Syntax/Semantics/Implementations*, 2d ed. Science Research Associates, Chicago.
28. Maurer, W. D.: 1979, 'Letter to the editor', *Commun. ACM 22*, 11, 625—629.
29. Michalos, A.: 1969, *Principles of Logic*, Prentice-Hall, Englewood Cliffs, N.J.
30. Moor, J. H.: 1988, 'The pseudorealization fallacy and the Chinese room', in *Aspects of Artificial Intelligence*, Fetzer, J. (Ed.) Kluwer, Dordrecht/Boston/London/Tokyo, pp. 35—53.
31. Pagels, H.: 1982, *The Cosmic Code*, Simon & Schuster, New York.
32. Popper, K. R.: 1965, *Conjectures and Refutations*, Harper & Row, New York.
33. Popper, K. R.: 1972, *Objective Knowledge*, Clarendon Press, Oxford.
34. Suppe, F. (Ed.): 1977, *The Structure of Scientific Theories*, 2d ed. University of Illinois Press, Urbana, Ill.
35. Teller, P.: 1980, 'Computer proof', *J. Philos. 77*, 12, 797—803.
36. Tymoczko, T.: 1979, 'The four-color theorem and its philosophical significance', *J. Philos. 76*, 2, 57—83.
37. van den Bos, J.: 1979, 'Letter to the editor', *Commun. ACM 22*, 11, 623.

AVRA COHN

THE NOTION OF PROOF IN HARDWARE
VERIFICATION

1. INTRODUCTION

The verification of hardware systems has recently become an attractive application area for theorem provers for several reasons. First, hardware verification is in many ways a more tractable problem than software (program) verification — it is often easier to write a clear specification that captures the functionality of a system of hardware than of software — and hardware proofs tend to have a certain uniformity of structure which is well suited to mechanical treatment. Second, compelling economic reasons exist for trying to get hardware correct early on; correcting errors in a chip can involve expensive refabrication, not merely the exiting of text. Finally, it is becoming increasingly important to invest time and effort in the verification of hardware that is intended for *safety-critical* applications.

Several issues pertaining to the scope and limitations of verification have recently been raised by the project to formally verify aspects of the Viper microprocessor at the University of Cambridge. In this article, some of the issues are discussed; although this discussion is presented in the context of Viper, the remarks are more general. The intention is to encourage intelligent understanding of the sense in which a piece of hardware can be called 'verified', and not to undermine research in verification or to discredit this vital work. As devices begin to appear on the market purporting to be 'verified' or 'mathematically proved' — possibly with the implication that they cannot therefore fail — a sharp watch must be kept for unqualified claims and also for failures to convey the sense, extent, and nature of the verification effort. This observation applies particularly to very hazardous applications such as nuclear power plant control.

Various computing systems in recent years have been claimed to do verification or proof, or something akin, when in fact they were doing something distinct from (if no less valuable than) formal proof in an explicit and well-understood logic. That is, they have done simulation, or reasoning in *ad hoc* logical systems, or informal reasoning, etc. For

359

Timothy R. Colburn et al. (eds.), Program Verification, 359—374.
© 1993 *Kluwer Academic Publishers.*

present purposes, 'verification' is taken to mean formal proof in the usual mathematical sense of a sequence of valid inference steps.

2. THE VIPER MICROPROCESSOR

Viper [8—11, 20, 24] is a microprocessor designed by W. J. Cullyer, C. Pygott, and J. Kershaw at the Royal Signals and Radar Establishment of the U.K. Ministry of Defence (henceforth RSRE) for use in safety-critical applications such as civil aviation and nuclear power plant control. Viper chips are now commercially available. They are currently finding uses in military areas such as the deployment of weapons from tactical aircraft [12] and in civilian areas such as railway signalling systems. To support safety-critical applications, Viper has a particularly simple design; for example, interrupts in the usual sense are not permitted; the instruction set is kept to a minimum; and the machine is designed to stop if it detects itself in an error or illegal configuration. (The stopping feature is intended to support the running of several Vipers simultaneously for increased reliability.) The simplicity of the design makes it amenable to formal analysis using current techniques.

Aspects of the formal, mechanical verification of Viper were sub-contracted to the Hardware Verification Group at the University of Cambridge from early 1986 to late 1987. The results of this project are reported in [6] and [7]. A pilot study for the main proof is reported in [5].

3. THE HOL VERIFICATION SYSTEM

The verification of Viper has been approached within the HOL (Higher-Order Logic) system [2, 14, 15], a theorem-proving system derived from R. Milner's LCF (Logic for Computable Functions) system [13, 23] and based on the version of higher-order logic formulated by A. Church [3]. HOL was implemented by M. Gordon at the University of Cambridge and is currently in use by the Hardware Verification Group at Cambridge and at several sites throughout the world. 'Verification' was understood by Viper's designers at RSRE (as by the LCF and HOL communities) to mean complete, formal proof in an explicit and well-understood logic. Proofs in HOL are normally constructed interactively, combining machine assistance with user

NOTION OF PROOF IN HARDWARE VERIFICATION   361

guidance, and not fully automatically (although the extent to which this
is so is a function of user-designed proof strategies).

4. THE MODELS OF VIPER

The designers of Viper, who deserve a great deal of credit for the
promotion of formal methods, intended from the start that Viper be
formally verified. Their approach was to specify Viper in a hierarchy of
decreasingly abstract levels, each of which concentrated on some
specific aspects of the design. That is, each level was to be a specification of the more abstract level above it (if any) and an implementation of the one below (if any). The verification effort would then be
simplified by being structured according to the abstraction levels. These
levels of description were characterized by the design team at RSRE.
The first two levels, and part of the third, were written by them in a
logical language amenable to reasoning and proof (a predecessor of
HOL's higher-order logic). (The systematic study of abstraction
hierarchies and mechanisms in the modeling of hardware is discussed
by T. Melham [21, 22].)

The highest level specification of Viper is a simple state-transition
function describing the way in which an abstract state (representing
Viper's memory and its visible registers) changes as Viper executes
each of its possible instruction types (see [8] and [6] for details). The
specification is thus an operational semantics of the instruction set. It
characterizes no more than the fetch-decode-execute cycle of Viper; it
does not specify all of the possible behaviors of the actual microprocessor. In particular, the capacity of Viper to be *reset* externally by
an operator (i.e. to have its registers cleared from the outside) is not
covered, nor is its capacity for 'timing out' after some fixed number of
clock cycles because of memory failure. As will be discussed, any verification of a more concrete model relative to this top-level specification
must consequently be limited to the behavior manifest at the top level.
That is, no such proof can establish that Viper resets or times-out in an
acceptable way.

The next (more concrete) level is called the *major-state* level. At this
level, an instruction is processed via a sequence of events rather than in
a single step. An event may affect the visible registers or the memory of
the top-level specification, *or* any of several internal registers (which

constitute the *internal state*). These internal registers are still part of an abstract view of Viper and do not necessarily correspond to parts of the actual Viper chip. The 'next' event in a sequence is determined according to the current event, the visible state, and the internal state; some events are recognizably terminal and some initial, in the sequences. From all of this, a new state-transition function can be extracted and its properties established by proof. In particular, the cumulative result of the sequence of events that processes each of the instruction types can be inferred, and then compared to the result of the corresponding high-level state-transformation function.

The 'block model' is the most concrete level considered in the formal verification project (although the RSRE design continues down to the gate-level circuit design, which could in principle also be formally verified). The block model was presented by the designers in a form that was partly pictorial and partly textual (expressed in a functional style). The model consists of 'blocks', that is, computational units such as Viper's instruction decoder, its arithmetic-logic unit (ALU), and its memory. Information passes between blocks, and to/from the outside world at fixed clock cycles. The textual specifications describe only the internal combinational logic of the various blocks. Neither their behavior over time (e.g. the delay units — registers — which give them memory) nor the connections between separate blocks are covered by the textual specifications: the pictorial specification fills in the rest of that information. The block definitions and the pictorial information were supplied by the designers at RSRE; a fully formal specification had to be constructed from these sources — a complicated process — and the block machine's behavior had to be logically inferred from that specification.

The block model isolates the computational behavior of Viper; to relate it to either of the more abstract specifications (the top-level specification or the major-state model), computational behaviors such as additions, shifts, negations, and comparisons had to be considered in detail. The block model also specifies more of the actual behaviors of Viper (e.g. the behavior of resets and time-outs) than appear in the top-level formal specification. At the block level, one begins to approach the functional units and connectivity of the actual circuit, though still in a rather abstract way.

5. THE VIPER VERIFICATION PROJECT

The correctness proof of the major-state level of Viper relative to its top-level specification was straightforward (if lengthy) in HOL, since the possible execution sequences of the model were explicitly given. That is, the conditions under which one event followed another were explicitly defined. The proof consisted, therefore, of a number of cases (one for each instruction type) in which the cumulative effects of the sequence of events processing that instruction type were inferred from the definition of the model. In each case, the effects were then proved equivalent to the effects specified at the top level for components appearing at both levels. It also had to be proved that every possible execution sequence had been considered, to justify the case analysis.

The correctness of the block model is more difficult to establish. The first task in the verification effort was to derive a mathematical function representing the behavior of the block over a single cycle. This had to be expressed in a formal logic suitable for reasoning and proof, since it is not obvious how to reason formally about a schematic diagram indicating the transfer of information to and from its subunits simultaneously.

The second task (analogous to, but more complex than that in the major-state level proof) was to infer, using the mathematical function for each instruction type, the accumulated effects on the registers of the block model after all of the steps that process that instruction had been performed. This involved extracting from the formal representation (i) the conditions under which one step led to another, and (ii) any assumptions that had to be made about initial states and 'normal' behavior in order to resolve the state transitions. (These were implicitly determined by the mathematical representation of the block model; at the major-state level, the conditions were given explicitly, and no assumptions were required.) Normal behavior means behavior that is within the scope of the high-level specification. For example, as mentioned, it had to be assumed in the verification of the Viper block model that the machine was not reset at any time during the course of processing an instruction, and that the block machine's time-out facility was never invoked. The initial conditions, for example, had to include the assumptions that at the start of processing each instruction (i) the time-out counter was not set at its maximum value, and (ii) no errors were being signalled. It also had to be shown, as before, that the state-transition conditions covered all logical possibilities, to ensure that no possible instruction types had been omitted from the analysis.

The third task would be to verify the results of the block model relative to the results of the top-level transformation function at each instruction type. The first and second tasks of the block model verification have been completed to date, giving a provably correct and complete description of the behavior of the formal representation of the block model (under the assumptions mentioned above); but for practical reasons, the third task has not been completed. All three tasks are discussed in [7].

6. LIMITATIONS OF PROOF IN HARDWARE VERIFICATION

The notion of formal proof began to receive serious attention in its own right just before the age of computing. Since computers have been used to assist with formal proofs, there has been renewed discussion of what proof is and what it actually ensures. This may be in part because there is no prior reason to insist that machines construct proofs in the *way* that mathematicians do; nor is there yet any well-agreed 'standard of evidence' that a proof has been successfully completed by a machine, of the sort that mathematicians are required to supply. In this section, attention is drawn to some of the fresh concerns that have been raised by the Viper verification project.

6.1. *Chips And Intentions Cannot Be Verified*

Ideally, one would like to prove that a chip such as Viper correctly implemented its intended behavior in all circumstances; we could then claim that the chip's behavior was predictable and correct. In reality, neither an actual *device* nor an *intention* is an object to which logical reasoning can be applied. The intended behavior rests in the minds of the architects and is not itself accessible. It can, of course, be reported in a formal language — but not with checkable accuracy. Similarly, a material device can only be observed and measured; it cannot be verified. Again, a device can be described in a formal way, and the description verified; but as with intentions, there is no way to assure the accuracy of the description. Indeed, any description is bound to be inaccurate in *some* respects, since it cannot be hoped to mirror an entire physical situation even at an instant, much less as it evolves through time; a model of a device is necessarily an abstraction (a simplification). In short, verification involves a pair of *models* that bear an uncheckable and possibly imperfect relation to the intended design and to the actual device.

Although these points seem obvious, they are not merely philosophical quibbles. Errors were found both in the top-level specification of Viper and in its major-state model, none of which was either intended by the designers *or* evident in the manufactured Viper chips. (These errors are discussed in [6].) The errors were fairly minor and quickly repaired, but their presence highlights the rather limited sense in which a formal specification can be said to have been verified against the architects' intended design or against the actual chip; there remains the danger that — secure as the proof may be — the models themselves may be wrong.

There is no complete solution to this problem, but there are avenues of approach to be explored. In particular, as we produce clearer and more concise and readable abstract specifications, their intuitive plausibility should be increased. At the other extreme, as we devise more realistic and detailed models, their correspondence with actual devices should become more convincing. Attention has been drawn to these points by T. Melham [21].

6.2. *Links Between Designer, Verifier, And Manufacturer*

That the actual Viper chips appear not to suffer from the errors found in the models also illustrates the still quite abstract nature of the research described in [6] and [7]. The chips were already in the process of being built by the time the subcontracted verification work began on the major-state model at Cambridge; and they had been built and were being advertised by the time the work on the block model was undertaken. While it is possible in theory that an error in an abstract specification had been reflected in the circuit design given by RSRE to the manufacturers — the abstract specifications were no doubt in the architects' minds while they designed the circuit — it seems more likely, because of the indirect links between the designers' abstract specifications, the circuit design process, the manufacturers, and the verifiers, that problems in the specification would *not* propagate down to chip problems. In fact, it would seem to be the case that the manufacturers worked from different 'design texts' than the verifiers. Until common models in a common language are adopted, we are only studying models that bear an informal connection to the devices they are modelling. In this respect too, there is good reason to hope that a common language will be agreed on and an integrated approach taken in the future.

6.3. *The Lack Of A Fully Formal Description*

At more concrete levels of description, the situation may be further complicated by not beginning with fully formal descriptions. For example, Viper's top-level specification and its major-state level were both supplied in a logical language; but its block-level model was given partly formally and partly pictorially (as was natural). Combining these two parts required both ingenuity and some guesswork. The guesses were based on the coincidence of line names, on the names of bound variables in function definitions, and on annotations in the text of the definitions. None of these notational devices can be regarded as a formal specification. Before verification can be meaningfully applied in such cases, a fully formal description must be produced. Once again, however, accuracy cannot be checked; the new formal description may be a flawed translation of the pictorial specification, or a flawed combination of picture and text, but this cannot be rigorously tested. One may therefore end up proving properties of a formal description bearing an imperfect relation to the intended design — and possibly never know it.

In fact, this *was* a problem in the block-level representation of Viper; in the author's first attempt at a formal representation of the Viper block diagram, there was a pair of interchanged line names. This flawed description was subsequently used to deduce plausible-looking block results. The error in the representation was discovered rather late in the proof and only by an unsystematic inspection. The problem of the accuracy of a representation could appear at the gate level, the transistor level, or any other level at which a linguistic description has to be constructed creatively from a pictorial one, i.e. at which diagrams are the usual and natural model of specification. This situation further limits the sense in which a system can be called verified.

This problem is at least partly addressed by the preceding section; if the designers, for example, are in a position to read and scrutinize the formal description derived from the informal specification, they may well be able to spot mistakes, particularly those that require a deep understanding of the design. For example, a typographical error in the formal representation caused some results to be deduced that the designers at RSRE queried. They were then able to locate the author's error.

6.4. *The Level And Completeness Of The Model*

As verification relates a less abstract implementation to a more abstract specification, it is important to be explicit about the level of abstraction and the degree of completeness of the models in question. We say that a device has been verified 'at the major-state level' or 'at the register-transfer level', and so on — it is not enough to say simply 'verified'. For example, Viper's major-state machine has been fully verified with respect to its top-level specification; but the proof establishing the equivalence of these two sets of results depends only on the flow of control in the two models, and does not depend on any of the computational behaviors of Viper. (That is, the same formal expression represents the arithmetic—logic unit at both levels, so that expression is never evaluated.) Therefore, the fact that Viper has been verified 'at the major-state level' does not actually ensure very much; the essence of the microprocessor (the behavior of its ALU) has not, at that stage, been treated. Viper certainly could not, on that basis alone, be usefully called 'verified'.

The block model of Viper does concern itself with Viper's arithmetic and logical operations and with the transfer of information between registers and memory. Thus, verifying Viper to the block level would be a significant step towards a 'verified' microprocessor. (In any case, the proof has not been fully completed at this level.) However, the block model does not concern itself with gate layout, transistors, electrical effects, timing problems, or many similar areas in which unsuspected errors would seem particularly likely to appear. (In those areas, enormous amounts of research remain to be done on finding useful, tractable models, even before we begin to verify them.) Thus, again, the term 'verified' cannot properly be used without an indication of the levels of the models involved. At *every* level of abstraction, some properties are included and some ignored.

In addition, the models involved may be incompletely specified. For example, Viper's highest level specification is complete only for the processing of instructions, and does not cover such features as the resetting or timing out of the machine, or other possible behaviors specified at the block level. This factor, from the outset, restricts any analysis to the high-level behavior alone, again missing the more subtle and perplexing issues.

6.5. *Normality Assumptions*

In discussing what was proved in the Viper verification project, it was indicated that certain assumptions had to be made (about initial and normal behaviors) in order to infer the cumulative effects of processing instructions. These assumptions are perfectly natural, and reflect the fact that devices are intended to operate only under certain conditions. The only cause for concern here arises if these assumptions are ignored when claims are made about what was proved. In the formal correctness statement, of course, any persistent assumptions will appear explicitly as the antecedents of an implication. It is in informal summaries (marketing material and so on) that the assumptions can easily be overlooked.

In the end, for example, the effect of each of Viper's instructions on the registers of its block model was deduced. This was done by assuming that the machine was initialized in a reasonable way, and assuming that it was run under certain ideal conditions. The effect was not deduced, say, of assuming that a reset operation could occur — it could have been, but to no useful end, since that effect is not specified at the top level. Thus, even a fully verified block design could remain incorrect in its resetting behavior, and the error could propagate, despite the proof, down to the chip itself. This illustrates the importance of knowing the conditions under which the block model has been analyzed.

6.6. *Putting Formal Proof In Context*

Finally, the correctness of an abstract representation of a system must be placed in context when we talk about its reliability in safety-critical applications. The author claims no expertise in the field of reliability, but this much is obvious: that an abstract and limited sense of correctness (for example, for Viper, the equivalence of a register-transfer level specification to a functional specification of the fetch—decode—execute cycle) is only one of many issues that have to be considered collectively.

Aside from possible problems at more concrete levels of description, which have already been discussed, safety will also depend on factors as yet outside the world of formal description: these range from issues of social administration and communication, as well as staff training and group behavior, at one end, to the performance of mechanical and chemical parts, and so on, at the other. One has only to contemplate the mass catastrophes of the last ten years or so to perceive the predominant role played by these extra-logical factors.

It is the author's guess (although, again, not an expert opinion) that the sort of abstract design correctness discussed here, though of undoubted importance, is still a relatively minor contribution to the overall reliability of real systems. This seems so at least at the present state of research into representation and proof, and with the present weak links between designer, verifier, and manufacturer. That is, using a hardware design verified only at a fairly abstract level — and only under idealized operating conditions — as part of the control system in hazardous applications (say, in which large populations are at risk) does not *yet* seem significantly safer than using any other design. If only because of the number of extra-logical factors involved, the use of the word 'verified' must under no circumstances be allowed to confer a false sense of security.

7. CONCLUSIONS

Various of the limitations on the use of the word 'verified' are obscured in claims such as the following (both taken from promotional material):

Viper is the first commercially available microprocessor with . . . a formal specification and a proof that the chip conforms to it. [26, 27]

One unique feature of Viper is that the instruction set is specified mathematically . . . and the gate-level logic design has been proven to conform to this specification. [16]

As discussed, a chip as such cannot be verified — but this is perhaps just an imprecise use of words. The second example, depending on one's interpretation of 'proved', could be called a false claim; no formal proofs of Viper (to the author's knowledge) have thus far been obtained at or near the gate level. The gate-level design of Viper *has* been checked by C. Pygott using an innovative simulation method called intelligent exhaustion [25], but it has not yet been formally verified.

Such assertions as those quoted, taken as assurance of the impossibility of design failure in safety-critical applications, could have catastrophic results. To summarize:

- Neither an intended behavior nor a physical chip is an object to which the word 'proof' meaning fully applies. Both an intention and a chip may themselves be inadequately represented in formal language, and this is not itself verifiable.
- Because of the present weak links between designer, verifier, and manufacturer, it is not at all obvious that errors deduced in very abstract specifications are likely to manifest themselves in actual products, or *vice versa*. We must then ask how much extra security verification currently affords. (This is an argument for continued research, not an argument against verification!)
- Any verification effort is necessarily limited to those behaviors specified at the most abstract level. It should be clearly stated when a system is called 'verified' which actual features are not covered.
- It should also be clearly stated to what level of concreteness the specifications extend. It seems fair to expect that the more concrete the models, the greater is the likelihood of finding errors in the design, particularly errors that would propagate through to the actual product. Since any model is an abstraction of a material device, it is *never* correct to call a system 'verified' without reference to the level of the models used.
- Any working assumptions about initial states or normal behaviors should also appear in verification claims. Particularly in informal descriptions, the assumptions may not always be evident.
- A proof that one specification implements another — despite being completely rigorous, expressed in an explicit and well-understood logic, and even checked by another system — should still be viewed in context of the many other extra-logical factors that affect the correct functioning of hardware systems. In addition to the abstract design, everything from the system operators to the mechanical parts must function correctly — and correctly together — to avoid catastrophe.

For a long time, automated theorem proving was sufficiently difficult that researchers frequently drew upon simple (or occasionally less simple) mathematical problems on which to exercise their automated proof systems. Advances in theory as well as in technology have now made proof efforts feasible that once appeared impossibly large, uneconomic, or labor intensive. Sophisticated theorem-proving environments, together with modern workstations, operating systems, and editors, have supported this progress. The proofs, for example, of the basic theorems of arithmetic [1] or of the correctness of schematic compiling algorithms [4] — to choose two examples — were challenging problems in their time, yet current verification efforts are focusing on properties of realistic (and sometimes commercial) hardware designs. Besides the Viper microprocessor, examples include the verification by W. Hunt in the Boyer—Moore system of the FM8501 [18], a computer designed (by Hunt) for the purpose of verification; the verification in HOL by J. Joyce [19] of Tamarack, a computer designed by M. Gordon, also for the purpose of verification; the verification in HOL by J. Herbert of the ECL chip [17], a network interface designed by A. Hopper as part of the Cambridge Fast Ring; and the verification in HOL by T. Melham of the T-Ring [22], a very simple ring network designed by D. Gaubatz and M. Burrows.

It would seem, in conclusion, that we are now beginning to be able to verify real hardware designs to useful levels of detail. None of the remarks in this article should be taken as pessimistic — just cautious. As 'verified' hardware begins to be used in life-critical applications (which could include fly-by-wire aircraft, bomb deployment systems, nuclear power stations, medical equipment, automotive braking systems, railway signalling systems, and so on), it will become increasingly important to insist that the word 'verified' and its synonyms are modified, qualified, and explained so that we know exactly what claims are being made, and can assess them intelligently.

8. FUTURE WORK

At the beginning of Section 6, the problem was mentioned of establishing a standard of evidence for having achieved a proof in an automated theorem prover. In this capacity, neither the long chains or primitive inferences that proofs comprise nor the particular procedures that have constructed these proofs, have so far found much favor. The Viper block model proofs consist of several million primitive inference steps, for example; and the procedures that generate them comprise dozens of pages of code in the functional programming language ML. The question of proof evidence is typical of a variety of fundamental issues that have not been broached in this article, but that at some point must also be addressed. For example, the consistency of any abstruse or special-purpose logic has to be established; this is a standard problem but not always easy. Worse, it could be asked on what basis we place our confidence in the implementation of a theorem-proving methodology (and the operating system on which it runs, the hardware of which the host machine is built, and so on).

One pragmatic answer (which is a topic of planned research at Cambridge) is to try to reduce the number of systems in which we must trust by agreeing on a standard for 'proof deliverables'. That is, we could agree on a proof output format such that the proofs produced at one site could be independently (and mechanically) checked at another. This idea, in the context of hardware verification, is due to K. Hanna. Part of its attraction is that proof checking is generally much less difficult than proof construction.

Another research goal is to find a uniform representation language for everyone involved in producing a hardware device: designers, verifiers, fabricators, etc. This would help to integrate the various communities, and thus reduce the danger, for example, that the models that are verified differ from the plans used by the manufacturers. It also would increase the chances that the errors turned up by verification were actual errors in the physical devices. Higher-order logic has been proposed for this purpose, but any standard logic could be a candidate.

A very large step toward reliable systems would be a verification effort extending all the way from the software level down to the gate level. Research is currently being planned in this area (i) jointly at the University of Cambridge, SRI International, and INMOS, and (ii) at Computational Logic Inc. in Texas.

Finally, research is continuing at various places into models for more realistic levels of representation of hardware, in the hope of expressing and locating the more subtle and worrisome errors that beset digital systems. Once the models are found, there appears to be no shortage of theorem-proving tools with which to verify them.

ACKNOWLEDGEMENTS

Many thanks to Tom Melham and Mike Gordon for helpful comments and discussions. Thanks also to Robin Milner, Thomas Forster, and Jeff Joyce. The opinions expressed here are the author's alone. The Viper verification work at Cambridge was supported by a grant from RSRE.

The preparation of this article was suggested by Larry Wos and was supported by a grant from the U.K. Science and Engineering Research Council. An earlier version of part of this article appears in [7].

Computer Laboratory,
University of Cambridge, U.K.

REFERENCES

1. Boyer, R. S. and Moore, J. S.: 1979, *A Computational Logic*, Academic Press.
2. Camilieri, A., Gordon, M., and Melham, T.: 1987, 'Hardware Verification Using Higher-Order Logic', Proceedings of the IFIP WG 10.2 Working Conference: *From H.D.L. Descriptions to Guaranteed Correct Circuit Designs*, Grenoble, September, 1986, (Ed.) D. Borrione, North-Holland, Amsterdam.
3. Church, A.: 1940, 'A Formulation of the Simple Theory of Types', *Journal of Symbolic Logic* 5.
4. Cohn, A.: 1979, 'Machine Assisted Proofs of Recursion Implementation', Ph.D. Thesis, Dept. of Computer Science, University of Edinburgh.
5. Cohn, A., and Gordon, M.: 1986, 'A Mechanized Proof of Correctness of a Simple Counter', University of Cambridge, Computer Laboratory, Tech. Report No. 94.
6. Cohn, A.: 1987, 'A Proof of Correctness of the Viper Microprocessor: The First Level', *VLSI Specification Verification and Synthesis*, (Eds.) G. Birtwistle and P. A. Subrahmanyam, Kluwer, 1987; Also University of Cambridge, Computer Laboratory, Tech. Report No. 104.
7. Cohn, A.: 1988, 'Correctness Properties of the Viper Block Model: The Second Level', *Current Trends in Hardware Verification and Automated Deduction*, (Eds.) G. Birtwistle and P. A. Subrahmanyam, Springer-Verlag, 1988; Also University of Cambridge, Computer Laboratory, Tech. Report No. 134.
8. Cullyer, W. J.: 1985, 'Viper Microprocessor: Formal Specification', RSRE Report No. 85013, Oct.
9. Cullyer, W. J.: 1986, 'Viper — Correspondence between the Specification and the "Major State Machine",' RSRE report No. 86004, Jan.
10. Cullyer, W. J.: 1987, 'Implementing Safety-Critical Systems: The Viper Micro-processor', *VLSI Specification, Verification and Synthesis*, (Eds.) G. Birtwistle and P. A. Subrahmanyam, Kluwer.
11. Cullyer, W. J., Kershaw, J., and Pygott, C.: forthcoming book on Viper.
12. Gane, C. (Computing Devices Company Ltd.): 1988, 'Computing Devices, Hastings' VIPER-VENOM Project: VIPER in Weapons Stores Management, Safety Net: Viper Microprocessors in High Integrity Systems, Enq. No. 021', Issue 2, July—August—September, Viper Technologies Ltd., Worcester, England.
13. Gordon, M., Milner, R., and Wadsworth, C. P.: 1979, 'Edinburgh LCF', *Lecture Notes in Computer Science*, No. 78, Springer-Verlag.
14. Gordon, M.: 1985, 'HOL: A Machine Oriented Formulation of Higher-Order Logic', University of Cambridge, Computer Laboratory, Tech. Report No. 68.

15. Gordon, M.: 1987, 'HOL: A Proof Generating System for Higher-Order Logic', University of Cambridge, Computer Laboratory, Tech. Report No. 103, 1987; Revised version in *VLSI Specification, Verification and Synthesis*, (Eds.) G. Birtwistle and P. A. Subrahmanyam, Kluwer.

16. Halbert, M. P. (Cambridge Consultants Ltd.): 1988, 'Selfchecking Computer Module Based on the Viper 1A Microprocessor, Safety Net: Viper Microprocessors in High Integrity Systems', Enq. No. 017, Issue 2, July—August—September, Viper Technologies Ltd., Worcester, England.

17. Herbert, J. and Gordon, M. J. C.: 1985, 'A Formal Hardware Verification Methodology and its Application to a Network Interface Chip', *IEEE Proceedings, Computers and Digital Techniques*, Special issue on Digital Design Verification, Vol. 133, Part E, No. 5, 1986; Also in draft version: University of Cambridge, Computer Laboratory, Tech. Report No. 66.

18. Hunt, W. A. Jr.: 1985, 'FM8501: A Verified Microprocessor', University of Texas, Austin, Tech. Report 47.

19. Joyce, J. J.: 1987, 'Formal Verification and Implementation of a Microprocessor', *VLSI Specification, Verification and Synthesis*, (Eds.) G. Birtwistle and P. A. Subrahmanyam, Kluwer.

20. Kershaw, J.: 1985, 'Viper: A Microprocessor for Safety-Critical Applications', RSRE Memo. No. 3754, Dec.

21. Melham, T.: 1987, 'Abstraction Mechanisms for Hardware Verification', *VLSI Specification, Verification and Synthesis*, (Eds.) G. Birtwistle and P. A. Subrahamanyam, Kluwer.

22. Melham, T., forthcoming Ph.D. Thesis, University of Cambridge, Computer Laboratory.

23. Paulson, L.: *Logic and Computation*, Cambridge, University Press.

24. Pygott, C. H.: 1986, 'Viper: The Electronic Block Model', RSRE Report No. 86006, July.

25. Pygott, C. H.: 1986, 'Formal Proof of a Correspondence between the Specification of a Hardware Module and its Gate Level Implementation', RSRE Report No. 85012, Nov.

26. Viper Microprocessor: Verifiable Integrated Processor for Enhanced Reliability: Development Tools, Charter Technologies Ltd., Publication No. VDT1, Issue 1, Dec. 1987.

27. Application for Admission and Registration Form, Second VIPER Symposium, RSRE, Malvern, England, 6—7 September, 1988.

TIMOTHY R. COLBURN

PROGRAM VERIFICATION, DEFEASIBLE REASONING, AND TWO VIEWS OF COMPUTER SCIENCE

1. INTRODUCTION

The current debate over the limits of program verification (Fetzer, 1988; Letters, 1989; Technical, 1989; Fetzer *et al.*, 1990) seems to have its roots in an apparent fundamental philosophical difference concerning the methods and goals of computer science. On the one hand, there is the view that computer science is, as its name implies, a science, but more importantly, an *empirical* science in the sense which de-emphasizes pure mathematics or logic. This sense is meant to cover all and only those experimental disciplines included in the 'natural' and 'social' sciences. This view is expounded implicitly and explicitly in many current standard computer science texts, for example, by Thomas L. Naps *et al.*:

Perhaps nothing is as intrinsic to the scientific method as the formulation and testing of hypotheses to explain phenomena. This same process plays an integral role in the way computer scientists work. [Naps *et al.* (1989) p. 5]

This view is also exemplified by the curricula and attitudes of many current academic computer science departments, in which computer science is put forth as the science of problem solving using computers, and not as 'mere computer programming'.

On the other hand, there is the view that computer science is, again as its name implies, a science, but an *exact* science in the sense which emphasizes the primacy of mathematics or logic. In this sense, there is no role for the testing of hypotheses to explain phenomena, since formal disciplines like mathematics or logic are not concerned with explanation of observable phenomena; they are *a priori*, or prior to experience, rather than seeking to explain experience. This view has as one of its most outspoken proponents C. A. R. Hoare:

Computer programming is an exact science in that all of the properties of a program and all of the consequences of executing it in any given environment can, in principle, be found out from the text of the program itself by means of purely deductive reasoning. (Hoare, 1969, p. 576)

375

Timothy R. Colburn et al. (eds.), Program Verification, 375—399.
© 1993 *Kluwer Academic Publishers.*

As James H. Fetzer has shown (Fetzer, 1991), citing other examples of writers comparing programming to formal mathematics or holding programming up to standards of perfection, this view is not a straw man and actually represents "several decades of commitment to formal methods by influential computer scientists".

These two views would indeed seem to be antithetical if they were both purporting to describe the same thing. However, a closer look at the second view at least as exemplified above by Hoare shows that he is referring to computer *programming* as an exact science, presumably leaving it open whether computer programming is to be identified with computer *science*. This paper attempts to reconcile the two views through a casting of computer programming within a rather broader context of computer science. If successful, it will provide an interpretation under which in some circumstances computer science is best viewed as the *engineering*, possibly using mathematical methods, of problem solutions on a computer, while in other circumstances it is best viewed as concerned with the experimental *testing of hypotheses* and the subsequent recording of observations in the real world. As I shall argue, computer science 'in the large' can be viewed as an experimental discipline which holds plenty of room for mathematical methods within theoretical limits of the sort emphasized by Fetzer (1988).

2. THE PROGRAM VERIFICATION DEBATE

My point in giving such an interpretation of the methodology of computer science is primarily to defuse the so-called program verification debate that has raged on several fronts during the past two years. The debate is over the proper way to go about the process of *program verification*, or the attempt to show that a program's behavior corresponds to its specified requirements. To clarify the debate and put it in perspective, place yourself in the following (admittedly simplified) scenario:

You are responsible for purchasing software for a large institution which has a need to automate a critical task. You specify, in a fairly rigorous way, the requirements that a program must meet, in terms of the kind of input it will accept, the kind of output it will produce, how long it will take for the program to run, etc., in solving your problem. You find that two software companies have anticipated your need and claim to have programs which meet your requirements, so you invite their sales representatives to your office.

Salesman A: We are offering, for a limited time only, a program that meets your requirements. We know that it meets your requirements because we have pain-stakingly tested it over many months and on many machines. Although it is not possible to test every combination of inputs, we have failed to uncover any errors. To the extent possible through testing, this is a verified product.

Salesman B: We are also offering a program that meets your requirements. But we know that the program is correct because we have formally *proven* it correct with the most sophisticated mathematical and logical techniques available. We are so sure of its correctness that we did not even need to test it on any specific machine. This is a completely verified product.

Assuming that the programs are comparable in price, which do you buy?

This scenario, besides being simplified, is also hypothetical since there are no large, critical programs that have been claimed to be formally proven correct. Still, it is a caricature of a scenario the possibility of which some researchers in software reliability are working to bring about. It tends to suggest that formal program proving methods always provide an alternative to program testing, as though the goals of these two activities were always the same. By taking a closer look at what it means for a program to be run as a test, I shall argue that this is not a realistic portrayal. But before turning to this, we must better understand the limits of both program testing and formal program verification.

3. THE DEFEASIBLE NATURE OF PROGRAM TESTING

In practice, non-trivial programs are verified by running them and observing their behavior with varying sets of representative input. It has long been recognized that for most non-trivial programs, it is either impossible or impractical to test every possible input configuration. Raymond D. Gumb, for example, writes: "... testing a program does not always establish its correctness; crucial test cases may be over-looked or there may be too many cases to test" (Gumb, 1989, p. 2). This inherent incompleteness of program testing has led to two dif-ferent responses. One is to admit that program testing is incomplete but to deny that it must necessarily be *ad hoc*, and to develop rigorous testing methodology which maximizes test case coverage through both static and dynamic program analysis (cf. Myers, 1979). The other response is to consider the incompleteness of program testing as grounds for eschewing it in favor of another approach to program verification, namely, the mathematical approach. In the same passage, for example, Gumb writes: "Program correctness must be understood in the context of a mathematical theory." Since there are alternative approaches to the problem of the incompleteness of testing, the 'must' in this quote cannot be taken to be a logical 'must', but a way of advocating the primacy of mathematical methods in both the construction and verification of programs.

But for an advocate of the experimental science approach to computer science, the incompleteness of testing is not grounds for eschewing it, but simply being more careful about it. E. Dijkstra's famous remark that testing can only show the presence, not the absence of program errors, is damning to the whole platform of program testing only if understood in the context of mathematical theorem proving. In this context, it is of course true that the fact that *some* specific inputs to a program issue in correct output in no way proves the theorem that *every* input would issue in correct output. But if interpreted in the context of experimental science, the fact that testing does not show the absence of errors is no more damning than the fact that observing only black crows does not show that all crows are black, where the intent is to inconclusively *confirm* rather than conclusively *prove*.

The reasoning involved in program testing is a species of inductive reasoning, since it argues from the observed (the input sets that are actually tested) to the unobserved (those that are not). For example, consider the procedure in Figure 1 written in the programming language Pascal. It attempts to complete the factorial of a positive integer supplied in the input variable N, leaving the result in the output variable M. In testing such a procedure, one typically supplies a value in N, runs the procedure, and checks to see whether the value in M is correct.

```
procedure factorial (N: integer; var M: integer);
begin
  M := N;
  while N > 1 do
    begin
      N := N - 1;
      M := M * N
    end
end;
```

Fig. 1. Procedure to compute factorial.

After repeating this process for various input values, say, integers 0 through 5, and being satisfied of the correctness of the output, one infers that the unchecked input values *would* also issue in correct output *were* they tested.[1] As such, this reasoning is no more 'incorrect' than that involved in any experimental endeavor attempting to confirm a generalization from observed instances. It does, however, depend in some sense on an assumption of the *typicality* of the unobserved instances in relation to the observed. That is, if an unobserved instance has a property distinguishing it significantly from the observed instances, it may *defeat* the claim that all inputs issue in correct output.

The fact that reasoning can involve possibly false assumptions does not thereby make it faulty or incorrect. It does, however, change the nature of the warrant that such reasoning provides. To describe that warrant, we must look at the distinction between *defeasible* and *indefeasible* reasoning, a distinction first studied by philosophers (cf. Pollock, 1974) and now also by researchers in artificial intelligence and cognitive science (Nute, 1988; Kyburg *et al.*, 1990). This distinction is based upon the concept of a *defeater*, or piece of evidence which can make an ordinarily justified inference no longer justified. The possibility of defeaters pervades empirical reasoning. For example, I may infer that a statue is red on the basis that it looks red to me, but if I later discover that there are red lights shining on it I am no longer entitled to this inference, since the normal connection between the color a statue *looks* and the color it *is* is defeated by this new knowledge. Besides examples of perception, the possibility of defeaters also accounts for the non-absolute (or 'fallible') nature of reasoning in other areas of empirical inquiry, including reasoning on the basis of memory, testimony, and inductive evidence.[2]

Defeasible reasoning is also called *non-monotonic*, since the consequence relation involved in the reasoning fails to have the property of monotonicity. Suppose R is a consequence relation and A is a set of sentences. Let us represent the fact that sentence p is a consequence of A by $R(A, p)$. Then the set of consequences of A using R can be represented by $\{x : R(A, x)\}$. R is monotonic if and only if whenever $A_1 \subseteq A_2, \{y : R(A_1, y)\} \subseteq \{z : R(A_2, z)\}$.[3] Thus if I can infer *the statue is red* from a set of sentences A which includes *the statue looks red to me* but not from the set $A \cup \{red\ lights\ are\ shining\ on\ the\ statue\}$, then *red lights are shining on the statue* is a defeater and the consequence relation in my logic is *non*-monotonic.

Reasoning is *indefeasible* if it is not defeasible, i.e., it employs a monotonic consequence relation. The difference between defeasible and indefeasible reasoning is often masked by the terms in which they are couched, where it is not immediately obvious whether a monotonic or non-monotonic consequence relation is involved. Consider, for example, the syntactically similar argument

(A)   1. Birds are vertebrates.

        2. Fred is a bird.

        3. Therefore, Fred is a vertebrate.

and

(B)   1. Birds are flyers.

        2. Fred is a bird.

        3. Therefore, Fred is a flyer.

Despite the syntactic similarities, arguments (A) and (B) employ different consequence relations and thus provide different warrants for their conclusions. In argument (A), the predicate calculus model is entirely adequate in characterizing the consequence relation as monotonic and thus the reasoning as indefeasible:

(C)   1. $(\forall x)(Bird(x) \rightarrow Vertebrate(x))$

        2. $Bird(Fred)$

        3. $\vdash Vertebrate(Fred)$

In argument (B), however, a predicate calculus model is not as straight-forward because (i) since the conclusion is defeasible (e.g., suppose Fred is a penguin), the consequence relation employed is nonmono-tonic, and (ii) the warrant given by the argument is not truth-preserving, as provided by $\vdash$, but only justification-preserving. Any model of the reasoning involved in (B) must provide representation for at least the following premises in the expanded argument:

        1. Typically, birds are flyers.

        2. Fred is a bird.

        3. There is no reason to suppose that Fred is not typical, i.e., that Fred cannot fly.

But even if these premises are all true — in particular, even if I have no reason to think that Fred is a non-flying bird (I have it on good evidence that Fred is a falcon) — they cannot guarantee that Fred can fly (Fred might, unbeknownst to me, have clipped wings). Thus there is no deductive warrant for the conclusion

4. Therefore, Fred is a flyer.

since the premises can be true while the conclusion is false. The fact that I am reasoning about a property which not all birds have assures that my reasoning is defeasible. I can at best conclude the *justifiedness* of the intended result:

4'. Therefore, 'Fred is a flyer' is justified.

One might wonder, from a purely logical point of view, what warrants even this inference. In this argument and in the argument schemata which follow, we shall assume an implicit conditional premise simply relating the explicit premises to the conclusion. In this case the implicit premise is,

3'. If birds are typically flyers and Fred is a bird and there is no reason to suppose that Fred is not typical, then 'Fred is a flyer' is justified.

These implicit premises can be regarded as instances of general epistemic principles without which we would have no hope of acquiring empirical knowledge. (Roderick M. Chisholm, for example, describes the nature of some of these principles in Chisholm (1977), Ch. 4.) Thus, a full rendering of the reasoning involved in argument (B) is captured in:

(D)   1. Typically, birds are flyers.

2. Fred is a bird.

3. There is no reason to suppose that Fred is not typical, i.e., that Fred cannot fly.

4. If birds are typically flyers and Fred is a bird and there is no reason to suppose that Fred is not typical, then 'Fred is a flyer' is justified.

5. Therefore, 'Fred is a flyer' is justified.

Note that the inclusion of premise 4 does give argument (D) a deductive form: the conclusion is a modal statement in a meta-language about another statement's justifiedness in the object-language.

The reasoning underlying program testing is also defeasible, due, one might say, to the assumption of *typicality* of the unobserved test inputs. For example, after testing my factorial procedure with the integers 0 through 5 and being satisfied with the output, I assume that unobserved test inputs also work correctly because the test cases using integers 0 through 5 are typical of the (unrealized) test cases using 6 on up. This is not to say that the *integers* 0 through 5 are typical of the *integers* 6 on up, but that the test cases using them in procedure factorial are. That is, the integers 0 through 5 *exercise* the procedure in the same way that larger integers would. Although the simple factorial example does not display much complexity in its flow of control, in more complex programs the set of possible inputs divides naturally into subsets which characterize the different paths of control flow through the program. Rigorous testing attempts to choose test inputs from each of the subsets, reducing the danger of leaving untested any cases that are not typical of those tried, that is, that exercise different parts of the program. Still, for large programs there often is no guarantee that every path of control flow has been anticipated. Reasoning about their untested cases, like reasoning about Fred and flying, carries only a guarantee of justification, not truth:

(E) 1. Program **P** works on machine **M** for the test cases observed.

2. There is no reason to suppose that the unobserved cases are not typical, i.e., that they exercise **P** differently than the observed cases.

3. Therefore, 'Program **P** works on machine **M** for the unobserved cases' is justified.

The defeasible nature of this pattern of reasoning in program testing, viewed as a method whereby generalizations are made in empirical science, should not be a problem. But if the goal of program verification is to reason about a program in such a way that arguments about unrealized behavior are truth-preserving, i.e., absolute guarantees of program correctness, then testing is not adequate. One approach with this goal in mind is *formal* program verification, but as we shall see, although its methods are entirely different, the reasoning behind it is also defeasible by nature.

4. THE DEFEASIBLE NATURE OF FORMAL PROGRAM VERIFICATION

As a representative example of techniques in research on formal program verification, consider the method of *inductive assertions*. Briefly, this method attempts to prove *output assertions* about the program on the basis of *input assertions* as well as assertions concerning the relationships that constantly obtain among variables while the program runs. Most often the program executes a loop, so these latter assertions are sometimes called *loop invariants*. The method of proof is called 'inductive' because mathematical induction, a special kind of deductive reasoning, is used on the number of iterations in the loop to prove that some loop invariant holds throughout the loop, establishing the output assertions.

For example, consider again the procedure in Figure 1. By restricting N to positive integers we have the input assertion:

$$N \in \text{positive integers}$$

The output assertion we wish to prove is:

$$M = N!$$

To prove this, we must prove something about the loop, a loop invariant, from which it follows that $M = N!$. Loop invariants typically state a general relationship among variables after an arbitrary number of iterations through the loop. We can let N_i and M_i refer to the contents of N and M, respectively, after i iterations through the loop. The invariant we would like to prove, for all i, is:

(a) $\qquad M_i \times (N_i - 1)! = N!$

To see this, suppose the loop requires k iterations. Then when the loop is complete $N_k = 1$ so $(N_k - 1)! = 0! = 1$ and therefore the invariant reduces to

$$M_k = N!$$

This asserts that at the end of the loop M contains $N!$, our original output assertion. To prove the invariant (a), we use induction on i.

TIMOTHY R. COLBURN

(F)   1. (*Base step*:) When $i = 0$, $M_i = N$ and $N_i = N$. Thus $M_i \times (N_i - 1)! = N \times (N - 1)! = N!$

2. (*Inductive step*: From $M_i \times (N_i - 1)! = N!$ show that $M_{i+1} \times (N_{i+1} - 1)! = N!$. By observing the program, we see that after any iteration of the loop, M contains what it had before the iteration multiplied by the contents of N after the iteration. Thus $M_{i+1} = M_i \times N_{i+1}$. We can also see that $N_{i+1} = N_i - 1$. Making use of these equalities and the inductive assumption, we can prove the inductive conclusion:

$$M_{i+1} \times (N_{i+1} - 1)! =$$
$$M_i \times (N_i - 1) \times (N_i - 2)! =$$
$$M_i \times (N_i - 1)! = N!$$

(F) appears to be an ironclad proof of the loop invariant (a). However, (F) is unusual for mathematical proofs in that what we mean by M and N makes a difference in the type of reasoning (F) involves. That is, as stated, (F) harbors an ambiguity of the type that Fetzer has pointed out in Fetzer (1988). To see this, consider the crucial premise in the inductive step that

(b)   ... after any iteration of the loop, M contains what it had before the iteration multiplied by the contents of N after the iteration.

The ambiguity is over the ontological status of M and N. If M and N are assumed to be abstract memory locations in an *abstract* machine, the proof is indeed ironclad since abstract machines are *ideal* in the sense that their operation is simply assumed never to fail. Abstract machines are represented and conceived of purely functionally, and not in terms of physical control units, registers, adders, and memory elements. Thus if M and N are abstract memory locations, (b) can be conclusively established simply by observing the program and understanding the language.

However, if M and N are assumed to be physical memory locations in a given *physical* machine, then (b) can be established only by making a host of assumptions, including the proper functioning of the hardware units mentioned above, as well as all the supporting software which issues in a program executing in physical memory. So insofar as proof

(F) refers to physical memory locations, these assumptions render the verification provided, in Fetzer's words, as 'relative' rather than 'absolute' (Fetzer, 1988, pp. 1050–1051). That is, the significance of proof (F) for the physical behavior of procedure factorial on an actual machine is entirely dependent upon the proper functioning of the underlying software and hardware supporting the program which manipulates M and N.

Thus, reasoning about physical machine behavior on the basis of (F) is defeasible. Just as, in the program testing case, I must make the possibly false assumption that the unobserved cases are in some sense typical in order to infer that they would issue in the correct output, in formal program verification I must also make possibly false assumptions about the typicality (proper functioning) of the underlying software and hardware. Thus, the truth of (a) cannot be inferred by (F), though its justifiedness can. So the overall method of reasoning by formal program verification, as applied to predicted behavior on a physical machine, also conforms to a general schema of defeasible reasoning:

(G)   1. Program **P** has been formally verified.

2. There is no reason to suppose that the underlying hardware and software for machine **M** are not typical, i.e., that they are not functioning correctly.

3. Therefore, 'Program **P** works correctly on machine **M**' is justified.

It is, of course, an epistemic obligation to justify as far as is practical the beliefs in one's underlying assumptions. For me to commit to the belief that Fred can fly, the amount of effort I put into ruling out any abnormalities concerning Fred's flying abilities is proportional to the amount of importance I attach to not being wrong. The amount of importance attached to the correctness of computer programs, of course, can reach enormous levels, in matters of medicine, financial transactions, and national security, so that the obligation to minimize assumptions becomes not merely an epistemic but an ethical one. In the case of formal program verification, the desire to minimize assumptions has led to the movement toward *system* verification, in which the goal is to apply formal methods not only to individual application programs, but also to the system software (operating systems, compilers, assem-

blers, loaders, etc.) and hardware designs of the components on which
they will run.

5. THE DEFEASIBLE NATURE OF
FORMAL HARDWARE VERIFICATION

As programs manipulating abstract memory locations, we might allow
that the programs comprising system software can theoretically be
proven correct, ignoring the practical limitations on proving programs
of this size and complexity. But as programs manipulating physical
memory locations on a given machine, arguments about them are
subject to the same assumptions concerning the 'typicality' (correctness
and proper functioning) of the integrated circuitry in memory and
processor as application programs. Thus significant efforts in 'hardware
verification' are underway. Here the object of scrutiny in the proving
process is not a program but a formal specification of some device.
Typically, the proof process proceeds by levels, each level being
concerned with a more specific description of the device than that
provided in the level before it. At each level relevant assertions about
the specification are proved, possibly using mechanical tools (theorem-
proving programs), the goal being to show that the behavior of the
device modeled at some level is the same as that modeled at the next
higher level of abstraction. The ultimate objective is to continue this
process down to a formal proof of the gate-level circuit design.

It is important to note that in this process of hardware verification
we are still talking about the behavior of *abstract* memory locations and
processing elements, and not actual devices *per se*. Just as it is a
mistake to claim that program proof (F) conclusively establishes the
correctness of a program executing on some physical machine, it is a
mistake to claim that hardware proofs conclusively establish the
correctness of behavior of physical devices. This latter point is even
made by the hardware verification practitioner Avra Cohn:

... a material device can only be observed and measured; it cannot be verified ... a
device can be described in a formal way, and the description verified; but ... there is
no way to assure accuracy of the description. Indeed, any description is bound to be
inaccurate in *some* respects, since it cannot be hoped to mirror an entire physical
situation even at an instant, much less as it evolves through time; a model of a device is
necessarily an abstraction (a simplification). In short, verification involves a pair of
models that bear an uncheckable and possibly imperfect relation to the intended design
and to the actual device. (Cohn, 1989, pp. 131–132)

Cohn is conceding, in the hardware verification realm, the same point introduced by Fetzer in the program verification realm, namely, that there is a fundamental difference between the nature of the abstract and the nature of the physical, preventing conclusions about the physical from having the same warrant as those in the abstract. Thus even if we were to grant the possibility of proving complicated system software, equally complicated theorem-proving software, and the gate-level circuit designs of the devices that make programs run, we still do not have a mathematical proof of any physical behavior of a given application program.

One might argue that this is quibbling, and that the difference between gate designs and gates themselves is so trifling that the program is as good as formally verified. But to conclude from the formal proof of an abstract hardware specification that the corresponding physical device, say, a chip, will work correctly involves the making of a host of assumptions concerning the conformance of the chip to the specification, including assumptions about the fabrication of the chip, the properties of its materials, the environment in which it runs, etc. Thus, no matter how rigorous its proofs about formal, abstract designs, for a hardware verification argument to be saying anything about an actual physical device, it must be regarded as defeasible and not conclusive in the sense reserved for arguments in mathematics or logic. As a general schema:

(H)   1. The hardware design **D** of physical device **P** has been formally verified.

2. There is no reason to suppose that the fabrication of **P** from **D** is not typical, i.e., that **P** does not conform to **D**.

3. Therefore, 'The hardware of **P** works correctly' is justified.

So even if a program and its supporting system software 'superstructure' can be formally proven correct, the defeasibility of reasoning about the program's physical behavior is insured by the defeasibility of reasoning about the behavior of the chip on which it runs.

Since reasoning by program testing, formal program verification, and formal hardware verification are all defeasible, it would seem that a combination of them in an overall attempt at verifying program correctness would be desirable. The defeasible assumptions required by the testing point of view might be mitigated by the formalist's methods, while those required by the formalist might be mitigated by testing. On one hand, if you start with the testing point of view, you cannot have the right to be sure about untested cases until you have reasoned correctly and *a priori* about them. On the other hand, if you start with the formalist point of view, you cannot have the right to be sure about the physical behavior of a program until you have tested it. But in what sense of testing: merely to check the underlying functioning of hardware, or to verify the program itself, or for some other purpose? Can *all* program testing be replaced, in principle, by a combination of *a priori* reasoning about both the application program and supporting system software, *a priori* reasoning about the gate level designs of the chip on which it runs, and *a posteriori* observations of the actual chip in pre-release testing? To answer this requires a closer examination of the role of program testing in computer science.

6. THE ROLE OF PROGRAM TESTING IN COMPUTER SCIENCE

With respect to a simple program like factorial, it seems that we might be prepared to admit that, on the basis of proof (F), one can be justified in assuming that the program is correct, in the sense that it will issue in correct physical memory contents given certain inputs, *without actually running it*, provided that (i) we are satisfied that the supporting software has — mathematically or not — been 'proved' correct, and (ii) the physical chip on which it and the supporting software runs has been extensively tested — whether or not its gate-level design was mathematically proved — to the point where we are satisfied that it carries out instructions without error. Still, owing to the fact that the chip could fail for a number of physical reasons, our justification is not conclusive, so we are in a position of being *defeasibly justified* in predicting what the behavior of the program would be were it run on this chip.

Although we have a (somewhat idealized) situation in which program testing can be foregone in favor of a formal proof, we have seen that, although the proof is mathematical, its warrant is defeasible. The focus has now shifted from the correctness of procedure factorial to the operability of the chip on which it runs. The running of the procedure would not be a test of its correctness at all, but of whether the chip was working. So we can distinguish at least two ways in which a program can be run as a test:

Testing Sense 1: We might run a program to determine whether its behavior conforms to its specifications.

Testing Sense 2: We might run a program to determine whether the hardware/software superstructure supporting it is functioning properly.

In the first case the program tests itself, while in the second case it tests its supporting superstructure. These are not independent notions, however, for two reasons. First, one cannot run a program as a test in sense 2 unless one has already determined how the program is supposed to behave. Thus, unless the program's intended behavior has been completely understood and formally verified, it must be tested in sense 1 in advance. Second, the failure of a program test in sense 1, without independent confirmation of the operability of the superstructure, is equally explained by program incorrectness or superstructure failure; the same superstructure malfunctioning that can undermine a formal proof of program correctness may also undermine the running of a program to test itself. Thus, testing a program in sense 1 presupposes the proper functioning of the superstructure, which involves testing in sense 2 (though not, of course, via the same program).

Normally, a typical application program is run primarily as a test in sense 1, while only very specialized programs are run to test the superstructure: programs like self-diagnosing routines for memory boards that execute when the machine is booted, or program test suites used in compiler validation. Still, we regard the *positive* outcome of a test in sense 1 as also a positive outcome of a test in sense 2. (This must allow, however, for the highly unlikely possibility that the program as written is *in*correct, but a subtle defect in the superstructure causes their interaction to produce correct behavior — a false positive result!)[4]

It is tempting to conclude that any problems involved in the interdependence of testing in senses 1 and 2 above could be avoided by replacing testing in sense 1 with formal program verification and leaving testing in sense 2 to hardware technicians and hardware testing equipment. But even if we perfected mathematical program proving techniques, developed more sophisticated mechanical theorem provers, and wrote all of our application software in a way amenable to analysis and formal proof, program testing would still be indispensable, not as a method of demonstrating programs' correctness, but in the broader, general role that testing plays in any empirical science. The example we have considered here — the computing of the factorial of a number, not considered in the context of some greater problem — is a mathematical activity whose modeling on a computer is very straightforward and whose verification can easily give way to the same mathematical approach. But in many cases the role of testing cannot be so easily supplanted.

Suppose I am a geneticist trying to understand a process that I suspect involves combinatorial analysis, so I devise a number of methods of modeling it, one of which involves the computing of factorial. Now I may be able to convince myself of the correctness of my factorial procedure, if not through a proof like (F) then through some mental walk-through, without testing, but that is not what is being tested anyway; I have coded a particular genetic model and *that model* is what I endeavor to test when I run a program, not the factorial procedure, or even the totality of the procedure I have written to implement the model.

Note that this example casts the role of computers in the solving of a problem in a very different light than that exemplified in the opening scenario above. In that scenario, as the data processing manager, I have a problem which I suspect can be solved with a computer program and I provide a complete specification of the program requirements for competing software producers. Once I choose one and satisfy myself that its program satisfies my requirements, the primary purpose in running it is to implement a solution to my original problem.

In the genetics example, I also must produce requirements for a program, but rather than implement an already understood solution to a problem, the program will implement a genetic model. Once I am satisfied that the program meets my input/output and performance requirements, the primary purpose in running it is not to implement a problem soltuion, but to *test* whether the model accurately depicts a genetic process. I shall now contend that these examples correspond to two different but complementary conceptions of computer science.

7. COMPUTER SCIENCE AS SOLUTION ENGINEERING

In the original scenario, a well-designed solution to a problem in terms of rigorous requirements is outlined, and an implementation of the solution is engineered via a computer program. The principal objective is to meet the requirements in a program solution, so the method can be characterized as *solution engineering* with computers. It is tempting to associate this paradigm with the mathematical view of computer programming as an exact science, with formal program verification playing the role (in theory if not in current practice) of defeasibly warranting a program's correct behavior, testing only being necessary to confirm the proper functioning of underlying software and hardware. But even within the restricted paradigm of computer science as solution engineering, with formal program verification playing its role of verifying that a program meets its requirements, there still seems to be a role for program testing.

Our treatment of computer science methodology has so far glossed over the use of algorithms in problem solving, but in practice they play an important role in the process of designing programs. Ordinarily conceived, an algorithm is not a program, but an abstraction of a problem solution, in the form of a terminating sequence of steps, which when implemented in a programming language issues in a list of computer instructions which carry out the solution steps. The role of an algorithm in computer problem solving can be elucidated through an analogy with natural science.

Natural science seeks explanations of observable phenomena through hypothesis construction and subsequent hypothesis testing via experiments. For example, the hypothesis that *aurora borealis* are caused by interaction of the earth's ionosphere with solar atoms can be tested by creating laboratory conditions which imitate this interaction and observing for the *aurora* effect. Thus, the method of natural science can be described by:

(I)   1. Formulate a HYPOTHESIS for explaining a phenomenon.

        2. TEST the hypothesis by conducting an EXPERIMENT.

        3. CONFIRM or DISCONFIRM the hypothesis by evaluating the results of the experiment.

Similarly, we might understand the role of an algorithm in the method of computer science via an analogy between explaining phenomena in natural science and problem solving in computer science. We might say that in computer science we:

(J)   1. Formulate an ALGORITHM for solving a problem.

2. TEST the algorithm by writing and running a PROGRAM.

3. ACCEPT or REJECT the algorithm by evaluating the results of running the program.

So we can add the following to the ways in which a program can be run as a test:

Testing Sense 3: We might run a program to determine whether an algorithm solves a problem.

If this analogy holds up, then just as what is on trial in scientific reasoning is not an experiment but a hypothesis giving rise to an experiment; what is on trial in computer science is not a program, but a problem solving method or algorithm for which a program is produced. To continue the analogy, writing a program is to computer science what experiment construction is to natural science. Just as test implications need to be produced from a scientific hypothesis, one needs to produce a program from an algorithm. Program verification, as an endeavor to make sure that input/output specifications required by an algorithm are enforced by the program, can be viewed as the process of making sure that the program actually does implement the algorithm being tested. This would be analogous to that part of the scientific method which is the process of making sure that the experiment actually does set up the conditions required for a correct testing of the hypothesis. Under this conception, then, program testing is indispensable since an algorithm, and not a program, is on trial.

All analogies inevitably break down, however, and an advocate of the primacy of formal methods in computer science will emphasize a disanalogy between the hypothesis/experiment relationship in natural science and the algorithm/program relationship in computer science. The short history of the field has seen programming languages get more and more high-level and algorithm description languages more and more formal, so that the distinction between an algorithm and a program gets smaller and smaller. The trend toward higher-level languages has one current thrust in the direction of the declarative style of programming, as opposed to the traditional imperative style, in which programs do not direct a computer as to how to solve a problem, but rather simply state what is the case in a particular computational world in terms of executable facts and rules, leaving to the language itself the details of working out the solution. (Cf. Robert Kowalski's treatment of logic programming in Kowalski (1979).) The possibility of declarative languages has also led to the movement toward the automatic generation of programs from formal specification languages. Since a program specification tells *what* a program is supposed to do and not *how* it will do it, the possibility of success in this area threatens to obliterate not only the distinction between algorithms and programs, but also the need for human programmers to program in a traditional sense at all. (Cf. Robert Balzer's work in automatic programming: Balzer, 1985.) The vision is one of program specification writers replacing programmers, with programs produced mechanically by other programs.

Models of the software development process have changed over the years, from the earliest linear phase models, to the waterfall model, to the rapid phototyping model. But whatever the current model, the trends and research areas just mentioned, including program verification, endeavor either to automate the process of going from program specification to actual program or to make this transition totally straightforward through formal methods. For when a process can be modeled formally, it is more likely able to be achieved mechanically. The desideratum, then, is for the software development process to become a "black box", with program specifications as input and totally correct program as output, with as little need for human involvement as possible. (See Figure 2.) The underlying paradigm of computer science as *solution engineering* is upheld in this vision, since the objective is to achieve a computer implementation of a problem whose solution is so well designed and conceived as to admit of formal treatment. But a case can still be made for the necessity of testing, despite the mechanical fabrication of programs from specifications. For *unless there is some sort of independent guarantee that the program specifications, no matter how formally rendered, actually specify a program which solves the problem*, one must run the program to determine whether the solution design embodied by the specification is correct. So there is another sense in which a program can be run as a test:

> *Testing Sense 4*: We might run a program to determine whether a program specification solves a problem.

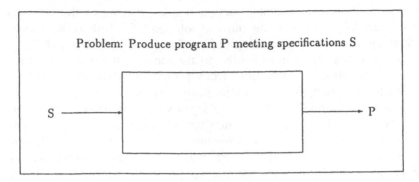

Fig. 2. Software development as a black box.

In this sense the method of computer science can be captured by:

(K)   1. Formulate a SOLUTION DESIGN, in terms of a program specification, for solving a problem.

2. TEST the solution design by generating and running a PROGRAM.

3. ACCEPT or REJECT the solution design by evaluating the results of running the program.

When a problem application domain is mathematical, as in computing the factorial of an integer, it may be easy to see that a program specification is correct for a problem and obviously solves it. Limiting problem solving to mathematical areas will reinforce the conception of computer science as an exact science. But in other problem areas, for example, involving trial and error or search, it is not at all obvious that a particular program specification issues in a program that solves a problem, even if the program is mechanically generated. This point is emphasized even more if the problem includes constraints on the amount of resources the program is allowed to use to solve the problem. To perform computer problem solving in these areas requires a conception of computer science in which its method is held not merely to be *analogous* to that of experimental science but a *species* of it.

8. COMPUTER SCIENCE AS EXPERIMENTAL SCIENCE

In the genetic modeling example above, the point of writing the program can also be described, albeit innocuously, as implementing a solution to a problem, but it is more appropriately characterized as testing a model of a phenomenon. Here, the method of writing and running the program is best conceived not as implementing a problem solution, but as a *bona fide* case of hypothesis testing. The only difference between this method and that of experimental science is that experimentation is carried out via a *model* of reality in a program. Testing the program in this sense is not at all to make sure it or an algorithm works. So we have yet another way in which a program can be run as a test:

> *Testing Sense 5*: We might run a program to test a hypothesis in a computer model of reality.

In this sense program testing *cannot* be supplanted by formal program verification since it is essential to the whole method of experimental science. (See Figure 3.)

This is not to say that formal program verification cannot play any role in programming as hypothesis testing. On the contrary, any or all of the methods of computer science as solution engineering come into play within the broader context of computer science as experimental science. The problem of producing a program with certain specifications must still be solved, but it can be seen as a subproblem of the overall problem of explaining a phenomenon. Figure 4 shows the

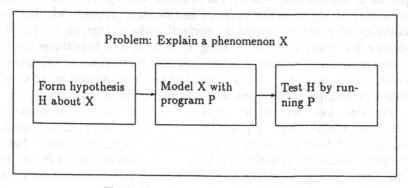

Fig. 3. Hypothesis testing using a program.

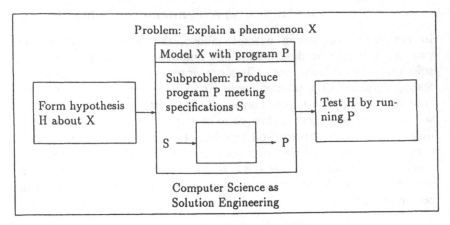

Fig. 4. Computer science as experimental science.

relationship of computer science as experimental science and computer science as solution engineering.

The method of 'doing' computer science as experimental science is exemplified in many applications involving modeling and simulation. But perhaps the field in which the experimental approach is most pervasive is artificial intelligence, in which human problem solving capabilities are not just aided by programs, but emulated by them. Invariably, the attempt to emulate problem solving capability in a program reveals a dearth of understanding as to the nature of that problem solving capability, as the program's behavior displays inadequacies in comparison with human problem solving. There is no way, obviously, to tell from an artificial intelligence program whether its modeling of problem solving is 'correct' without running it. This is because the main point of running it is to test a hypothesis about human problem solving, not to check whether it meets its specifications. Thus research in artificial intelligence is often characterized as the study of human intelligence through attempts to model it in a program.

The question then might be raised, is the artificial intelligence researcher really doing computer science or something else, say, cognitive science? Similarly, is the genetic model programmer really doing computer science or genetics? These questions, being matters of how to label fields of inquiry, should not be troubling, arising from the unique multi-disciplinary nature of computer science. No matter what we choose to call these marriages involving computer methods, they are united in their reliance upon the testing of programs to test hypotheses and sharply distinguished from the 'solution engineering' view of computer science, which sees programs as solutions to well specified problems.

9. CONCLUSION

The two views of computer science described here should not be perceived as contraries, since the view of computer science as the engineering of program solutions can be seen as being properly contained in the view of computer science as experimental science (or computer science 'in the large'). Formal program verification, as a method of computer science as solution engineering, might, at least in theory if not in current practice, replace program testing as a method of guaranteeing that abstract devices such as memory locations change state correctly in the running of a program. But even combined with formal hardware verification, formal program verification can only defeasibly guarantee the correctness of state changes in physical devices.

Formal program verification may be able to show (defeasibly) that a program meets its specifications, but as such formal program verification only (theoretically) replaces one of the ways in which programs are run as tests. Programs are also run as tests of whether a program specification solves a problem, and, in many non-mathematical problem areas, this sort of testing will continue to be indispensable. In particular, when computer programs are used for the modeling and explaining of phenomena, it is only through testing that the correctness of a program can be ascertained. For in these cases the testing of the program also doubles as the testing of the hypothesis.

Department of Computer Science,
University of Minnesota, Duluth, U.S.A.

NOTES

[1] Actually, in this example one *could* check all of the possible input values, since typical implementations of Pascal cannot hold integers larger than the factorial of 12. However, other languages, for example Common Lisp, place no practical limit on the size of integers representable.

398   TIMOTHY R. COLBURN

² Questions about the relationship of defeasible and inductive reasoning, e.g., whether one is a subspecies of the other, are open to philosophical debate, the outcome of which does not affect the points made here. See Rankin (1988) for a relevant discussion.
³ Much of the work in artificial intelligence in the 1980's and today is devoted to formalizing and modeling non-monotonic consequence relations (cf. Ginsberg, 1987; Reinfrank, 1989).
⁴ This situation brings to mind an analogy with the so-called 'Gettier' problem in epistemology, in which bizarre counterexamples to the traditional definition of knowledge as justified true belief were proposed (cf. Roth *et al.*, 1970, Part 2).

REFERENCES

Balzer, R.: 1985, 'A 15 Year Perspective on Automatic Programming', *IEEE Transactions on Software Engineering* **SE-11**(11), 1257–1267.
Chisholm, R. M.: 1977, *Theory of Knowledge*, Englewood, NJ: Prentice-Hall.
Cohn, A.: 1989, 'The Notion of Proof in Hardware Verification', *Journal of Automated Reasoning* **5**(2), 127–139.
Fetzer, J. H.: 1988, 'Program Verification: The Very Idea', *Communications of the ACM* **31**(9), 1048–1063.
Fetzer, J. H.: 1991, 'Philosophical Aspects of Program Verification', *Minds and Machines* **1**(2), forthcoming.
Fetzer, J. H. and Martin, C. R.: 1990, ' "The Very Idea", Indeed! An Intellectual Brawl in Three Rounds (For Adults Only)', National Biomedical Simulation Resource, Technical Report no. 1990-2.
Ginsberg, M. L. (Ed.): 1987, *Readings in Nonmonotonic Reasoning*, Los Altos, CA: Morgan Kaufmann.
Gumb, R. D.: 1989, *Programming Logics: An Introduction to Verification and Semantics*, New York, NY: Wiley and Sons.
Hoare, C. A. R.: 1969, 'An Axiomatic Basis for Computer Programming', *Communications of the ACM* **12**, 576–580, 583.
Kowalski, R.: 1979, *Logic for Problem Solving*, New York, NY: North Holland.
Kyburg, H. E., Jr., Loui, R. P., and Carlson, G. N. (Eds.): 1990, *Knowledge Representation and Defeasible Reasoning*, Dordrecht, The Netherlands: Kluwer Academic Publishers.
Letters to the Editor: 1989, 'ACM Forum', *Communications of the ACM* **32**(3), 287–290.
Myers, G. J.: 1979, *The Art of Software Testing*, New York, NY: Wiley and Sons.
Naps, T. L., Nance, D. W., and Singh, B.: 1989, *Introduction to Computer Science: Programming, Problem Solving, and Data Structures*, Alternate Edition, St. Paul, MN: West Publishing Company.
Nute, D.: 1988, 'Defeasible Reasoning: A Philosophical Analysis in Prolog', in J. Fetzer (Ed.), *Aspects of Artificial Intelligence*, Dordrecht, The Netherlands: Kluwer Academic Publishers, pp. 251–288.
Pollock, J. L.: 1974, *Knowledge and Justification*, Princeton, NJ: Princeton University Press.

Rankin, T.: 1988, 'When Is Reasoning Nonmonotonic?', in J. Fetzer (Ed.), *Aspects of Artificial Intelligence*, Dordrecht, The Netherlands: Kluwer Academic Publishers, pp. 289—308.

Reinfrank, M. (Ed.): 1989, *Proceedings of the 2nd International Workshop on Non-Monotonic Reasoning*, New York, NY: Springer-Verlag.

Roth, M. D. and Galis, L. (Eds.): 1970, *Knowing: Essays in the Analysis of Knowledge*, New York, NY: Random House.

Technical Correspondence: 1989, *Communications of the ACM* **32**(3), 287—290.

EPILOGUE

JAMES H. FETZER

PHILOSOPHICAL ASPECTS OF
PROGRAM VERIFICATION

Not least among the fascinating issues confronted by computer science is the extent to which purely formal methods are sufficient to secure the goals of the discipline. There are those, such as C. A. R. Hoare and Edsgar Dijkstra, who maintain that, in order to attain the standing of a science rather than of a craft, computer science should model itself after mathematics. Others, including Richard DeMillo, Richard Lipton, and Alan Perlis, however, deny that the goals of the discipline can be gained by means of purely formal methods.

Much of the debate between adherents to these diverse positions has revolved about the extent to which purely formal methods can provide a guarantee of computer system performance. Yet the ramifications of this dispute extend beyond the boundaries of the discipline itself. The deeper questions that lie beneath this controversy concern the paradigm most appropriate to computer science. The issue not only influences the way in which agencies disburse funding but also the way in which the public views this discipline.

Some of the most important issues that arise within this context concern questions of a philosophical character. These involve 'ontic' (or ontological) questions about the kinds of things computer and programs are as well as 'epistemic' (or epistemological) questions about the kind of knowledge we can possess about thing of these kinds. They also involve questions about crucial differences between 'pure' and 'applied' mathematics and whether the performance of a system when it executes a program can be guaranteed.

The purpose of this essay is to explore the similarities and differences between mathematical proofs, scientific theories, and computer programs. The argument that emerges here suggests that, although they all have the character of syntactic entities, scientific theories and computer programs possess semantic significance not possessed by mathematical proofs. Moreover, the causal capabilities of computer programs distinguish them from scientific theories, especially with respect to the ways they can be tested.

Timothy R. Colburn et al. (eds.), Program Verification, 403—427.
© 1993 *Kluwer Academic Publishers.*

1. THE FORMAL APPROACH TO PROGRAM VERIFICATION

The phrase 'program verification' occurs in two different senses, one of which is broad, the other narrow. In its broad sense, 'program verification' refers to any methods, techniques, or procedures that can be employed for the purpose of assessing software reliability. These methods include testing programs by attempting to execute them and constructing prototypes of the systems on which they are intended to be run in an attempt to discover possible errors, mistakes, or 'bugs' in those programs that need to be corrected.

In its narrow sense, 'program verification' refers specifically to formal methods, techniques, or procedures that can be employed for the same purpose, especially to 'proofs' of program correctness. This approach seeks to insure software reliability by utilizing the techniques of deductive logic and pure mathematics, where the lines that constitute the text of a program are subjected to formal scrutiny. This approach has inspired many members of the community (cf. Linger *et al.*, 1979; Gries, 1979; and Berg *et al.*, 1982).

Thus, while 'program verification' in its broad sense includes both formal and non-formal methods for evaluating reliability, in its narrow sense 'program verification' is restricted to formal methods exclusively. The use of these methods tends to be driven by the desire to put computer science on a sound footing by means of greater reliance on mathematics in order to "define transformations upon strings of symbols that constitute a program, the result of which will enable us to predict how a given computer would behave when under the control of that program" (Berg *et al.*, 1982, p. 1).

The conception of programming as a mathematical activity has been eloquently championed by Hoare, among others, as the following reflects:

Computer programming is an exact science in that all of the properties of a program and all of the consequences of executing it in any given environment can, in principle, be found out from the text of the program itself by means of purely deductive reasoning. (Hoare 1969, p. 576)

Thus, if this position is well-founded, programming ought to be viewed as a mathematical activity and computer science as a branch of mathematics. If it is not well-founded, however, some other paradigm may be required.

2. COMPUTER PROGRAMS AND FORMAL PROOFS
OF CORRECTNESS

No doubt, the conception that underlies Hoare's position exerts an immense intuitive appeal. A computer M, after all, can be viewed abstractly as a set of transformation functions T for effecting changes in its states S:

(I) $M = \langle T, S \rangle.$

A program P in turn can be viewed as a function that transforms a computer from an initial state si to a final state sf when that program is executed E:

(II) $E\langle P, si \rangle = sf.$

(II) thus represents a 'special purpose' instance of a universal machine (I). (The account in this section follows that of Berg $et\ al.$, 1982, Ch. 2 and 3.)

Since every program is intended to effect a change of state from an initial state si before execution to a final one sf after execution, Hoare (1969) introduced a notation that is equivalent to '$\{si\}P\{sf\}$', where '$\{si\}$' denotes the state prior to and '$\{sf\}$' the state after the execution of program P (or, in general, to '$\{X\}P\{Y\}$', where 'X' describes some property satisfied by M prior to and 'Y' another property satisfied by M after the execution of P).

As Berg $et\ al.$ (1982, pp. 20–21) explain, a distinction has to be drawn between $proof\ of\ correctness$ and $proofs\ of\ partial\ correctness$, where the difference depends upon whether or not the execution of program P terminates. When P terminates, then a formal proof that $E\langle P, si \rangle = sf$ becomes possible, where $E\langle P, si \rangle = sf$ if and only if $\vdash \{si\}P\{sf\}$ (that is, if and only if $\{si\}P\{sf\}$ is syntactically derivable on the basis of the axioms defining M).

Given a formal specification of the state transformations that are desired of a program P when it is executed in the general form $\{X\}P\{Y\}$, the aim of a (complete) correctness proof is to demonstrate $\vdash\{X\}P\{Y\}$, where:

(III) $\vdash\{X\}P\{Y\}$ if and only if
            $(si)\,[\vdash X(si) \rightarrow \vdash(P \text{ terminates}) \text{ and } \vdash Y(E\langle P, si \rangle)],$

(employing '$\ldots \rightarrow \_$' as the material conditional 'if ... then $\_$' sign);

and the aim of a (partial) correctness proof is to demonstrate $\vdash \{X\} P \{Y\}$ *if P terminates*, where termination is something that may or may not occur:

(IV) $\vdash \{X\} P \{Y\}^*$ if and only if
         $(si) [\vdash X(si)$ and $\vdash (P$ terminates$) \rightarrow \vdash Y(E \langle P, si \rangle)]$,

where the asterisk "*" attached to $\vdash \{X\} P \{Y\}$ indicates partial correctness. Proofs of correctness are therefore not precluded by the halting problem.

3. THE IMPLIED ANALOGY WITH PURE MATHEMATICS

One need not be a student of the history of mathematics to appreciate the implied analogy with pure mathematics. The model of formal proofs of program correctness immediately brings to mind Euclidean geometry with its formal proofs of geometrical theorems. Indeed, it is a plausible analogy to suppose that formal proofs of program correctness in computer science are precisely analogous to formal proofs of geometrical theorems in Euclid (see Figure 1).

	Mathematics	*Programming*
Objects of Inquiry:	Theorems	Programs
Methods of Inquiry:	Proofs	Verifications

Fig. 1. A plausible analogy.

Thus, by employing deductive reasoning that involves the application of formal rules of inference to the premises of an argument, various assertions could be shown to be valid formulae of Euclidean geometry (that is, their status as theorems could be established). Analogously, by employing deductive reasoning that involves the application of formal rules of inference to the premises of an argument, various assertions might be shown to be valid formulae about computer programs (they could be theorems too!).

Indeed, the realization that programs can be viewed as functions from initial states (or 'input') to final states (or 'outputs'), as (I) and especially (II) above imply, provides an even more persuasive foundation for conceding the force of this comparison. For programs in computer science can be viewed as functions from inputs to outputs and rules of inference in mathematics can be viewed as functions from premises to conclusions (theorems) (see Figure 2).

	Mathematics	*Programming*
Domain:	Premises	Input
Function:	Rules of Inference	Programs
Range:	Theorems	Output

Fig. 2. A more plausible analogy.

As though these analogies were not convincing enough, the use of mathematical demonstrations seems to be warranted on the basis of at least two further benefits that accrue from adopting the mathematical paradigm. One is that the abstract characterization of computing machines represented by (I) and (II) attains a generality that transcends the special characteristics of specific machines. Another is that abstract characterizations also emphasize the propriety of adopting formal methods within formal domains. The use of deductive methodology thus appear appropriate (cf. Berg *et al.*, 1982, p. 9).

4. RECENT ADVOCATES OF PURELY FORMAL METHODS

Hoare has not been alone in a advocating the position that computer programming should model itself after mathematics. A fascinating illustration of a similar perspective has been advanced by William Wulf, who contends:

The *galling* thing about the generally poor quality of much current software is that there is no extrinsic reason for it; perfection is, in principle, possible. Unlike physical devices: (1) There are no natural laws limiting the tolerance to which a program can be manufactured; it can be built *exactly* as specified. (2) There is no Heisenberg uncertainty principle operative; once built, a program will behave exactly as prescribed. And (3) there is no friction or wear; the correctness and performance of a program will not decay with time. (Wulf, 1979, p. 40, original italics)

This conception of perfection in programming where, once built, a program will behave exactly as prescribed, could be called 'Wulfian Perfectionism'.

Similar views have been elaborated more recently by Dijkstra, among others, who has definite views on "the cruelty of really teaching computer science", especially when it is properly taught as a branch of mathematics:

Finally, in order to drive home the message that this introductory programming course is primarily a course in formal mathematics, we see to it that the programming language in question has *not* been implemented on campus so that students are protected from the temptation to test their programs. (Dijkstra, 1989, p. 1404, original italics)

Hoare (1969), Wulf (1979), and Dijkstra (1989) thus represent several decades of commitment to formal methods by influential computer scientists.

It should come as no surprise, of course, that belief in formal methods tends to go hand in hand with the denigration of testing and prototyping. Consider recent remarks by J. Strother Moore of Computation Logic, Inc., in support of the extension of formal methods to whole computer systems:

System verification grew out of dissatisfaction with program verification in isolation. Why prove software correct if it is going to be run on unverified systems? In a verified system, one can make the following startling claim: *if the gates behave as formally modeled, then the system behaves as specified.* (Moore, 1989, p. 409, italics added)

This claim is indeed 'startling'. (We shall explore later whether it is true.)

And consider some recent messages sent out by Hal Render on USENET. On 17 January 1990, for example, Render transmitted the following claims:

The process of proving a program correct should either indicate that there are no errors or should indicate that there are (and often what and where they are). Thus, *successful program proving methods should eliminate the need for testing.* In practice, things are not so straightforward, because verification is tough to do for many kinds of programs, and one still has to contend with erroneous specifications. (Render, 1990a, italics added)

Approximately two weeks later, Render returned to this theme in his response to a USENET critic, whom he dismissed with the following remarks:

No one (except maybe you) thinks that "proving a program correct" means proving absolutely, positively that there is not a single (no, not even one) error in a program.

Since the specification and the verifications can be in error, there is NO (not even one) way to infallibly prove a program correct. I know this, all informed proponents of verification know this, and you should know this (enough people have told you). (Render 1990b, original emphasis; this was not directed toward me.)

This position thus implies that only inadequate specifications, on the one hand, or mistaken reasoning, on the other, can generate program errors (cf. Smith, 1985). Taken together, there can be little doubt that passages such as these from Hoare, Wulf, Dijkstra, Moore, and Render, represent a coherent position. The question thus becomes whether it can be justified.

5. IS THE ANALOGY WITH MATHEMATICS PROPER?

Arguments by analogy compare two things (or kinds of things) with respect to certain properties, contending that, because they share certain properties in common and one of them has a certain additional property, the other has it too. When there are more differences than similarities or few but crucial differences or these arguments are taken to be conclusive, however, arguments by analogy can go astray. Some analogies are faulty. Perhaps programming and mathematics are very different kinds of things.

There is clearly some foundation for a comparison between mathematics and programming, especially since mathematical theorems and computer programs are both syntactical entities that consist of sequences of lines. Computer programs seem to differ from mathematical theorems, however, insofar as they are intended to possess a semantical significance that mathematical theorems (within pure mathematics) do not possess, a difference arising because programs, unlike theorems, are instructions for machines.

Indeed, a comparison between mathematical theorems and computer programs becomes more striking when scientific theories (classical mechanics, special relativity, quantum mechanics, and so forth) are considered as well. For scientific theories, like computer programs, have semantical significance that mathematical proofs do not possess. The lines that make up a program, like the sentences that make up a theory, after all, tend to stand for other things for the users of those programs and those theories.

The specific commands that constitute a program, for example, stand for corresponding operations by means of computing machines, while the generalizations that constitute a theory stand for lawful properties of the physical world. Yet even scientific theories do not possess the causal capabilities of computer programs, which can affect the performance of those machines when they are loaded and then executed. For a more adequate comparison of their general features see Figure 3.

	Mathematics Proofs:	Scientific Theories:	Computer Programs:
Syntactic Entities	Yes	Yes	Yes
Semantic Significance:	No	Yes	Yes
Causal Capability:	No	No	Yes

Fig. 3. A more general comparison.

The comparison that is reflected by Figure 3 suggests rather strongly that the differences between theorems and programs may outweigh their similarities. Whether a difference should make a difference in relation to an analogical argument, however, depends on how strongly it is weighted. If their existence as syntactical entities is all that matters, then computer programs, scientific theories, and mathematical theorems would appear to be exactly on a par. Why should differences such as these matter at all?

6. ARE OTHER IMPORTANT DIFFERENCES BEING OVERLOOKED?

The sections that follow are intended to explain why differences like these are fundamental to understanding programming as an activity and computer science as a discipline. Before pursuing this objective, however, it should be observed that yet another difference has sometimes been supposed to be the fundamental difference between the construction of proofs in mathematics and the verification of programs in computing. This arises from a difference in dependence upon social processing in these disciplines.

DeMillo, Lipton, and Perlis (1979), in particular, have suggested that the success of mathematics is crucially dependent upon a process of social interaction between various mathematicians. Without this social process, they contend, mathematics could not succeed. Because formal proofs of program correctness are complex and boring, however, they doubt there will ever be a similar process of social interaction between programmers. Such comparisons with mathematics thus depend upon a faulty analogy.

Even advocates of formal methods have been willing to acknowledge the significance of social processes within mathematical contexts. Some have accented the subjective character of mathematical proof procedure:

A mathematical proof is an agenda for a repeatable experiment, just as an experiment in a physics or chemistry laboratory. But the main subject in each experiment is another person instead of physical objects or material. *The intended result of the experimenter is a subjective conviction on the part of the other person that a given logical hypothesis leads to a given logical conclusion.* . . . A successful experiment ends in a subjective conviction by a listener or reader that the hypothesis implies the conclusion. (Linger *et al.*, 1979, p. 3, italics added)

While they concede that subjective conviction provides no guarantee of the validity of a proof or the truth of a theorem, Linger *et al.* maintain that this process of social interaction is essential to producing mathematical products.

Since even advocates of formal methods seem to be willing to grant this premise of DeMillo, Lipton, and Perlis' position, the extent to which mathematics is comparable to programming apparently depends on the extent to which programs are like theorems, especially with respect to features that might affect their accessibility to social processing. While these arguments have been highly influential, there are reasons to believe that social interaction should be far more important to mathematics than to programming.

7. ARE SUBJECTIVE 'PROOFS'
THE ONLY AVAILABLE EVIDENCE?

The ground on which Linger *et al.* stake their claim to the importance of proof procedures (within programming as well as within mathematics) — in spite of their subjectivity — is the absence of procedural alternatives:

Why bother with mathematics at all, if it only leads to subjective conviction? *Because that is the only kind of reasoned conviction possible*, and because the principal experimental subject who examines your program proofs is yourself! Mathematics provides language and procedure for your own peace of mind. (Linger *et al.*, 1979, p. 4, italics added)

Although this position may initially appear very plausible, it fails to take into account other features that may distinguish programs and theorems.

After all, if computer programs possess a semantical significance and a causal capability that mathematical theorems do not possess, then there would appear to be opportunities for their evaluation of kinds other than social processing. Scientific theories (such as classical mechanics, special relativity, and so on) are suggestive illustrations, because (almost) no one would want to maintain that their acceptability is exclusively a matter of subjective conviction. Science relies upon observations and experiments.

If scientific theories are viewed as conjectures, then observation and experimentation afford nature opportunities for their refutation. The results of these observations and experiments are not mere matters of subjective conviction. There thus appear to be other methods for evaluating scientific theories that go beyond those available for evaluating proofs of theorems in mathematics. If that is the case, however, then there would appear to be kinds of 'reasoned conviction' besides mathematical proofs.

If computer programs possess causal capability as well as semantical significance, then there should also be means for evaluating their correctness going beyond those available for scientific theories. Prototyping and testing offer opportunities for further kinds of experiments that arise from atempting to execute programs by machine: these are prospects over which we have (almost) complete control! But this indicates yet another kind of 'reasoned conviction' going beyond social processing alone.

8. ARE FORMAL PROOFS OF
PROGRAM CORRECTNESS NECESSARY?

If these reflections are well-founded, then the position of Linger *et al.* (1979), which suggests that mathematical proofs of program correctness are indispensable, appears to be very difficult to sustain. If those who advocate formal methods for evaluating programs want to insist upon the primacy of this methodology, they need to find better ground for their position. And, indeed, that support appears to arise from at least two quite different directions, one of which is ontic, the other epistemic.

The ontic defense consists in characterizing 'programs' as abstract objects to which physical machines have to conform. Consider Dijkstra:

What is a program? Several answers are possible. We can view the program as what turns the general-purpose computer into a special-purpose symbol manipulator, and it does so without the need to change a single wire. . . . I prefer to describe it the other way around. The program is an abstract symbol manipulator which can be turned into a concrete one by supplying a computer to it. After all, *it is no longer the purpose of programs to instruct our machines; these days, it is the purpose of machines to execute our programs.* (Dijkstra, 1989, p. 1401, italics added)

One of the attractions of this approach, no doubt, is its strong intimation that programming only involves reference to purely abstract machines.

The identification of 'program' with abstract symbol manipulators, however, warrants further contemplation. While 'computers' are sometimes characterized as physical symbol systems — by Alan Newell and Herbert Simon (1976), for example — 'programs' are typically supposed to provide instructions that are executed by machine. When he tacitly transforms 'programs' from instructions into machines, Dijkstra thereby distorts the traditional distinction between 'programs' and 'computers'.

The epistemic defense consists in maintaining that formal methods provide the only access route to the kind of knowledge that is required. The maxim that "testing can be used to show the presence of bugs, but never to show their absence" has been promoted, especially by Dijkstra (1972). Such a position seems to conform to Sir Karl Popper's conception of scientific methodology as a deductive domain, where scientific hypotheses can possibly be shown to be false but can never be shown to be true.

The epistemic defense, however, appears to be far weaker than the ontic defense, not least because it almost certainly cannot be justified. A successful compilation supports the inference that certain kinds of bugs are absent, just as a successful execution supports the inference that certain other kinds of bugs are absent. Even the conception of conjectures and (attempted) refutations, after all, must take into account the positive significance of *unsuccessful* attempted refutations as well as the negative significance of *successful* attempted refutations (Fetzer, 1981, esp. Ch. 7).

9. PROGRAMS AS ABSTRACT ENTITIES AND AS EXECUTABLE CODE

The principal defense of the primacy of formal methods in evaluating computer programs, therefore, appears to depend upon the adequacy of the conception of *programs* as "abstract symbol manipulators". Dijkstra offers several alternative accounts that reinforce the ontic defense. Thus,

... Another way of saying the same thing is the following one. A programming language, with its formal syntax and with the proof rules that define its semantics, is a formal system for which program execution provides only a model. It is well-known that formal systems should be dealt with in their own right and not in terms of a specific model. And, again, the corollary is that *we should reason about programs without even mentioning their possible "behavior"*. (Dijkstra, 1989, p. 1403, italics added)

Observe in particular that Dijkstra translates a claim about formal systems into a claim about computer programs without justifying this identification.

Dijkstra further obscures the distinction between 'programs' as abstract objects and 'programs' as executable entities by asserting the benefits of working with the definition of a set rather than with its specific members:

... the statement that a given program meets a certain specification amounts to a statement about *all* computations that could take place under control of that given program. And since this set of computations is defined by the given program, our recent moral says: deal with all computations possible under control of a given program by ignoring them and working with the program. *We must learn to work with program texts while (temporarily) ignoring that they admit the interpretation of executable code.* (Dijkstra, 1989, p. 1403, italics added)

The differences between sets and programs, however, are such that there are several good reasons to doubt that even this position can be justified.

Consider, for example, that sets are completely extensional entities, in the sense that two sets are the same when and only when they have the same members. The set $\{x, y, z\}$, for example, is the same as the set $\{x, x, y, z, z\}$, precisely because they have all and only the same elements. Programs, by comparison, are quite different, because two iterations of the same command qualifies as two commands, even if two iterations of the same element qualify as one member of the corresponding set. The principles that govern extensional entities do not govern causal relations.

Thus, although Dijkstra wants to view programs as "abstract symbol manipulators", he thereby obfuscates fundamental differences between things of two different kinds. 'Programs' understood as *abstract entities* (to which individual concrete machines have to conform) and 'programs' as *executable code* (that causally interacts with physical machines) have enormously different properties. Indeed, their differences are sufficiently great as to bring into doubt the adequacy of the mathematical analogy.

10. THE AMBIGUITY OF 'PROGRAM VERIFICATION'

Part of the seductive appeal of formal methods within this context, I surmise, arises from a widespread reliance upon abstract models. These include abstract models that represent the *program specification*, which is the problem to be solved. (A program is shown to be 'correct' when it is shown to satisfy its specification.) They also include abstract models that represent the *programming language*, which is a tool used in solving such a problem. (Pascal, LISP, and other high-level languages simulate the behavior of abstract machines rather than describe those of target machines.)

Enormous benefits accrue from the use of these abstract models. The construction of programs in Pascal, LISP, and so forth, for example, is far easier than the composition of programs in assembly language, where there is a one-to-one correspondence between program instructions and machine behavior. And, as was observed in Section 3, the abstract characterization of computing machines attains a generality that transcends the special characteristics of specific machines. But there is the correlated risk of mistaking properties of the abstract models for properties of the machines themselves.

Some machines, for example, have 8-bit words, while others have 16-bit (32-bit, ...) words. And while register size makes no difference with respect to many operations, it can affect others [see Tompkins, 1989 and Garland, 1990). The capacity to represent real numbers by numerals obviously depends upon the characteristics of specific machines. While it is clearly possible to construct abstract models that successfully model specific machines, these abstract models no longer possess inherent generality.

The formal approach to program verification thus appears to trade upon an equivocation. While it is indeed possible to construct a formal proof of program correctness with respect to an abstract model of a machine, the significance of that proof depends upon the features of that model. When the model is an abstract model (the axioms of which can be given as stipulations) that does not represent a specific physical machine, a formal proof of correctness can then guarantee that a program will perform as specified (unless mistakes have been made in its construction). Otherwise, it cannot.

11. DISAMBIGUATING 'PROGRAM VERIFICATION'

The differences at stake here can be made explicit by using the term 'PROGRAM' to refer to the operations that can be executed by an abstract machine for which there is no physical counterpart and by using the term "program" to refer to the operations that may be executed by an abstract machine for which there is some physical counterpart. It should then become apparent that, although the properties of abstract machines with no physical counterpart can be stipulated by definition, those of abstract machines with physical counterparts have to be discovered by investigation.

This difference can be illustrated by means of rules of inference such as Hoare has introduced. Consider, for example, the following illustrations:

$Consequence$ 1: If '$\{X\}P\{Y\}$' and '$Y \Rightarrow Z$',
    then infer '$\{X\}P\{Z\}$';
$Consequence$ 2: If '$X \Rightarrow Y$' and '$\{Y\}P\{Z\}$',
    then infer '$\{X\}P\{Z\}$'; and,
$Conjunction$: If '$\{X\}P\{Y\}$' and '$\{X'\}P\{Y'\}$',
    then infer '$\{X \& X'\}P\{Y \& Y'\}$',

(employing '$\ldots \Rightarrow \_$' as the semantical entailment 'if ... then __' sign) (cf. Marcotty and Ledgard, 1986, p. 118). $Consequence$ 1 asserts that, if a line of the form '$\{X\}P\{Y\}$' is true and '$Y \Rightarrow Z$' is true, then a line of the form '$\{X\}P\{Z\}$' will also be true, necessarily. And likewise for $Consequence$ 2, etc.

The problem that arises here is how it is possible to know when a line of the form '$\{X\}P\{Y\}$' is true. Even on the assumption that semantic entailments of the form '$Y \Rightarrow Z$' are knowable *a priori* (on the basis of grammar and vocabulary), that does not provide any foundation for ascertaining the truth of lines that describe the state of the machine **M** before and after the execution of program P. There appear to be only two alternatives. If **M** is an abstract machine with no physical counterpart, axioms such as '$\{X\}P\{Y\}$' can be true of **M** as a matter of stipulation, but not if **M** has a counterpart.

When **M** is an abstract machine with a physical counterpart, then the axioms that are true of **M** are only knowable *a posteriori* (on the basis of observation and experiment). But this means that two different kinds of 'program verification' are possible, only one of which can be conclusive:

(D1)   (conclusive) *absolute verification* can guarantee the behavior of an abstract machine, because its axioms are true by definition;

(D2)   (inconclusive) *relative verification* cannot guarantee the behavior of a counterpart, because its axioms are not definitional truths.

Verifications that are conclusive are only significant for abstract machines, while those that are significant for physical machines are never conclusive.

12. THE DISTINCTION BETWEEN 'PURE' AND 'APPLIED' MATHEMATICS

The differences that have been uncovered here are parallel to those between 'pure' and 'applied' mathematics, where theorems of applied mathematics, unlike those of pure mathematics, run the risk of observational and experimental disconfirmation. A theorem about the natural numbers, say,

(a) $2 + 2 = 4,$

for example, cannot possibly be false within the context of that theory, since it follows from axioms that are true by definition. The application of those same numerical relations to physical phenomena, however, might be false.

Sometimes the results of mathematical descriptions of physical phenomena are surprising, indeed. Consider, for example, the following sentence:

(b)   2 units of water + 2 units of alcohol = 4 units of mixture.

This certainly sounds plausible. Yet it turns out that, as an effect of their atomic structure, molecules of water and molecules of alcohol have the capacity to partially occupy the same volume when they are mixed together. This apparently true sentence, therefore, turns out to be empirically false.

Other examples afford equally interesting illustrations of the problem we have discovered. What truth-value should be assigned to these claims:

(c)   2 drops of water + 2 drops of water = 4 drops of water?

(d)   2 lumps of plutonium + 2 lumps of plutonium = 4 lumps of plutonium?

(e)   2 gaggle of geese + 2 gaggle of geese = 4 gaggle of geese?

Even if the assumed behavior of abstract machines can be known with deductive certainty, the real behavior of physical things can only be known with the empirical uncertainty that accompanies scientific investigations.

But surely this is something that we ought to have expected all along. The example of Euclidean geometry affords an instructive illustration; for, when 'points' are locations without extension and 'lines' are the shortest distances between them, their relations can be ascertained deductively on the basis of the axioms that define them. But as soon as 'lines' are identified with the paths of light rays in space and 'points' with their intersection, these relations can only be discovered empirically by means of observation and experimentation. (Space even turns out to be non-Euclidean!)

13. WHAT ABOUT HOARE'S 'FOUR BASIC PRINCIPLES'?

From this point of view, certain contentions whose truth-values may have been unclear tend to become obvious. When Hoare maintains that,

Computer programming is an exact science in that all of the properties of a program and all of the consequences of executing it in any given environment can, in principle, be found out from the text of the program itself by means of purely deductive reasoning. (Hoare, 1969, p. 576)

it should now be apparent that, while this may be true for PRO-GRAMS on abstract machines, it is certainly false for programs on physical machines.

Similarly, when Dijkstra asserts that, "We must learn to work with program texts while (temporarily) ignoring that they admit the interpretation of executable code" (Dijkstra, 1989, p. 1403), it should be evident that, while this attitude appears to be appropriate for PRO-GRAMS whose properties can be known by deduction and with certainty, it is not appropriate for programs whose properties can only be known by experience and with uncertainty. Indeed, it should be apparent that the very idea of the mathematical paradigm for computer science trades on ambiguity.

Consider the 'Four Principles' of Hoare (1986), which can be viewed as defining the "mathematical paradigm" in application to this discipline:
(1) computers are mathematical machines;
(2) computer programs are mathematical expressions;
(3) a programming language is a mathematical theory; and,
(4) programming is a mathematical activity.
Clearly, the structures and entities that satisfy these relations (as Hoare intended them to be understood) are PROGRAMS and abstract machines.

As soon as attention turns to programs and physical machines, it is evident that the proper comparison is with applied mathematics instead:
(1') computers are *applied* mathematical machines;
(2') computer programs are *applied* mathematical expressions;
(3') a programming language is an *applied* mathematical theory; and,
(4') programming is an *applied* mathematical activity.
To the extent to which computer science can properly be entertained as a mathematical discipline, it qualifies as an applied rather than a pure one.

14. ANOTHER ARGUMENT FOR THE SAME CONCLUSION

The differences between PROGRAMS and programs that have been elaborated here, however, would never have come to light had we failed to appreciate the differences between mathematical theorems, scientific theories, and computer programs. Advocates of formal methods in computer science tend to overlook these differences, especially because they are drawn toward the position "that the meaning or the semantics of a program are precisely equivalent to what the program causes a machine to do" (Berg *et al.*, 1982, p. 9; for a related discussion of 'procedural semantics', see Fodor, 1978). But an equivocation arises at this juncture.

The differences between the positions outlined in Hoare (1969) and in Fetzer (1988) can be formulated to address this point by distinguishing 'programs-as-texts' (unloaded), which consist of sequences of lines, from 'programs-as-causes' (loaded), which affect machine performance. Formal verification invariably involves the application of deductive techniques to programs-as-texts, where it does not even make sense to talk about the application of techniques of deduction to programs-as-causes (cf. Barwise, 1989, p. 848).

Hoare and I both assume (*) that programs-as-causes are represented by programs-as-texts, except that Hoare contends that verifying a program-as-text guarantees what will happen when a program-as-cause is executed, which I deny. Otherwise, in contending that even a successful verification of a program is never enough to guarantee what will happen when that program is executed, I would have simply changed the subject. In that case, we would not have joined issues and we might both be right.

Indeed, a key difference between Hoare's position and my own can be formulated in terms of different versions of thesis (*). For Hoare maintains,

(*')   that programs-as-causes are *appropriately* represented by programs-as-texts;

but I assert that all he is entitled to assume is the strikingly weaker thesis,

(*")   that programs-as-causes are *supposed-to-be* appropriately represented by program-as-texts.

For one of the major points I have sought to convey is that there might be two different ways in which this supposition could be grounded, namely: (a) when the program-as-text concerns an abstract machine for which there is no physical counterpart (where this relation can be true by definition); or, (b) when the program-as-text concerns as abstract machine for which there is a physical counterpart (where this relation must be empirically justified). Hoare's position — which, I believe, is tacitly adopted by the vast majority of those favoring the verificationist position — clearly overlooks this distinction.

15. ARE THERE OTHER ALTERNATIVE POSITIONS?

In their reflections on this debate as it has been carried in pages of the *Communications of the ACM*, John Dobson and Brian Randell (1989) raise an important question namely: Why do many proponents of formal methods in computer science deny that they hold the position that I have attacked? There may be more than one answer to this question, of course, since some of them may simply hold inconsistent positions (perhaps by persisting in the belief that purely formal methods *are* capable of providing guarantees concerning what would happen if a program were executed by a machine).

The situation is also complicated by differences regarding the ultimate importance of social processing as an aspect of computer methodology. If issues concerning social processing that DeMillo, Lipton, and Perlis (1979), among others, have addressed, are left aside, however, then the views of Avra Cohn (1989) on behalf of provers of PROGRAMS and Glenford Myers (1979) on behalf of testers of programs contribute to defining some of the most important alternative positions. The following flow chart (Figure 4) is intended as a summary of the debate's central course without exhausting its details.

Thus, Cohn maintains that formal proofs of correctnesss, even those that involve entire computer systems, cannot possibly guarantee the performance of any real system, even when verified programs are being executed:

Ideally, one would like to prove that a chip such as Viper correctly implements its intended behavior under all circumstances; we could then claim that the chip's behavior was predictable and correct. In reality, neither an actual *device* nor an *intention* is an object to which logical reasoning can be applied. . . . In short, verification involves a pair of *models* that bear an uncheckable and possibly imperfect relation to the intended design and to the actual device. (Cohn, 1989, pp. 131—132, original italics)

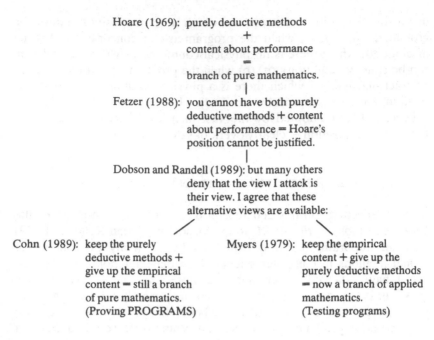

Hoare (1969): purely deductive methods
+
content about performance
=
branch of pure mathematics.
|
Fetzer (1988): you cannot have both purely
deductive methods + content
about performance = Hoare's
position cannot be justified.
|
Dobson and Randell (1989): but many others
deny that the view I attack is
their view. I agree that these
alternative views are available:

Cohn (1989): keep the purely   Myers (1979): keep the empirical
deductive methods +   content + give up the
give up the empirical   purely deductive methods
content = still a branch   = now a branch of applied
of pure mathematics.   mathematics.
(Proving PROGRAMS)   (Testing programs)

Fig. 4. The program verification debate.

Indeed, as she hastens to point out, these points "are not merely phi-losophical quibbles": errors were located even in the models assumed for Viper!

Strictly speaking, however, although specification models and machine models are *deductively* 'uncheckable', they are not therefore *inductively* 'uncheckable' as well, where 'induction' relies upon the kinds of observation and experimentation that computer systems permit. It would therefore be mistaken to assume that the relationship between a specification and the world, on the one hand, or an abstract machine and a counterpart, on the other, are matters about which only subjec-tive opinion is possible. Empirical evidence can even support reasoned convictions about models.

While Cohn disavows responsibility for what happens when a system executes a program, Myers embraces a Popperian conception of testing in which primary emphasis is placed upon and discovery of remaining errors. In his view, "*Testing is the process of executing a program with the intent of finding errors*" (Myers, 1979, p. 5). Yet it is no more necessary to suppose that testing can only discover errors than it is to suppose that only formal proofs of programs can establish their absence. These are two widely held but unjustifiable beliefs that qualify as myths of computer science.

16. WULFIAN PERFECTIONISM CANNOT BE SUSTAINED

It should be obvious by now that Wulfian Perfectionism, which maintains that, once written, a program will execute exactly as prescribed, cannot possibly be correct. The problems with which it cannot cope even extend to those of the Intel 80486 microprocessor, which turned out to have a (potentially fatal) flaw. When certain instructions were executed in one sequence, they executed properly, yet in another sequence, they failed to execute properly (Markoff, 1989, p. 39). Although this specific case was fixed, it is impossible to know which other cases remain to be discovered.

It should also be obvious that Moore's optimism in asserting that, if the gates behave as formally modeled, the system behaves as specified, is unwarranted even for the case of fully verified systems (Moore, 1989, p. 409). The verification of an abstract model, even in relation to an entire computer system, cannot guarantee system performance; even if the gates behave as formally modeled, unless there is no more to that system than the gates themselves! Moreover, unless gate performance can be guaranteed, this claim amounts to a promissory note that can never be cashed in.

And it should also be obvious that the conception of computer science as a branch of pure mathematics cannot be sustained. The proper conception is that of computer science as a branch of applied mathematics, where even that position may not go far enough in acknowledging the limitations imposed by physical devices. Remarks like these from Hamilton Richards,

... the conversion of programming from a craft into a mathematical discipline requires an unorthodox type of mathematics in which the traditional distinction between "pure" and "applied" need not appear. (Richards, 1990, p. viii)

should be recognized as nonsense. They are both misleading and untrue.

Perhaps the resolution of these difficulties is to concede a point made by David Parnas in response to Dijkstra's piece on teaching programming:

There is no engineering profession in which testing and mathematical validation are viewed as alternatives. It is universally accepted that they are complementary and that both are required. (Parnas, 1989, p. 1405)

Similar views are endorsed by Berg *et al.* (1982, p. 124) and by others (including Goodenough and Gerhart (1975) and Gerhart and Yelowitz (1976)). Computer science, after all, has both theoretical and experimental dimensions, where both formal methods and empirical procedures have a place.

17. WHAT POSITIONS SHOULD BE DISTINGUISHED?

As I have elsewhere emphasized (Fetzer and Martin, 1990), some of the issues at stake here involve questions of logic, others matters of methodology, and others issues of verifiability, which can be summarized here. In relation to questions of logic, for example, two positions are in dispute:

Positions of LOGIC:
(T1) Formal proofs of program correctness can provide an absolute, conclusive guarantee of program performance;
(T2) Formal proofs of program correctness can provide only relative, inconclusive evidence concerning program performance.

The purpose of Fetzer (1988) was to establish that (T1) is false but that (T2) is true. I would find it difficult to believe that anyone familiar with the course of this debate could continue to disagree. These are separate from three other positions that arise in relation to questions of methodology:

Positions of METHODOLOGY:
(T3) Formal proofs of program correctness should be the exclusive methodology for assessing software reliability;
(T4) Formal proofs of program correctness should be the principal methodology for assessing software reliability;
(T5) Formal proofs of program correctness should be one among various methodologies for assessing software reliability.

I maintain (T3) is false and, at the ACM Computer Science Conference 90, I argued further that (T4) is false, but I have no doubt that (T5) is true. My opponents on that occasion — David Gries and Mark Ardis — did not choose to defend the merits of (T4). In relation to verification in the broad sense, finally,

Positions on VERIFICATION:

(T6) Program verifications always require formal proofs of correctness;
(T7) Program verifications always require proof sketches of correctness;
(T8) Program verifications always require the use of deductive reasoning.

I maintain that (T6) is clearly false and that (T8) is clearly true but that the truth-value of (T7) is subject to debate. Much appears to hang on whether such proof sketches have to be written down, could be merely thought through, or whatever. If the former, (T7) becomes closer to (T6); if the latter, (T7) becomes closer to (T8). Charlie Martin believes 'hand proofs' may be a suitable standard (Fetzer and Martin, 1990). However these matters may ultimately be resolved, (T1) through (T8) seem to reflect the principal distinctions that must be drawn to understand the range of issues at stake.

18. WHAT IS THE APPROPRIATE ATTITUDE TO ADOPT?

To admit that formal methods have a place in computer science is not to grant that they have the capacity to guarantee program performance. If Berg *et al.* and others have appreciated the situation any more clearly than the Hoares, the Wulfs, the Dijkstras, the Moores, the Renders, and the Richards of the world, they are certainly to be commended. But it does not follow that Hoare, Wulf, Dijkstra, Moore, Render, Richards and their followers do not hold inadequate positions. The views they offer cannot be justified.

If the entire debate has brought into the foreground some indefensible assumptions that have been made by some influential figures in the field, then it will have been entirely worthwhile. It should be obvious by now that pure mathematics provides an unsuitable paradigm for computer science. At the very least, the discipline appears to have some characteristics that are more akin to those that distinguish empirical sciences and others indicating that it ought to be viewed as an engineering discipline instead.

Although the issues that have been addressed here concern questions of ontology and of epistemology, the most important implications of this debate are ethical. The greater the seriousness of the consequences that would ensue from making a mistake, the greater our obligation to insure that it is not made. This suggests that the role for prototyping and testing increases dramatically as life-critical tasks come into play (cf. Blum, 1989). We can afford to run a system that has been only formally assessed if the consequences of mistakes are relatively minor, but not if they are serious.

Formal proofs of program correctness provide one variety of assurance that a system will perform properly. Testing supplies another. Constructing prototypes yields a third. When we deal with abstract models, formal methods are available. When we want to know whether or not a physical system will perform as it is intended, however, there are no alternatives to prototyping and testing. The program verification debate has implications that are practical and ethical as well as theoretical and philosophical. And the future of our species may depend upon how well we understand them.

ACKNOWLEDGEMENT

Special thanks to David Nelson and Chuck Dunlop for their helpful criticism.

REFERENCES

Barwise, J.: 1989, 'Mathematical Proofs of Computer System Correctness', *Notices of the AMS* **36**, 844—851.
Berg, H. K. *et al.*: 1982, *Formal Methods of Program Verification and Specification*, Englewood Cliffs, NJ: Prentice-Hall.
Blum, B.: 1989, 'Formalism and Prototyping in the Software Process', RMI-89-011, Applied Physics Laboratory, Johns Hopkins University.
Cohn, A.: 1989, 'The Notion of Proof in Hardware Verification', *Journal of Automated Reasoning* **5**, pp. 127—139.
DeMillo, R., Lipton, R., and Perlis, A.: 1979, 'Social Processes and Proofs of Theorems and Programs', *Communications of the ACM* **22**, 271—280.
Dijkstra, E. W.: 1972, 'Notes on Structured Programming', in: O. Dahl *et al.* (Eds.), *Structured Programming*, New York, NY: Academic Press.
Dijkstra, E. W.: 1989, 'On the Cruelty of Really Teaching Computing Science', *Communications of the ACM* **32**, 1398—1404.
Dobson, J. and Randell, B.: 1989, 'Viewpoint', *Communication of the ACM* **32**, 420—422.

Fetzer, J. H.: 1981, *Scientific Knowledge*, Dordrecht, The Netherlands: D. Reidel.

Fetzer, J. H.: 1988, 'Program Verification: The Very Idea', *Communications of the ACM* **31**, 1048—1063.

Fetzer, J. H. and Martin, C. R.: 1990, ' "The Very Idea", Indeed!', Technical Report, Department of Computer Science, Duke University.

Fodor, J.: 1978, 'Tom Swift and His Procedural Grandmother', *Cognition* **6**, 229—247.

Garland, D.: 1990, 'Technical Correspondence Letter', *Communications of the ACM*, forthcoming.

Gerhart, S. and Yelowitz, L.: 1976, 'Observations of Fallibility in Applications of Modern Programming Methodologies', *IEEE Transactions on Software Engineering* **2**, 195—207.

Goodenough, J. and Gerhart, S.: 1975, 'Toward a Theory of Test Data Selection', *IEEE Transactions on Software Engineering* **1**, 156—173.

Gries, D. (Ed.): 1979, *Programming Methodology*, New York, NY: Springer-Verlag.

Hoare, C. A. R.: 1969, 'An Axiomatic Basis for Computer Programming', *Communications of the ACM* **12**, 576—580, 584.

Hoare, C. A. R.: 1986, 'Mathematics of Programming', *BYTE* (August), 115—149.

Linger, R. C., Mills, H., and Witt, B.: 1979, *Structured Programming: Theory and Practice*, Reading, MA: Addison-Wesley.

Marcotty, M. and Ledgard, H.: 1989, *Programming Language Landscape: Syntax/ Semantics/Implementations*, 2nd ed., Chicago, IL: Science Research Associates.

Markoff, J.: 1989, 'Top-of-Line Intel Chip Is Flawed', *The New York Times* (Friday, October 27), pp. 25 and 39.

Moore, J. Strother: 1989, 'System Verification', *Journal of Automated Reasoning* **5**, pp. 409—410.

Myers, G. J.: 1979, *The Art of Software Testing*, New York, NY: John Wiley & Sons.

Newell, A. and Simon, H.: 1976, 'Computer Science as Empirical Inquiry: Symbols and Search', *Communications of the ACM* **19**, 113—126.

Parnas, D.: 1989, 'Colleagues Respond to Dijkstra's Comments', *Communications of the ACM* **32**, 1405—1406.

Render, H.: 1990a, Article 755 (comp.software.eng), USENET, 17 January 1990, 20:01:00 GMT.

Render, H.: 1990b, Article 1413 (comp.software.eng), USENET, 1 February 1990, 01:31:30 GMT.

Richards, H.: 1990, 'Foreword', in E. W. Dijkstra (Ed.), *Formal Development of Programs and Proofs*, Reading, MA: Addison-Wesley, pp. vii—ix.

Smith, B. C.: 1985, 'The Limits of Correctness', *Computers and Society* **14**(4) (Winter), 18—28.

Tompkins, H.: 1989, 'Verifying Feature — Bugs', *Communications of the ACM* **32**, 1130—1131.

Wulf, W. A.: 1979, 'Introduction to Part I: Comments on "Current Practice",' in P. Wegner (Ed.), *Research Directions in Software Technology*, Cambridge, MA: MIT Press, pp. 39—43.

SELECTED BIBLIOGRAPHY

I. THE MATHEMATICAL PARADIGM

Bakker, J. W. de: 1968, 'Axiomatics of Simple Assignment Statements', M.R. 94, Mathematisch Centrum, Amsterdam, June 1968.

Burstall, R. M.: 1968, 'Proving Properties of Programs by Structural Induction', Experimental Programming Report No. 17 DMIP, Edinburgh, pp. 41—48.

Carnap, R.: 1953, 'Formal and Factual Science', in: *Readings in the Philosophy of Science*, H. Feigl and M. Brodbeck (Eds.), Appleton-Century-Crofts, New York, pp. 123—128.

Cerutti, E. and Davis, P.: 1969, 'Formac Meets Pappus', *Am. Math. Monthy* **76**, 895—904.

Church, A.: 1940, 'A Formulation of the Simple Theory of Types', *Journal of Symbolic Logic* **5**.

Church, A.: 1941, 'The Calculi of Lambda-Conversion', *Annals of Mathematics Studies* **6**, Princeton University Press, Princeton.

Church, A.: 1952, *Introduction to Mathematical Logic*, Princeton University Press, Princeton.

Courant, R.: 1961, 'In Applied Mathematics: What is Needed in Research and Education — A Symposium', *SIAM Rev* **4**, 297—320.

Davis, M.: 1958, *Computability and Unsolvability*, McGraw-Hill, New York.

Dijkstra, E. W.: 1968, 'A Constructive Approach to the Problem of Program Correctness', *BIT* **8**, 174—186.

Floyd, R.: 1967, 'Assigning Meanings to Programs', *Proceedings of Symposia in Applied Mathematics* **19**, 19—32.

Floyd, R. W.: 1967, 'The Verifying Compiler', *Computer Science Research Review*, Carnegie-Mellon University, Annual Report, pp. 18—19.

Hempel, C. G.: 1949, 'On the Nature of Mathematical Truth', in *Readings in Philosophical Analysis*, Feigl, H. and Sellars, W. (Eds.), Appleton-Century-Crofts, New York, pp. 222—237.

Hempel, C. G.: 1949, 'Geometry and Empirical Science', in *Readings in Philosophical Analysis*, Feigl, H. and Sellars, W. (Eds.), Appleton-Century-Crofts, New York, pp. 238—249.

Hoare, C. A. R.: 1969, 'An Axiomatic Basis for Computer Programming', *Communications of the ACM* **12**, 576—583.

Igarashi, S.: 1968, 'An Axiomatic Approach to Equivalence Problems of Algorithms with Applications', Ph.D. Thesis, Rep. Compt. Centre, U. Tokyo, pp. 1—101.

Kaplan, D. M.: 1967, 'Correctness of a Compiler for Algol-Like Programs', Stanford Artificial Intelligence Memo No. 48, Department of Computer Science, Stanford University.

429

Timothy R. Colburn et al. (eds.), Program Verification, 429—443.
© 1993 *Kluwer Academic Publishers.*

King, J. C.: 1969, 'A Program Verifier', Ph.D. thesis, Computer Sciences Department, Carnegie-Mellon University, Pittsburgh, PA.

King, J. C.: 1971, 'A Program Verifier', IFIPS Congress, Ljubljana, Yugoslavia, August 1971.

Kleene, S. C.: 1953, 'Recursive Predicates and Quantifiers', *Transactions of the American Mathematical Society* **53**, 41.

Knuth, D. E.: 1968, *The Art of Computer Programming*, Addison-Wesley, Reading, MA.

Landin, P. J. and Burstall, R. M.: 1969, 'Programs and Their Proofs: An Algebraic Approach, *Machine Intelligence* **4**.

Laski, J.: 1968, 'Sets and Others Types', *ALGOL Bull.* **27**.

McCarthy, J.: 1960, 'Recursive Functions of Symbolic Expressions and Their Computation by Machine, Part I', *Communications of the ACM* **3**(April), 184—195.

McCarthy, J.: 1960, *The Lisp Programmer's Manual*, MIT Computation Center.

McCarthy, J.: 1961, 'Checking Mathematical Proofs by Computer', *Proc. American Mathematical Society Symposium on Recursive Function Theory*, New York.

McCarthy, J.: 1962, 'Towards a Mathematical Science of Computation', *Proceedings of the IFIP Congress 62*, pp. 21—28.

McCarthy, J.: 1964, 'A Formal Description of a Subset of ALGOL', *Proc. Conference on Formal Language Description Languages*, Vienna.

McCarthy, J.: 1965, 'Problems in the Theory of Computation', *Proc. IFIP Congress 65*, Vol. 1, pp. 219—222.

McCarthy, J. and Painter, J. A.: 1967, 'Correctness of a Compiler for Arithmetic Expressions', in *Mathematical Aspects of Computer Science*, American Mathematical Society, Providence, RI, pp. 33—41.

Naur, P.: 1966, 'Proof of Algorithms by General Snapshots', *BIT* **6**, 310—316.

Naur, P.: 1969, 'Programming by Action Clusters', *BIT* **9**(3), 250—258.

Naur, P., *et al.*: 1960, 'Report on the Algorithmic Language ALGOL 60', *Communications of the ACM* **3**(May).

Painter, J. A.: 1967, 'Semantic Correctness of a Compiler for an Algol-Like Language', Stanford Artificial Intelligence Memo No. 44, Department of Computer Science, Stanford University.

Yanov, Y. I.: 1958, 'Logical Operator Schemes', *Kybernetika* **1**.

Yanov, Y. I.: 1960, 'The Logical Schemes of Algorithms', from *Problems of Cybernetics I*, Pergamon Press Ltd., New York, pp. 82—140.

II. ELABORATING THE PARADIGM

Abrial, J. R.: 1980, 'The Specification Language Z: Syntax and Semantics', Oxford University Computing Laboratory, Programming Research Group, Oxford.

Abrial, J. R. and Schuman, S. A.: 1979, 'Specification of Parallel Processes', in *Semantics of Concurrent Computation* (Proc. Int'l Symp., Evian, France, July 2—4, 1979), G. Kahn (Ed.), Springer-Verlag, Berlin, New York.

Abrial, J., Schuman, S. A. and Meyer, B.: 1980, 'A Specification Language', in *On the Construction of Programs*, R. McNaughten and R. C. McKeag (Eds.), Cambridge University Press.

Aho, A. V., Hopcroft, J. E. and Ullman, J. D.: 1974, *The Design and Analysis of Computer Algorithms*, Addison-Wesley, Reading, Mass.

Alagic, S. and Arbib, M. A.: 1978, *The Design of Well-Structured and Correct Programs*, Springer-Verlag.

Alford, M. W.: 1977, 'A Requirements Engineering Methodology for Real-Time Processing Requirements', *IEEE Trans. Software Engineering* SE-3(1), 60—69.

Backhouse, R. C.: 1987, *Program Construction and Verification*, Prentice-Hall International, U.K.

Backus, J. W.: 1959, 'The Syntax and Semantics of the Proposed International Algebraic Language of the Zurich ACM-GAMM Conference', *Proc. International Conf. on Information Processing*, UNESCO, pp. 125—132.

Balzer, R.:1981, 'Transformational Implementation: An Example', *IEEE Transactions on Software Engineering* SE-7(1), 3—14.

Balzer, R.: 1985, 'A 15 Year Perspective on Automatic Programming', *IEEE Transactions on Software Engineering* SE-11(11), 1257—1267.

Barstow, D. R.: 1980, 'The Roles of Knowledge and Deduction in Algorithm Design', Yale Research Report 178.

Barstow, D. R.: 1985, 'Domain-Specific Automatic Programming', *IEEE Trans. Software Engineering* SE-11, 1321—1336.

Bates, J. L.: 1979, 'A Logic for Correct Program Development', Ph.D. Thesis, Cornell University.

Bates, J. L. and Constable, R. L.: 1982, 'Proofs as programs', Cornell University Technical Report.

Bauer, F. L.: 1971, 'Software Engineering', IFIP congress 71, North-Holland, I, pp. 267—274.

Bauer, F. L.: 1982, 'From Specifications to Machine Code: Program Construction Through Formal Reasoning', Sixth International Conference on Software Engineering.

Bauer, F. L., *et al*: 1981, 'Programming in a Wide Spectrum Language: A Collection of Examples', *Science of Computer Programming* 1, 73—114.

Beckman, L., Haraldsson, A., Oskarsson, O. and Sandewall, E.: 1976, 'A Partial Evaluator and its use as a Programming Tool', *Artificial Intelligence* 7, 319—357.

Broy, M. and Pepper, P.: 1981, 'Program Development as a Formal Activity', *IEEE Transactions on Software Engineering* SE-7(1), 14—22.

Bruijn, N. G. de: 1987, 'The Mathematical Vernacular, A Language for Mathematics with Typed Sets', *Workshop on Programming Logic*, pp. 1—36.

Budde, R., K. Kuhlenkamp, L. Mathiassen, H. Zullighoven: 1984, 'Approaches to Prototyping', *Proceedings of the Working Conference on Prototyping*, Springer-Verlag Berlin, Heidelberg, New York, Tokyo.

Burstall, R. M. and Darlington, J.: 1977, 'A Transformation System for Developing Recursive Programs', *Journal of the ACM* 24(1), 44—67.

Burstall, R. M. and Goguen, J. A.: 1977, 'Putting Theories Together to Make Specifications', *Proc. Fifth Int'l Joint Conf. Artificial Intelligence*, Cambridge, Mass., pp. 1045—1058.

Buxton, J. N. and B. Randell (Eds.): 1969, 'Software Engineering Techniques: Report on a conference sponsored by the NATO Science Committee', Brussels, Scientific Affairs Division, NATO, p. 5.

Cadiou, J. and Levy, J.: 1973, 'Mechanizable Proofs about Parallel Processes', *Proc. 14th IEEE Symposium on Switching Theory and Automata*, pp. 34—48.

Caracciolo di Forino, A.: 1968, 'String Processing Languages and Generalized Markov Algorithms', in *Symbol Manipulation Languages and Techniques*, D. G. Bobrow (Ed.), North Holland, Amsterdam.

Cheatham, T. E. and Wegbreit, B.: 1972, 'A Laboratory for the Study of Automatic Programming', AFIPS Spring Joint Computer Conference, Vol. 40.

Cheatham, T. E., Townley, J. A. and Holloway, G. H.: 1979, 'A System for Program Refinement', Fourth International Conference on Software Engineering, pp. 53—63.

Clark, K. and Darlington, J.: 1980, 'Algorithm Classification Through Synthesis', *Computer Journal* **23**(1).

Clint, M. and Hoare, C. A. R.: 1972, 'Program Proving: Jumps and Functions', *Acta Informatica* **I**, 214—224.

Cohn, A.: 1979, 'Machine Assisted Proofs of Recursion Implementation', Ph.D. Thesis, Dept. of Computer Science, University of Edinburgh.

Cristian, F.: 1985, 'On Exceptions, Failure and Errors', *Technology and Science of Informatics* **4**(1).

Cullyer, W. J.: 1985, 'Viper Microprocessor: Formal Specification', Royal Signals and Radar Establishment Report No. 85013.

Cullyer, W. J.: 1986, 'Viper — Correspondence between the Specification and the "Major State Machine",' Royal Signals and Radar Establishment Report No. 86004.

Cullyer, W. J.: 1987, 'Implementing Safety-Critical Systems: The Viper Microprocessor', in *VLSI Specification, Verification and Synthesis*, G. Birtwistle and P. A. Subrahmanyam (Eds.), Kluwer.

Detlefsen, M. and Luker, M.: 1980, 'The Four-Color Theorem and Mathematical Proof', *The Journal of Philosophy* **77**(12), 803—820.

Dijkstra, E. W.: 1971, 'Notes on Structured Programming', in *Structured Programming*, O. J. Dahl, E. W. Dijkstra, C. A. R. Hoare, (Eds.), Academic Press.

Dijkstra, E. W.: 1972, 'The Humble Programmer', *Communications of the ACM* **15**(10), 859—866.

Dijkstra, E. W.: 1976, *A Discipline of Programming*, Prentice-Hall, Englewood Cliffs, N.J.

Dijkstra, E. W.: 1977, 'A Position Paper on Software Reliability', *Software Engineering Notes, ACM SIGSOFT* **2**(5), 3—5.

Dijkstra, E. W.: 1982, *Selected Writings on Computing: A Personal Perspective*, Springer-Verlag, New York.

Dybjer, P., Nordstrom, B., Petersson, K., and Smith, J. M.: 1987, Workshop on Programming Logic, Programming Methodology Group, Univ. of Goteborg and Chalmers Univ. of Technology, Report 37.

Ershov, A. P.: 1978, 'On the Essence of Compilation', in *Formal Descriptions of Programming Concepts*, E. J. Neuhold (Ed.), North-Holland.

Feather, M. S.: 1982, 'A System for Assisting Program Transformation', *ACM Transactions on Programming Languages and Systems* **4**(1), 455—460.

Floyd, R. W.: 1979, 'The Paradigms of Programming', *Communications of the ACM* **22**(8), 455—460.

Gilb, T.: 1988, *Principles of Software Engineering Management*, Addison-Wesley, Reading, MA.

Goad, C.: 1982, 'Automatic Construction of Special-Purpose Programs', 6th Conference on Automated Deduction.

Goguen J.: 1980, 'Thoughts on Specification, Design and Verification', *Software Engineering Notes ACM SIGSOFT* **5**(3), 29—33.

Good, D.: 1970, 'Toward a Man-Machine System for Proving Program Correctness', Ph.D. thesis, University of Wisconsin.

Gordon, M.: 1985, 'HOL: A Machine Oriented Formulation of Higher-Order Logic', University of Cambridge, Computer Laboratory, Tech. Report No. 68.

Gordon, M., Milner, R., and Wadsworth, C. P.: 1979, *Edinburgh LCF*, Lecture Notes in Computer Science No. 78, Springer-Verlag.

Green, C. C. and Barstow, D. R.: 1978, 'On Program Synthesis Knowledge', *Artificial Intelligence* **10**, 241.

Green, C., et al.: 1981, 'Research on Knowledge-Based Programming and Algorithm Design', Kestrel Institute Technical Report.

Gries, D.: 1979, 'Current Ideas in Programming Methodology', in P. Wegner (Ed.), *Research Directions in Software Technology*, MIT Press, Cambridge, MA, pp. 254—275.

Gries, D.: 1981, *The Science of Computer Programming*, Springer, New York.

Gries, D. (Eds.): 1979, *Programming Methodology*, New York, NY: Springer Verlag.

Haken, W., Appel, K., and Koch, J.: 1977, 'Every Planar Map is Four-Colorable', *Illinois Journal of Mathematics* **21**(84), 429—567.

Hanna, K. and Daeche, N.: 1987, 'An Algebraic Approach to Computational Logic', Workshop on Programming Logic 1987, pp. 301—326.

Henhapl W. and Jones, C. B.: 1978, 'A Formal Definition of ALGOL 60 as described in the 1975 Modified Report', in *The Vienna Development Method: The Meta-Language*, D. Bjorner and C. B. Jones (Eds.), Springer Lecture Notes in Computer Science 61, Berlin, Heidelberg, New York, pp. 305—336.

Higman, B.: 1967, *A Comparative Study of Programming Languages*, American Elsevier, New York.

Hill, I. D.: 1972, 'Wouldn't It Be Nice If We Could Write Computer Programs In Ordinary Englsih — Or Would It?', *BCS Computer Bulletin* **16**(6), 306—312.

Hoare, C. A. R.: 1971, 'Proof of a Program: FIND', *Communications of the ACM* **14**m 39—45.

Hoare, C. A. R.: 1972, 'Proof of Correctness of Data Representations', *Acta Informatica* **1**(4), 271—281.

Hoare, C. A. R.: 1984, 'Programming: Sorcery or Science', *IEEE Software* **1**(2), 5—16.

Hoare, C. A. R.: 1986, 'Mathematics of Programming', *BYTE* (August 1986), 115—118, 120, 122, 124, and 148—149.

Hoare, C. A. R.: 1987, 'An Overview of Some Formal Methods for Program Design', *IEEE Computer* (Sept.), 85—91.

Hoare, C. A. R., et al.: 1987, 'Laws of Programming', *Communications of the ACM* **30**(8), 672—686.

Hoare, C. A. R., et al.: 1987, 'Corrigenda: Laws of Programming', *Communications of the ACM* **30**(9), 770—770.

Jackson, M. A.: 1975, *Principles of Program Design*, Academic Press, London.

Jones, C. B.: 1978, 'The Meta-Language: A Reference Manual', in *The Vienna Development Method: The Meta-Language*, D. Bjorner and C. B. Jones (Eds.), Springer Lecture Notes in Computer Science 61, Berlin, Heidelberg, New York, pp. 218—277.

Jones, C. B.: 1980, *Software Development: A Rigorous Approach*, Prentice-Hall, Englewood Cliffs, NJ.

Keen, P. G. W.: 1981, 'Information Systems and Organizational Change', *Communications of the ACM* **24**(1), 24—33.

Kernighan, B. W. and Plauger, P. J.: 1974, *The Elements of Programming Style*, McGraw-Hill, New York.

King, J. C.: 1975, 'A New Approach to Program Testing', *Proc. IEEE Intn'l. Conf. Reliable Software*, pp. 228—233.

King, J. C. and Floyd, R. W.: 1972, 'An Interpretation-Oriented Theorem Prover over Integers', *Journal of Computer and System Sciences* **6**(4), 305—323.

Knuth, D. E.: 1974, 'Structured Programming with Goto Statements', *Computing Surveys* **6**(4), 261—301.

Kriesel, G.: 1981, 'Neglected Possibilities of Processing Assertions and Proofs Mechanically: Choice of Problems and Data', in *University-Level Computer-Assisted Instruction at Stanford: 1968—1980*, Stanford University.

Ledgard, H. F.: 1973, *Programming Proverbs*, Hayden Book Company, Rochelle Park, NJ.

Linger, R. C., Mills, H. and Witt, B.: 1979, *Structured Programming: Theory and Practice*, Reading, MA: Addison-Wesley.

Liskov, B.: 1980, 'Modular Program Construction Using Abstractions', in *Abstract Software Specifications*, D. Bjorner (Ed.), Springer Lecture Notes in Computer Science 86, Berlin-Heidelberg, New York, pp. 354—389.

Liskov, B. and Zilles, S.: 1977, 'An Introduction to Formal Specifications of Data Abstractions', in *Current Trends in Programming Methodology* **1**, R. T. Yeh (Ed.), Prentice-Hall, Englewood Cliffs, NJ, pp. 1—32.

Locasso, R., Scheid, J., Schorre, V. and Eggert, P. R.: 1980, 'The Ina Jo Specification Language Reference Manual', Technical Report TM-(L)-/6021/001/00, System Development Corporation, Santa Monica, CA.

London, R. L.: 1970, 'Computer Interval Arithmetic: Definition and Proof of Correct Implementation', *Journal of the ACM* **17**.

Manna, Z. and Waldinger, R.: 1979, 'Synthesis: Dreams := Programs', *IEEE Transactions on Software Engineering* **SE-5**(4).

Manna, Z. and Waldinger, R.: 1987, 'The Deductive Synthesis of Imperative LISP Programs', *Workshop on Programming Logic*, pp. 39—45.

Martin-Lof, P.: 1979, 'Constructive Mathematics and Computer Programming', *Proc. 6th International Congress of Logic, Methodology and Philosophy of Science*, Hannover, North Holland, Amsterdam, pp. 153—175.

Meyer, B.: 1980, 'A Basis for the Constructive Approach to Programming', in *Information Processing 80*, Proc. IFIP World Computer Congress, (Tokyo, Japan, Oct. 6—9, 1980), S. H. Lavington (Ed.), North-Holland, Amsterdam, pp. 293—298.

Meyer, B.: 1984, 'A System Description Method', in *Int'l Workshop on Models and*

Languages for Software Specification and Design, R. G. Babb II and Ali Mili (Eds.), Orlando, Fla., pp. 42—46.

Meyer, B.: 1985, 'On Formalism in Specifications', *IEEE Software* (January), 6—26.

Mills, H. D.: 1972, 'Mathematical Foundations of Structured Programming', IBM Report FSC 72-6012.

Milner, R.: 1971, 'An Algebraic Definition of Simulation Between Programs', CS 205, Stanford University.

Moher, T. and Schneider, G. M.: 1982, 'Methodology and Experimental Research in Software Engineering', *Int. Journal of Man-Machine Studies* **16**, 65—87.

Moran, R. M. de, Hill I. D., and Wichman, B. A.: 1976, 'Modified report on the Algorithmic Language ALGOL 60', *Computer Journal* **19**, 364—379.

Morgan, C. and Sufrin, B.: 1984, 'Specification of the Unix File System', *IEEE Trans. Software Engineering* **SE-10**(2), 128—142.

Morris, F. L.: 1972, 'Correctness of Translations of Programming Languages', Technical Report CS-72-303, Stanford University Computer Science Department.

Musser, D. R.: 1980, 'Abstract Data Type Specification in the AFFIRM System', *IEEE Trans. Software Engineering* **SE-6**(1), 24—32.

Naur, P.: 1974, 'What Happens During Program Development — An Experimental Study', in *Systemering 75*, M. Lundberg and J. Bubenko (Eds.), Studentlitteratur, Lund, Sweden, pp. 269—289.

Naur, P.: 1974, *Concise Survey of Computer Methods*, Petrocelli Books, New York.

Naur, P.: 1982, 'Formalization in Program Development', *BIT* **22**, 437—453.

Naur, P. and Randell, B. (Eds.): 1970, 'Software Engineering: Report on a conference sponsored by the NATO Science Committee', Brussels, Scientific Affairs Division, NATO.

Oskarsson, O.: 1982, 'Mechanisms of Modifiability in Large Software Systems', Linkoping Studies in Science and Technology, Dissertations, No. 77, Linkoping.

Owicki, S. and Gries, D.: 1976, 'An Axiomatic Proof Technique for Parallel Programs', *Acta Informatica* **6**, 319—340.

Paulin-Mohring, C.: 1987, 'An Example of Algorithms Development in the Calculus of Constructions: Binary Search for the Computation of the Lambo Function', Workshop on Programming Logic 1987, pp. 123—135.

Popek, G. *et al.*: 1977, 'Notes on the Design of Euclid', *Proc. Conf. on Language Design for Reliable Software*, SIGPLAN Notices **12**(3), 11—18.

Proceedings of Conference on Proving Assertions about Programs (1972), SIGPLAN Notices 7.

Proctor, N.: 1985, 'Restricted Access Processor Message Block Processing System Formal Top-Level Specification', Technical Report TR-83002, Sytek.

Reif, J. and Scherlis, W. L.: 1982, 'Deriving Efficient Graph Algorithms', Carnegie-Mellon University Technical Report.

Rich, C. and Shrobe, H.: 1978, 'Initial Report on a Lisp Programmer's Apprentice', *IEEE Transactions on Software Engineering* **SE-4**(6), 456—467.

Rich, C. and Waters, R. C.: 1988, 'Automatic Programming: Myths and Prospects', *Computer* (August), 40—51.

Robinson L. and Roubine, O.: 1980, 'Special Reference Manual', Stanford Research Institute.

Ross, D. T. and Schoman, K. E., Jr.: 1967, 'Structured Analysis for Requirement Definitions', *IEEE Trans. Software Engineering* **SE-3**(1), 6—15.

Scherlis, W. L.: 1981, 'Program Improvement By Internal Specialization', *8th Symposium on Principles of Programming Languages*, pp. 41—49.

Scherlis, W. L. and Scott, D. S.: 1983, 'First Steps Towards Inferential Programming', *Information Processing 83*, pp. 199—212.

Schwartz, J. T.: 1973, 'On Programming, An Interim Report on the SETL Project', Courant Institute of Mathematical Sciences, New York University.

Schwartz, J. T.: 1977, 'On Correct Program Technology', Courant Institute of Mathematical Sciences, New York University.

Sintzoff, M.: 1980, 'Suggestions for Composing and Specifying Program Design Decisions', *International Symposium on Programming*, Springer-Verlag Lecture Notes in Computer Science.

Sufrin, B.: 1982, 'Formal Specification of a Display-Oriented Text Editor', *Science of Computer Programming* **1**(2).

Vuillemin, J. E.: 1973, 'Proof Techniques for Recursive Programs', Technical Report CS-73-393, Stanford University Computer Science Department.

Wand, M.: 1980, 'Continuation-Based Program Transformation Strategies', *Journal of the ACM* **27**(1), 164—180.

Wile, D. S.: 1981, 'Program Developments as Formal Objects', USC/Information Sciences Institute Technical Report.

Zemanek, H.: 1980, 'Abstract Architecture', in *Abstract Software Specifications*, D. Bjorner (Ed.), Springer Lecture Notes in Computer Science 86, Berlin, Heidelberg, New York, pp. 1—42.

III. CHALLENGES, LIMITS, AND ALTERNATIVES

Abramsky, S.: 1987, 'Domain Theory in Logical Form', *Workshop on Programming Logic*, pp. 191—208.

Agresti, W. W.: 1986, *New Paradigms for Software Development*, IEEE Computer Society Press, Order Number 707, Washington D.C.

Backhouse, R.: 1987, 'Overcoming the Mismatch Between Programs and Proofs', *Workshop on Programming Logic*, pp. 116—122.

Bauer, F. L.: 1985, 'Where Does Computer Science Come From and Where is It Going?', in: Neuhold, E. J. and Ghroust, G. (Eds.), *Formal Models in Programming*, North Holland.

Blum, B.: 1991, 'Formalism and Prototyping in the Software Process', to be published in *Information and Decision Technologies*.

Blum, B. I.: 1989, *TEDIUM and the Software Process*, MIT Press, Cambridge, MA.

Boehm, B. W.: 1976, 'Software Engineering', *IEEE Transactions on Computers* **C-25**(12), 1226—1241.

Boehm, B. W.: 1981, *Software Engineering Economics*, Prentice-Hall.

Boehm, B. W.: 1988, 'A Spiral Model of Software Development and Enhancement', *Computer* **21**(5), 61—72.

Brooks, F. P.: 1975, *The Mythical Man-Month*, Addison-Wesley, Reading, MA.

Brooks, F. P.: 1987, 'No Silver Bullet', *Computer* **20**(4), 10—19.

Brooks, R. E.: 1980, 'Studying Programmer Behavior Experimentally', *Communications of the ACM* **23**(4), 207—213.

Dreyfus, H. L.: 1972, *What Computers Can't Do — The Limits of Artificial Intelligence*, Harper and Row, New York.

Feyerabend, P.: 1978, *Against Method*, Verso Editions, London.

Fletcher, J. and McMillan, B.: 1984, 'Report of the Study on Eliminating the Threat Posed by Nuclear Ballistic Missiles', Vol. 5, Battle Management, Communications, and Data Processing, U.S. Department of Defense, February.

Floyd, C.: 1981, 'A Process-Oriented Approach to Software Development', in *Systems Architecture, Proceedings of the 6th European ACM Regional Conference*, Westbury House, pp. 285—294.

Floyd, C.: 1984, 'A Systematic Look at Prototyping', in: R. Budde, *et al.* (Eds.) *Approaches to Prototyping*, Springer-Verlag, pp. 1—18.

Floyd, C.: 1986, 'A Comparative Evaluation of System Development Methods', in: T. W. Olle, H. G. Sol, A. A. Verrijn-Stuart (Eds.), *Information Systems Design Methodologies: Improving the Practice*, North-Holland, pp. 163—172.

Floyd, C.: 1987, 'Outline of a Paradigm Change in Software Engineering', in *Computers and Democracy: A Scandinavian Challenge*, Gower Publishing Company, England, pp. 191—210.

Floyd, C. and Keil, R.: 1986, 'Adapting Software Development for Systems Design with the User', in U. Briefs, C. Ciborra, L. Schneider (Eds.), *System Design For, With and By the Users*, North-Holland, pp. 163—172.

Gerhart, S. and Yelowitz, L.: 1976, 'Observations of Fallibility in Applications of Modern Programming Methodologies', *IEEE Transactions on Software Engineering* **2**.

Gomma, H. and Scott, D. B. H.: 1981, 'Prototyping as a Tool in the Specification of User Requirements', *Proc. Int. Conf. S.E.*, 333—342.

Goodenough, J. B. and Gerhart, S.: 1975, 'Towards a Theory of Test Data Selection', *Proc. Third Int'l Conf. Reliable Software*, Los Angeles, 1975, pp. 493—510. Also published in *IEEE Trans. Software Engineering* **SE-1**(2), 156—173.

Goodenough, J. B. and Gerhart, S.: 1977, 'Towards a Theory of Test: Data Selection Criteria', in: *Current Trends in Programming Methodology* **2**, R. T. Yeh (Ed.), Prentice-Hall, Englewood Cliffs, N.J., pp. 44—79.

Hekmatpour, S. and Ince, D.: 1988, *Software Prototyping, Formal Methods and VDM*, Addison-Wesley, Reading, MA.

Lehman, M. M.: 1980, 'Programs, Life Cycles and Laws of Software Evolution', *Proceedings of the IEEE* **68**, 1060—1076.

Lehman, M. M., Stenning, V., and Turski, W. M.: 1984, 'Another Look at Software Design Methodology', *ACM SIGSOFT SEN* **9**(3).

Lientz, B. P. and Swanson, E. G.: 1980, *Software Maintenance Management*, Addison-Wesley, Reading, MA.

Maibaum, T. S. E. and Turski, W. M.: 1984, 'On What Exactly Is Going On When Software Is Developed Step-by-Step', Seventh Int. Conf. S.E., pp. 528—533.

Myers, G. J.: 1978, 'A Controlled Experiment in Program Testing and Code Walkthroughs/Inspections', *Communications of the ACM* **21**(9), 760—768.

Myers, G. J.: 1979, *The Art of Software Testing*, New York, NY: Wiley and Sons.

Naur, P.: 1975, 'Programming Languages, Natural Languages, and Mathematics', *Communications of the ACM* **18**(12), 676—683.

Naur, P.: 1983, 'Program Development Studies Based on Diaries', in: *Psychology of Computer Use*, Academic Press, London, pp. 159—170.

Naur, P.: 1984, 'Programming as Theory Building', in *Microprocessing and Microprogramming* **15**, North-Holland, pp. 253—261.

Naur, P.: 1989, 'The Place of Strictly Defined Notation in Human Insight', unpublished paper from 1989 Workshop on Programming Logic, Bastad, Sweden.

Newell, A. and Simon, H.: 1976, 'Computer Science as Empirical Inquiry: Symbols and Search', *Communications of the ACM* **19**, 113—126.

Nygaard, K. and Handlykken, P.: 1981, 'The System Development Process — Its Setting, Some Problems and Needs for Methods', in: H. Hunke (Ed.), *Proceedings of the Symposium on Software Engineering Environments*, North-Holland, pp. 157—172.

Parnas, D. L.: 1985, 'Software Aspects of Strategic Defense Systems', *Communications of the ACM* **28**(12), 1326—1335.

Parnas, D. L. and Clements, P. C.: 1986, 'A Rational Design Process: How and Why to Fake it', *IEEE Trans. Software Eng.* **SE-12**(2), 251—257.

Royce, W. W.: 1979, 'Managing the Development of Large Software Systems', *IEEE WESCON*, 1—9.

Smith, B. C.: 1985, 'Limits of Correctness in Computers', Center for the Study of Language and Information Report No. CSLI-85-36.

Smith, B. C.: 1985, 'The Limits of Correctness', *Computers and Society, ACM* **14**(4), 18—26.

Swartout, W. and Balzer, R.: 1982, 'On the Inevitable Intertwining of Specification and Implementation', *Communications of the ACM* **25**(7), 438—440.

Turski, W. M.: 1985, 'The Role of Logic in Software Enterprise', *Proc. Int. Conf. S.E.*, 400ff.

Turski, W. M. and Maibaum, T. S. E.: 1987, *The Specification of Computer Programs*, Addison-Wesley, Workingham, England.

Waldrop, M. M.: 1989, 'Will the Hubble Telescope Compute?', *Science* **243**, 1437—1439.

Weitzel, J. R. and Kerschberg, L.: 1989, 'Developing Knowledge-Based Systems: Reorganizing the System Development Life Cycle', *Communications of the ACM* **32**, 482—488.

Zave, P.: 1984, 'The Operational Versus the Conventional Approach to Software Development', *Communications of the ACM* **27**, 104—118.

IV. FOCUS ON FORMAL VERIFICATION

Abrahams, P.: 1990, 'Specifications and Illusions', *Communications of the ACM* **31**(5), 480—481.

Ardis, M. A., Gerhart, D. *et al.*: 1989, 'Editorial Process Verification', ACM Forum, *Communications of the ACM* **32**, 287—288.

Barwise, J.: 1989, 'Mathematical Proofs of Computer System Correctness', *Notices of the AMS* **36**(7), 844—851.

Barwise, J.: 1989, 'Mathematical Proofs of Computer System Correctness! A Response', *Notices of the ACM* **36**(10), 1352—1353.

Barwise, J.: 1990, 'More on Proving Computer Correctness', *Notices of the AMS* **31**(2), 123—124.

Barwise, J.: 1980, 'The Final Word on Program Verification', *Notices of the AMS* **37**(5), 562—563.

Bassett, P. G.: 1987, 'Brittle Software: A Programming Paradox', *J. Infor. Systs. Mgmt.* **4**(7).

Benacerraf, P. and Putnam, H. (Eds.): 1964, *Philosophy of Mathematics: Selected Readings*, Prentice-Hall, Englewood Cliffs, N.J.

Bentley, J.: 1986, *Programming Pearls*, Addison-Wesley.

Berg, H. K. *et al.*: 1982, *Formal Methods of Program Verification and Specification*, Englewood Cliffs, NJ: Prentice-Hall.

Bochmann, G. *et al.*: 1989, 'Trace Analysis or Conformance and Arbitration Testing', *IEEE Trans. Software Eng.* **15**(11), 1347—1356.

Bochner, S.: 1966, *The Role of Mathematics in the Rise of Science*, Princeton Univ. Press, Princeton, N.J.

Boyer, R. S. and Moore, J. S.: 1979, *A Computational Logic*, Academic Press.

Boyer, R. S. and Moore, J. S. (Eds.): 1981, *The Correctness Problem in Computer Science*, Academic Press, London.

Brilliant, S. S. *et al.*: 1989, 'The Consistent Comparison Problem in N-Version Software', *IEEE Trans. Software Eng.* **15**(11), 1481—1485.

Camilieri, A. Gordon, M., and Melham, T.: 1987, 'Hardware Verification Using Higher-Order Logic', *Proc. IFIP WG 10.2 Working Conference: From H. D. L. Descriptions to Guaranteed Correct Circuit Designs*, D. Borrione (Ed.), Grenoble, North-Holland, Amsterdam.

Cohn, A.: 1987, 'A Proof of Correctness of the Viper Microprocessor: the First Level', in *VLSI Specification, Verification and Synthesis*, G. Birtwistle and P. A. Subrahmanyam (Eds.), Kluwer.

Cohn, A.: 1988, 'Correctness Properties of the Viper Block Model: The Second Level', in *Current Trends in Hardware Verification and Automated Deduction*, G. Birtwistle and P. A. Subrahmanyam (Eds.), Springer-Verlag.

Cohn, A.: 1989, 'The Notion of Proof in Hardware Verification', *Journal of Automated Reasoning* **5**(2), 127—139.

Cohn, A. and Gordon, M.: 1986, 'A Mechanized Proof of Correctness of a Simple Counter', University of Cambridge, Computer Laboratory, Tech Report No. 94.

Colburn, T.: 1991, 'Program Verification, Defeasible Reasoning, and Two Views of Computer Science', *Minds and Machines* **1**(1), 97—116.

Davis, P. J.: 1972, 'Fidelity in Mathematical Discourse: Is One and One Really Two?', *The Amer. Math. Monthly* **79**(3), 252—163.

DeMillo, R., Lipton, R. and Perlis, A.: 1979, 'Social Processes and Proofs of Theorems and Programs', *Communications of the AMC* **22**, 271—280.

Demillo, R., Lipton, R. and Perlis, A.: 1978, 'Response', *ACM SIGSOFT, Software Engineering Notes* **3**(2), 16—17.

Demillo, R., Lipton, R. and Perlis, A.: 1979, 'The Authors' Response', *Communications of the ACM* **22**(11), 629—630.

Denning, P. J.: 1989, 'Reply from the Editor in Chief', *Communications of the ACM* **32**(3), 289—290.

Department of Defense: 1983, 'Trusted Computer Systems Evaluation Criteria', SCS-STD-001-83, August 15, 1983.

Dijkstra, E. W.: 1978, 'On a Political Pamphlet from the Middle Ages', *ACM SIGSOFT, Software Engineering Notes* **3**(2), 14—16.

Dijkstra, E. W.: 1990, 'On the Cruelty of Really Teaching Computing Science', *Communications of the ACM* **32**, 1398—1404.

Dijkstra, E. W.: 1990, *Formal Development of Programs and Proofs*, Addison-Wesley, Reading, MA.

Dobson, J. and Randell, B.: 1989, 'Program Verification: Public Image and Private Reality', *Communications of the ACM* **32**(4), 420—422.

Dudley, R.: 1990, 'Program Verification', *Notices of the AMS* **37**, 123—124.

Dybjer, P.: 1987, 'From Type Theory to LCF — A Case Study in Program Verification', *Workshop on Programming Logic*, pp. 99—109.

Farmer, W. M., Johnson, D. M. and Thayer, F. J.: 1986, 'Towards a Discipline for Developing Verified Software', in Burrows, J. H. and Gallagher, P. R. Jr. (Eds.), *Proceedings of the 9th National Computer Security Conference*, September 15—18, 1986, at the National Bureau of Standards, Gaithersburg, Maryland, pp. 91—98.

Fetzer, J. H.: 1988, 'Program Verification: The Very Idea', *Communications of the ACM* **31**(9), 1048—1063.

Fetzer, J. H.: 1989, 'Response From the Author', *ACM Forum, Communications of the ACM* **32**(3), 288—289.

Fetzer, J. H.: 1989, 'Program Verification Reprise: The Author's Response', *Communications of the ACM* **32**(3), 377—381.

Fetzer, J. H.: 1989, 'Author's Response', *Communications of the ACM* **32**(4), 510—512.

Fetzer, J. H.: 1989, 'Another Point of View', *Communications of the ACM* **32**(8), 920—921.

Fetzer, J. H.: 1990, *Artificial Intelligence: Its Scope and Limits*, Dordrecht, the Netherlands, Kluwer Academic Publishers.

Fetzer, J. H.: 1991, 'Philosophical Aspects of Program Verification', *Minds and Machines* **1**(2), 197—216.

Fetzer, J. H. and Martin, C. R.: 1990, '"The Very Idea", Indeed! An Intellectual Brawl in Three Rounds (For Adults Only)', National Biomedical Simulation Resource, Technical Report NO. 1990—2.

George, J. Alan: 1971, 'Computer Implementation of the Finite Element Method', Ph.D. Thesis, Stanford U., Stanford, CA.

Gerrity, G. W.: 1982, 'Computer Representation of Real Numbers', *IEEE Trans. Computers* **C-31**(8), 709—714.

Gordon, M.: 1987, 'HOL: A Proof Generating System for Higher-Order Logic', in *VLSI Specification, Verification and Synthesis*, G. Birtwistle and P. A. Subrahmanyam (Eds.), Kluwer.

Gries, D.: 1976, 'An Illustration of Current Ideas on the Derivation of Correctness Proofs and Correct Programs', *IEEE Trans. on Software Engineering* **2**, 238—243.

Gries, D.: 1976, 'Proof of Correctness of Dijkstra's On-the-fly Garbage Collector', Lecture notes in Computer Science 46, Springer-Verlag, pp. 57—81.

Gumb, R. D.: 1989, *Programming Logics: An Introduction to Verification and Semantics*, New York, NY: Wiley and Sons.

Herbert, J. and Gordon, M. J. C.: 1986, 'A Formal Hardware Verification Methodology and its Application to a Network Interface Chip', *IEEE Proc. Computers and Digital Technologies*, Vol. 133, Part E, No. 5.

Hill, R., Conte, P., *et al.*: 1989, 'More on Verification', *Communications of the ACM* 32(7), 790—792.

Hunt, W. A. Jr.: 1985, 'FM8501: A Verified Microprocessor', University of Texas, Austin, Tech Report 47.

Joyce, J. J.: 1987, 'Formal Verification and Implementation of a Microprocessor', in *VLSI Specification, Verification and Synthesis*, G. Birtwistle and P. A. Subrahmanyam (Eds.), Kluwer.

Kling, R.: 1987, 'Defining the Boundaries of Computing Across Complex Organizations', in *Critical Issues in Information Systems*, Boland, R. and Hirschheim, R. (Eds.), Wiley, New York.

Knight, J. C. and Leveson, N. G.: 1985, 'A Large Scale Experiment in N-Version Programming', *Fifteenth International Symposium on Fault-Tolerant Computing*, pp. 135—139.

Knight, J. C. and Leveson, N. G.: 1986, 'An Experimental Evaluation of the Assumption of Independence in Multi-Version Programming', *IEEE Transactions on Software Engineering* SE-12(1), 96—109.

Knight, J. C. and Leveson, N. G.: 1986, 'An Empirical Study of Failure Probabilities in Multiversion Software', in: *Proceedings of the 16th IEEE International Symposium on Fault-Tolerant Computing* (FTCS-16), pp. 165—170.

Kolata, G. B.: 1976, 'Mathematical Proof: The Genesis of Reasonable Doubt', *Science* 192, 989—990.

Kowalski, R.: 1979, *Logic for Problem Solving*, New York, NY: North Holland.

Kreisel, G.: 1965, 'Informal Rigour and Completeness Proofs', *Int. Colloq. Philos. Sci.*, pp. 138—186.

Lakatos, I.: 1976, *Proofs and Refutations: The Logic of Mathematical Discovery*, Cambridge University Press, England.

Larsen, K. G.: 1987, 'Proof Systems for Hennessy-Milner Logic with Recursion', *Workshop on Programming Logic*, pp. 444—481.

Leith, P.: 1986, 'Fundamental Errors in Legal Logic Programming', *Computer J* 29(6), 545—552.

London, R. L.: 1975, 'A View of Program Verification', *SIGPLAN Notices*, 534—545.

Maclennan, B. J.: 1982, 'Values and Objects in Programming Languages', *SIGPLAN Notices* 17(2), 70—79.

Marcotty, M. and Ledgard, H.: 1986, *Programming Language Landscape: Syntax, Semantics, Implementations*, 2d Ed. Science Research Associates, Chicago.

Markoff, J.: 1989, 'Top-of-Line Intel Chip is Flawed', *The New York Times* (Friday, October 27), pp. 25 and 39.

Maurer, W. D.: 1978, 'On the Correctness of Real Programs', *ACM SIGSOFT, Software Engineering Notes* 3(3), 22—24.

442   SELECTED BIBLIOGRAPHY

Melham, T.: 1987, 'Abstraction Mechanisms for Hardware Verification', in *VLSI Specification, Verification and Synthesis*, G. Birtwistle and P. A. Subrahmanyam (Eds.), Kluwer.

Meyer, A.: 1974, 'The Inherent Computational Complexity of Theories of Ordered Sets: A Brief Survey', *Int. Cong. of Mathematicians*. Aug.

Meyer, S.: 1978, 'Should Computer Programs be Verified?', *ACM SIGSOFT, Software Engineering Notes* 3(3), 18—19.

Mills, H. D., Dyer, M., and Linger, R.: 1987, 'Cleanroom Software Engineering', *IEEE Software*, 19—25.

Milner, R. and Weyrauch, R.: 1972, 'Proving Compiler Correctness in a Mechanized Logic', *Machine Intelligence* 7, Edinburgh University Press, pp. 65—70.

Moore, J. S.: 1989, 'System Verifications', *Journal of Automated Reasoning* 5, 409—410.

Muller, H., Holt, C., and Watters, A.: 1989, 'More on the Very Idea', *Communications of the ACM* 32(4), 506—510.

Nelson, D.: 1992, 'Deductive Program Verification (A Practitioner's Commentary)', *Minds and Machines* 2(3), forthcoming.

Parnas, D. L.: 1978, 'Another View of the Dijkstra—DeMillo, Lipton, and Perlis Controversy', *ACM SIGSOFT, Software Engineering Notes* 3(3), 20—21.

Parnas, D., *et al.*: 1990, 'Colleagues Respond to Dijkstra's Comments', *Communications of the ACM* 32, 1405—1406.

Paulson, L.: 1987, *Logic and Computation*, Cambridge University Press.

Platek, R.: 1990, 'Formal Methods in Mathematics', *Software Engineering Notes, ACM* 15(4), 100—103.

Pleasant, J., Paulson, L., *et al.*: 1989, 'The Very Idea', *Communications of the ACM* 32(3), 374—377.

Pomerantz, A., Glazier, D., *et al.*: 1979, 'Comments on Social Processes and Proofs', *Communications of the ACM* 22(11), 621—629.

Proctor, N.: 1985, 'Restricted Access Processor Verification Results Report', Technical Report TR-84002, Sytek.

Proctor, N.: 1985, 'The Restricted Access Processor: An Example of Formal Verification', *Proc. 1985 IEEE Symposium on Security and Privacy*, Oakland, CA.

Pygott, C. H.: 1985, 'Formal Proof of a Correspondence between the Specification of a Hardware Module and its Gate Level Implementation', Royal Signals and Radar Establishment Report No. 85012.

Render, H.: 1990a, Article 755 (Comp.software.eng), USENET, 17 January 1990, 20:01:00 GMT.

Render, H.: 1990b, Article 1413 (Comp.software.eng), USENET, 1 February 1990, 01:31:30 GMT.

Richards, H.: 1990, 'Foreword', in: E. W. DIjkstra (Ed.), *Formal Development of Programs and Proofs*, Reading, MA: Addison-Wesley, pp. vii—ix.

Suydam, W. E.: 1986, 'Approaches to Software Testing Embroiled in Debate', *Computer Design* 25(21), 49—55.

Swart, E. R.: 1980, 'The Philosophical Implications of the Four-Color Problem', *American Mathematical Monthly* 87(9), 697—707.

Teller, P.: 1980, 'Computer Proof', *Journal of Philosophy* 77(12), 803—820.

Tompkins, H.: 1989, 'Verifying Feature—Bugs', *Communications of the ACM* **32**, 1130—1131.

Tymoczko, T.: 1979, 'The Four-Color Problem and its Philosophical Significance', *Journal of Philosophy* **66**(2), 57—83.

Van Ghent, R.: 1978, 'Regarding the Commentary by Dijkstra and the Reply by DeMillo, Lipton, and Perlis', *ACM SIGSOFT, Software Engineering Notes* **3**(3), 20—21.

Williams, G. B., *et al.*: 1989, 'Software Design Issues: A Very Large Information Systems Perspective', *Software Eng. Notes, ACM* **14**(3), 238—240.

Wulf, W. A.: 1979, 'Introduction to Part I: Comments on 'Current Practice'',' in: P. Wegner (Ed.), *Research Directions in Software Technology*, MIT Press, Cambridge, pp. 39—43.

Wulf, W. A., Shaw, M., Hilfinger, P. N. and Flon, L.: 1981, *Fundamental Structures of Computer Science*, Addison-Wesley.

INDEX OF NAMES

445

INDEX OF SUBJECTS

451

STUDIES IN COGNITIVE SYSTEMS

1. J. H. Fetzer (ed.): *Aspects of Artificial Intelligence.* 1988
ISBN 1-55608-037-9; Pb 1-55608-038-7
2. J. Kulas, J.H. Fetzer and T.L. Rankin (eds.): *Philosophy, Language, and Artificial Intelligence.* Resources for Processing Natural Language. 1988
ISBN 1-55608-073-5
3. D.J. Cole, J.H. Fetzer and T.L. Rankin (eds.): *Philosophy, Mind and Cognitive Inquiry.* Resources for Understanding Mental Processes. 1990
ISBN 0-7923-0427-6
4. J.H. Fetzer: *Artificial Intelligence: Its Scope and Limits.* 1990
ISBN 0-7923-0505-1; Pb 0-7923-0548-5
5. H.E. Kyburg, Jr., R.P. Loui and G.N. Carlson (eds.): *Knowledge Representation and Defeasible Reasoning.* 1990 ISBN 0-7923-0677-5
6. J.H. Fetzer (ed.): *Epistemology and Cognition.* 1991 ISBN 0-7923-0892-1
7. E.C. Way: *Knowledge Representation and Metaphor.* 1991
ISBN 0-7923-1005-5
8. J. Dinsmore: *Partitioned Representations.* A Study in Mental Representation, Language Understanding and Linguistic Structure. 1991 ISBN 0-7923-1348-8
9. T. Horgan and J. Tienson (eds.): *Connectionism and the Philosophy of Mind.* 1991 ISBN 0-7923-1482-4
10. J.A. Michon and A. Akyürek (eds.): *Soar: A Cognitive Architecture in Perspective.* 1992 ISBN 0-7923-1660-6
11. S.C. Coval and P.G. Campbell: *Agency in Action.* The Practical Rational Agency Machine. 1992 ISBN 0-7923-1661-4
12. S. Bringsjord: *What Robots Can and Can't Be.* 1992 ISBN 0-7923-1662-2
13. B. Indurkhya: *Metaphor and Cognition.* An Interactionist Approach. 1992
ISBN 0-7923-1687-8
14. T.R. Colburn, J.H. Fetzer and T.L. Rankin (eds.): *Program Verification.* Fundamental Issues in Computer Science. 1993 ISBN 0-7923-1965-6

KLUWER ACADEMIC PUBLISHERS – DORDRECHT / BOSTON / LONDON